T0335530

Alcohol and Its Biomarkers

Clinical Aspects and Laboratory Determination of Biomarkers Series

Series Editor: Amitava Dasgupta

Volume 1
Alcohol and Its Biomarkers: Clinical Aspects and Laboratory Determination

Alcohol and Its Biomarkers

Clinical Aspects and Laboratory Determination

Amitava Dasgupta, Ph.D.
Professor of Pathology and Laboratory Medicine,
University of Texas Medical School at Houston

AMSTERDAM • BOSTON • HEIDELBERG • LONDON • NEW YORK • OXFORD
PARIS • SAN DIEGO • SAN FRANCISCO • SINGAPORE • SYDNEY • TOKYO

Elsevier
525 B Street, Suite 1800, San Diego, CA 92101-4495, USA
32 Jamestown Road, London NW1 7BY, UK
225 Wyman Street, Waltham, MA 02451, USA

British Library Cataloguing-in-Publication Data
A catalogue record for this book is available from the British Library

Library of Congress Cataloging-in-Publication Data
A catalog record for this book is available from the Library of Congress

ISBN: 978-0-12-800339-8

For Information on all Elsevier books, visit our website at http://store.elsevier.com/

Printed and bound in the United States of America

15 16 17 18 19 10 9 8 7 6 5 4 3 2 1

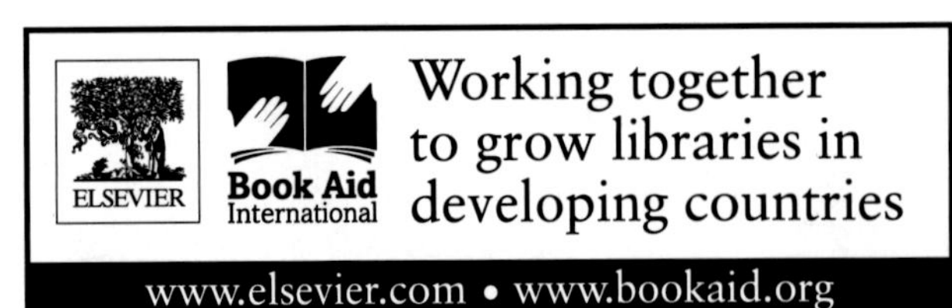

Contents

PREFACE .... **xi**

CHAPTER 1 Alcohol 1

1.1 Introduction 1
1.2 Alcohol Consumption: Historical Perspective 2
1.3 Alcohol Content of Various Alcoholic Beverages 3
1.4 Guidelines for Alcohol Consumption 4
1.5 Benefits of Drinking in Moderation 6
1.5.1 Moderate Alcohol Consumption and Reduced Risk of Cardiovascular Disease 7
1.5.2 Is Red Wine More Effective than other Alcoholic Beverages for Protecting the Heart? 10
1.5.3 Moderate Consumption of Alcohol and Reduced Risk of Stroke 10
1.5.4 Moderate Consumption of Alcohol and Reduced Risk of Developing Metabolic Syndrome and Type 2 Diabetes 11
1.5.5 Moderate Alcohol Consumption and Reduced Risk of Dementia/Alzheimer's Disease 13
1.5.6 Association between Moderate Alcohol Consumption and Reduced Cancer Risk 13
1.5.7 Can Moderate Alcohol Consumption Prolong Life? 14
1.5.8 Moderate Alcohol Consumption and Reduced Risk of Arthritis 15
1.5.9 Moderate Alcohol Consumption and Reduced Chance of Getting the Common Cold 15
1.6 Adverse Heath Effects Related to Alcohol Dependence 16
1.6.1 Liver Diseases and Cirrhosis of the Liver Associated with Alcohol Abuse 16
1.6.2 Alcohol Abuse and Neurological Damage 18
1.6.3 Alcohol Abuse and Increased Risk of Cardiovascular Disease and Stroke 21
1.6.4 Alcohol Abuse and Damage to the Immune System 21

1.6.5 Alcohol Abuse and Damage to the Endocrine System and Bone ... 22
1.6.6 Alcohol Abuse Increases the Risk of Certain Cancers ... 22
1.6.7 Fetal Alcohol Syndrome ... 23
1.6.8 Alcohol Abuse and Reduced Life Span ... 24
1.6.9 Alcohol Abuse and Violent Behavior/Homicide ... 25
1.6.10 Alcohol Poisoning ... 26
1.7 Blood Alcohol Level ... 27
1.7.1 Alcohol Odor on Breath and Endogenous Alcohol Production ... 29
1.8 Conclusions ... 30
References ... 30

CHAPTER 2 Genetic Aspects of Alcohol Metabolism and Drinking Behavior ... 37
2.1 Introduction ... 37
2.2 Alcohol Absorption: Effect of Food ... 37
2.3 First-Pass Metabolism of Alcohol ... 39
2.4 Alcohol Metabolism ... 40
2.4.1 Non-Oxidative Pathways of Alcohol Metabolism ... 42
2.4.2 Factors Affecting Alcohol Metabolism ... 43
2.5 Genes Encoding Alcohol Dehydrogenase ... 45
2.5.1 Polymorphism of Alcohol Dehydrogenase Genes ... 46
2.6 Genes Encoding Aldehyde Dehydrogenase ... 48
2.6.1 Polymorphisms of Alcohol Dehydrogenase and Aldehyde Dehydrogenase Genes that Protect Against the Development of Alcohol Use Disorder ... 51
2.6.2 Polymorphism of Alcohol Dehydrogenase and Aldehyde Dehydrogenase Genes that may Increase the Risk of Developing Alcohol Use Disorder ... 56
2.7 Polymorphism of the *CYP2E1* Gene ... 59
2.8 Conclusions ... 59
References ... 60

CHAPTER 3 Measurement of Alcohol Levels in Body Fluids and Transdermal Alcohol Sensors ... 65
3.1 Introduction ... 65
3.2 Breath Alcohol Determination ... 68
3.2.1 Chemical Principle of Breath Alcohol Analyzers ... 69
3.2.2 Effect of Breathing Pattern on Breath Alcohol Test Results ... 71
3.2.3 Interference in Various Breath Alcohol Analyzers ... 71

3.3 Blood Alcohol Determination 74
3.3.1 Enzymatic Alcohol Assays and Limitations 75
3.3.2 Gas Chromatography in Blood Alcohol Determination 78
3.3.3 Stability of Alcohol in Blood During Storage 79
3.3.4 Correlation between Blood and Breath Alcohol 80
3.4 Endogenous Production of Alcohol 81
3.5 Urine Alcohol Determination 82
3.6 Saliva Alcohol Determination 84
3.7 Transdermal Alcohol Sensors 85
3.8 Conclusions 87
References 87

CHAPTER 4 Alcohol Biomarkers 91
4.1 Introduction 91
4.2 State Versus Trait Alcohol Biomarkers 92
4.3 Liver Enzymes as Alcohol Biomarkers 94
4.4 Mean Corpuscular Volume as Alcohol Biomarker 97
4.5 Carbohydrate-Deficient Transferrin as Alcohol Biomarker 98
4.5.1 Combined CDT–GGT as Alcohol Biomarker 100
4.6 β-Hexosaminidase as Alcohol Biomarker 101
4.7 Ethyl Glucuronide and Ethyl Sulfate as Alcohol Biomarkers 102
4.8 Fatty Acid Ethyl Ester as Alcohol Biomarker 104
4.9 Phosphatidylethanol as Alcohol Biomarker 106
4.10 Total Plasma Sialic Acid as Alcohol Biomarker 108
4.11 Sialic Acid Index of Apolipoprotein J 109
4.12 5-HTOL/5-HIAA as Alcohol Biomarker 109
4.13 Other Alcohol Biomarkers 110
4.14 Clinical Application of Alcohol Biomarkers 111
4.14.1 Diagnosis Using *DSM-IV* and *DSM-5* 111
4.14.2 Self-Assessment of Alcohol Use 111
4.14.3 Alcohol Biomarkers and AUDIT 112
4.14.4 Application of Alcohol Biomarkers 113
4.14.5 Combining Alcohol Biomarkers 114
4.15 Conclusions 116
References 116

CHAPTER 5 Liver Enzymes as Alcohol Biomarkers 121
5.1 Introduction 121
5.2 Factors Affecting Liver Function Tests 122
5.3 Effect of Moderate Alcohol Consumption on Liver Enzymes 125
5.4 GGT as Alcohol Biomarker 126
5.4.1 Elevated GGT in Various Diseases and as a Risk Factor for Mortality and Certain Illnesses 130
5.4.2 GGT Fraction as Alcohol Biomarker 132

5.5 Laboratory Determinations of Liver Enzymes 133
5.6 Conclusions 134
References 134

CHAPTER 6 Mean Corpuscular Volume and Carbohydrate-Deficient Transferrin as Alcohol Biomarkers 139
6.1 Introduction 139
6.2 Mean Corpuscular Volume as Alcohol Biomarker 139
6.2.1 Mechanism of Increased MCV in Alcoholics 141
6.2.2 Other Causes of Macrocytosis 141
6.3 Carbohydrate-Deficient Transferrin 142
6.3.1 Mechanism of Formation of CDT 144
6.3.2 Cutoff Values, Sensitivity, and Specificity of CDT 144
6.3.3 CDT and GGT as Combined Alcohol Biomarker 147
6.3.4 Application of CDT 148
6.3.5 Limitations of CDT as Alcohol Biomarker 151
6.4 Laboratory Determination of CDT 154
6.5 Conclusions 158
References 158

CHAPTER 7 β-Hexosaminidase, Acetaldehyde—Protein Adducts, and Dolichol as Alcohol Biomarkers 163
7.1 Introduction 163
7.2 β-Hexosaminidase Isoforms 163
7.3 β-Hexosaminidase as Alcohol Biomarker 164
7.3.1 Pathophysiological Conditions that Cause Elevated Levels of β-Hexosaminidase 168
7.4 Laboratory Methods for Measuring β-Hexosaminidase 171
7.5 Acetaldehyde—Protein Adducts as Alcohol Biomarkers 172
7.5.1 Acetaldehyde—Hemoglobin Adducts 173
7.5.2 Acetaldehyde—Erythrocyte Protein Adducts 174
7.5.3 IgA Antibody Against Acetaldehyde-Modified Bovine Serum Albumin 175
7.6 Dolichol as Alcohol Biomarker 175
7.7 Conclusions 177
References 177

CHAPTER 8 Direct Alcohol Biomarkers Ethyl Glucuronide, Ethyl Sulfate, Fatty Acid Ethyl Esters, and Phosphatidylethanol 181
8.1 Introduction 181
8.2 Ethyl Glucuronide and Ethyl Sulfate 182
8.3 Ethyl Glucuronide and Ethyl Sulfate as Alcohol Biomarkers 186
8.3.1 Ethyl Glucuronide and Ethyl Sulfate Observed Due to Incidental Exposure to Alcohol 188
8.3.2 Ethyl Glucuronide and Ethyl Sulfate Cutoff Concentrations in Urine 191

8.3.3 Ethyl Glucuronide and Ethyl Sulfate Cutoff Concentrations in Hair, Meconium, and other Matrices ... 192
8.3.4 False-Positive/False-Negative Results with Ethyl Glucuronide ... 194
8.3.5 Application of Ethyl Glucuronide and Ethyl Sulfate as Alcohol Biomarkers ... 196
8.3.6 Laboratory Methods for Determination of Ethyl Glucuronide and Ethyl Sulfate ... 199
8.4 Fatty Acid Ethyl Esters as Alcohol Biomarkers ... 201
8.4.1 Fatty Acid Ethyl Esters in Hair ... 202
8.4.2 Fatty Acid Ethyl Esters in Meconium ... 204
8.4.3 Laboratory Analysis of Fatty Acid Ethyl Esters ... 206
8.5 Phosphatidylethanol as Alcohol Biomarker ... 207
8.5.1 Cutoff Concentration of Phosphatidylethanol ... 210
8.5.2 Laboratory Analysis of Phosphatidylethanol ... 211
8.6 Sensitivity and Specificity of Direct Alcohol Biomarkers ... 213
8.7 Conclusions ... 215
References ... 215

CHAPTER 9 Less Commonly Used Alcohol Biomarkers and Proteomics in Alcohol Biomarker Discovery ... 221
9.1 Introduction ... 221
9.2 Total Sialic Acid in Serum as Alcohol Biomarker ... 221
9.2.1 Other Causes of Elevated Plasma Sialic Acid Concentrations ... 224
9.2.2 Laboratory Determination of Total Sialic Acid ... 225
9.3 Sialic Acid Index of Apolipoprotein J as Alcohol Biomarker ... 227
9.3.1 Laboratory Methods for the Determination of the Sialic Acid Index of Plasma Apolipoprotein J ... 229
9.4 5-Hydroxytryptophol as Alcohol Biomarker ... 230
9.4.1 Laboratory Methods for Determining 5-HTOL and 5-HIAA ... 235
9.5 Other Alcohol Biomarkers ... 236
9.6 Proteomics in Alcohol Biomarker Discovery ... 237
9.6.1 Specific Proteins Identified as Alcohol Biomarkers Using the Proteomics Approach ... 238
9.7 Conclusions ... 241
References ... 241

CHAPTER 10 Genetic Markers of Alcohol Use Disorder ... 245
10.1 Introduction ... 245
10.2 Heredity, Environment, and Alcohol Use Disorder ... 246
10.2.1 Effect of Nongenetic Factors on the Development of Alcohol Use Disorder ... 247
10.3 Genes and Alcohol Use Disorder: An Overview ... 248

10.4 Polymorphisms in Genes Encoding Alcohol Dehydrogenase and Aldehyde Dehydrogenase 250
10.4.1 Polymorphisms that Protect from Alcohol Use Disorder 251
10.4.2 Polymorphisms that may Increase the Risk of Alcohol Use Disorder 253
10.5 Neurobiological Basis of Alcohol Use Disorder 253
10.6 Polymorphisms of Genes in Dopamine Pathway and Alcohol Use Disorder 254
10.6.1 Dopamine Receptors 255
10.6.2 Dopamine Transporters and Dopamine-Metabolizing Enzymes 258
10.6.3 Monoamine Oxidase 259
10.6.4 Catechol-*O*-Methyltransferase 259
10.7 Polymorphisms of Genes in the Serotonin Pathway and Alcohol Use Disorder 260
10.8 Polymorphisms of Genes in the Gaba Pathway and Alcohol Use Disorder 264
10.9 Polymorphisms of Genes Encoding Cholinergic Receptors and Alcohol Use Disorder 267
10.10 Polymorphisms of Genes in the Glutamate Pathway and Alcohol Use Disorder 269
10.11 Polymorphisms of Genes Encoding Opioid Receptors and Alcohol Use Disorder 272
10.12 Polymorphisms of Genes Encoding Cannabinoid Receptors and Alcohol Use Disorder 274
10.13 Adenylyl Cyclase and Alcohol Use Disorder 275
10.14 Neuropeptide Y and Alcohol Use Disorder 277
10.15 Possible Association of Polymorphisms of other Genes With Alcohol Use Disorder 278
10.16 Epigenetics and Alcohol Use Disorder 278
10.17 Conclusions 281
References 281

INDEX 289

Preface

Alcohol use by humans can be traced back to 10,000 BC. Consuming alcohol in moderation has many health benefits, including increased longevity. Some of these benefits are directly attributable to alcohol, whereas others are due to the combined effect of both alcohol and many beneficial phytochemicals present in beer and wine. However, all health benefits of alcohol disappear with heavy alcohol consumption, and continued excessive intake may lead to alcohol abuse disorder. Alcohol abuse is a serious public health concern worldwide and is a leading cause of mortality and morbidity internationally. The World Health Organization's (WHO) "Global Status Report on Alcohol and Health 2014" estimated that in 2012, 3.3 million deaths worldwide were attributable to alcohol abuse. WHO estimates that there are 140 million alcoholics worldwide, and an estimated 18 million alcohol-dependent individuals live in the United States. Moreover, it is estimated that alcohol dependence and alcohol abuse result in an estimated annual cost of $220 billion annually. Therefore, it is important for primary care physicians to be able to identify individuals who may be consuming a high amount of alcohol and intervene early so that these individuals do not develop alcohol use disorder. Alcohol biomarkers play an important role not only in identifying such individuals but also in monitoring the progress of alcohol rehabilitation therapy in alcohol-dependent patients.

The aim of this book is to provide a comprehensive overview of alcohol and alcohol biomarkers from both a clinical and a laboratory standpoint. Chapter 1 discusses alcohol use and abuse along with the benefits of consuming alcohol in moderation. The adverse effects of alcohol abuse are also discussed. In addition, issues regarding driving under the influence of alcohol are explored. Also in Chapter 1, the "drunken monkey hypothesis" is presented, which deals with the potential genetic basis of affinity toward alcohol by humans due to dependence of monkeys on ripe fruit as a major dietary source millions of years ago. In Chapter 2, the genetic aspects of alcohol metabolism are addressed, along with genetic polymorphisms of genes encoding alcohol dehydrogenase and aldehyde dehydrogenase that may protect individuals from alcohol abuse disorder. Chapter 3 discusses methods of detecting alcohol in various biological fluids (blood, urine, and saliva) and breath as well as transdermal alcohol determination, along with issues of interference in various methods, if applicable. A brief overview of state alcohol biomarkers is provided in Chapter 4. The various state alcohol biomarkers are discussed in detail in Chapters 5–9. In Chapter 10, the genetic aspect of alcohol abuse is addressed. One unique feature of this book is that both the clinical aspects of alcohol biomarkers and the various methods for detecting such alcohol biomarkers in the clinical laboratory are addressed, including issues of false-positive results due to interference. Moreover, at the end of each chapter, an extensive list of references is provided for further reading, if desired.

This book is aimed at clinicians, including primary care physicians, pathologists, clinical chemists, clinical toxicologists, and laboratory scientists, who deal with either the clinical aspect of alcohol abuse or the laboratory aspect of determining blood alcohol or alcohol biomarkers in biological fluids. Moreover, resident physicians and advanced medical students may find this book useful. Physicians and scientists who practice forensic medicine and students of forensic medicine may also find this book to be of use because alcohol is responsible for many deaths every year, including unnatural deaths related to alcohol use.

I thank my wife Alice for putting up with me during the long evening and weekend hours I spent preparing the manuscript. I also thank our department chair Robert L. Hunter, MD, PhD, for his support and encouragement. Finally, if readers find this book useful, my hard work will be rewarded.

Amitava Dasgupta

CHAPTER 1

Alcohol

Use, Abuse, and Issues with Blood Alcohol Level

1.1 INTRODUCTION

CONTENTS

1.1 Introduction 1
1.2 Alcohol Consumption: Historical Perspective 2
1.3 Alcohol Content of Various Alcoholic Beverages 3
1.4 Guidelines for Alcohol Consumption 4
1.5 Benefits of Drinking In Moderation 6
1.5.1 Moderate Alcohol Consumption and Reduced Risk of Cardiovascular Disease 7
1.5.2 Is Red Wine More Effective than other Alcoholic Beverages for Protecting the Heart?. 10
1.5.3 Moderate Consumption of Alcohol and Reduced Risk of Stroke 10

Alcohol use can be traced back to 10,000 BC. The "drunken monkey hypothesis," originally proposed by Professor Robert Dudley of the University of California at Berkeley, speculates that the human attraction to alcohol may have a genetic basis due to the high dependence of primate ancestors of *Homo sapiens* on fruit as a major source of food. Ethanol produced by yeast from fructose diffused out of the fruit and the alcoholic smell helped primates identify fruits as ripe and ready to consume. In tropical forests where monkeys lived, competition for ripe fruits was intense, and hungry monkeys capable of identifying ripe foods and eating them rapidly survived better than others. Eventually, "natural selection" favored monkeys with a keen appreciation for the smell and taste of alcohol. By the time humans evolved from apes approximately 1 to 2 million years ago, fruit consumption was mostly replaced by the consumption of roots, tubers, and meat. Although human ancestors stopped relying mainly on fruits as diet, it is possible that humans' taste for alcohol arose during our long-shared ancestry with other primates. Anecdotally, humans often consume alcohol with food, suggesting that this is a natural instinct. For millions of years, the amount of alcohol consumed by our ancestors was strictly limited, and the situation did not change even 10,000 years ago when humans had knowledge of agriculture and could produce plenty of barley and malt, the raw material for fermentation. The ancient beers and wines probably contained only 5% alcohol. After alcohol distillation was invented in Central Asia in approximately AD 700, drinks with higher alcoholic content became available, and the history of alcohol abuse by humans began. Alcohol abuse can also be considered as a disease of nutritional excess [1].

A. Dasgupta: Alcohol and Its Biomarkers. DOI: http://dx.doi.org/10.1016/B978-0-12-800339-8.00001-8

1.5.4 Moderate Consumption of Alcohol and Reduced Risk of Developing Metabolic Syndrome and Type 2 Diabetes....................... 11
1.5.5 Moderate Alcohol Consumption and Reduced Risk of Dementia/Alzheimer's Disease 13
1.5.6 Association between Moderate Alcohol Consumption and Reduced Cancer Risk 13
1.5.7 Can Moderate Alcohol Consumption Prolong Life?............... 14
1.5.8 Moderate Alcohol Consumption and Reduced Risk of Arthritis........................ 15
1.5.9 Moderate Alcohol Consumption and Reduced Chance of Getting the Common Cold.............................. 15
1.6 Adverse Heath Effects Related to Alcohol Dependence......... 16
1.6.1 Liver Diseases and Cirrhosis of the Liver Associated with Alcohol Abuse 16
1.6.2 Alcohol Abuse and Neurological Damage. 18
1.6.3 Alcohol Abuse and Increased Risk of Cardiovascular Disease and Stroke................... 21
1.6.4 Alcohol Abuse and Damage to the Immune System........................ 21
1.6.5 Alcohol Abuse and Damage to the Endocrine System and Bone............................ 22

1.2 ALCOHOL CONSUMPTION: HISTORICAL PERSPECTIVE

The first historical evidence of alcoholic beverages was the archeological discovery of Stone Age beer jugs from approximately 10,000 years ago. Egyptians probably consumed wine approximately 6000 years ago. The first beer was probably brewed in ancient Egypt, and Egyptians used alcoholic beverages (both beer and wine) for pleasure, rituals, and medical and nutritional purposes. The earliest evidence of alcohol use in China dates back to 5000 BC, when alcohol was produced mainly from rice, honey, and fruits. A Chinese imperial edict from approximately 1116 BC made it clear that the use of alcohol in moderation was the key and was prescribed from the heavens. In ancient India, alcohol beverages were known as "sura," a favorite drink of Indra, the king of all gods and goddesses. Use of such drinks was known in 3000–2000 BC, and ancient Ayurvedic texts concluded that alcohol was a medicine if consumed in moderation but a poison if consumed in excess. Beer was known to Babylonians as early as 2700 BC. In ancient Greece, wine making was common in 1700 BC. Hippocrates identified numerous medicinal properties of wine but was critical of drunkenness [2].

In ancient civilization, alcohol was used primarily to quench thirst because water was contaminated with bacteria. Hippocrates specifically cited that water from only springs and deep wells and from rainfall was safe for human consumption. Beer was a drink for common people, whereas wine was the preferred drink of elites. Drinking alcoholic beverages for thirst quenching was less common in ancient Eastern civilizations than in Western civilizations because drinking tea was very popular in Asian countries. During boiling of water to prepare tea, all pathogens die, thus making tea drinking a safe and healthy practice [3].

Yeast can be used to produce alcoholic beverages with up to 15% alcohol content. To prepare alcoholic beverages with higher alcohol content, distillation was needed, which probably originated in Asia. The distillation process became common in Europe only during the 11th century and later. During early American history, colonialists showed little concern about drunkenness, and the production of alcoholic beverages was a major source of commerce. In 1791, however, the "whiskey tax" was introduced, which was a tax on both privately and publicly brewed distilled whiskey. The whiskey tax was repealed by President Thomas Jefferson in 1802, but a new alcohol tax was temporarily imposed between 1814 and 1817 to pay for the War of 1812. In 1862, President Abraham Lincoln introduced an alcohol tax to pay for Civil War expenses. The act also created the office of Commissioner of Internal Revenue. In 1920, alcohol was prohibited in the United States, but Congress repealed the law in 1933. In 1978, President Jimmy Carter signed a bill legalizing home brewing of beer for personal use—the first time since prohibition [4].

1.3 ALCOHOL CONTENT OF VARIOUS ALCOHOLIC BEVERAGES

Alcohol content of alcoholic beverages varies widely; for example, beer contains approximately 4–7% alcohol, whereas the average alcohol content of vodka is 40–50%. However, due to wide differences in serving sizes of various alcoholic beverages, one drink (often called one standard drink) is considered to contain approximately 0.6 oz of alcohol, which is equivalent to 14 g. In the United States, a standard drink is defined as a bottle of beer (12 oz) containing 5% alcohol; 8.5 oz of malt liquor containing 7% alcohol; a 5-oz glass of wine containing 12% alcohol; 3.5 oz of fortified wine such as sherry or port containing approximately 17% alcohol; 2.5 oz of cordial or liqueur containing 24% alcohol; or one shot of distilled spirits such as gin, rum, vodka, or whiskey (1.5 oz). In general, the average bottle of beer contains 0.56 oz of alcohol, whereas a standard wine drink may contain 0.66 oz and distilled spirits may contain up to 0.89 oz of alcohol [5]. The alcohol content of various popular beverages is given in Table 1.1.

Historically, the alcohol content of various drinks was expressed as "proof," a term that originated in the 18th century when British sailors were paid with money as well as rum. To ensure that the rum was not diluted with water, it was "proofed" by dousing gunpowder with it and setting it on fire. If the gunpowder failed to ignite, this indicated that the rum was diluted with excess water. A sample of rum that was 100 proof contained approximately 57% alcohol by volume. In the United States, proof to alcohol by volume is defined as a ratio of 1:2. Therefore, a beer that has 4% alcohol by volume is defined as 8 proof. In the United Kingdom, alcohol by volume to proof is a ratio of 4:7. Therefore, multiplying alcohol by volume content by a factor of 1.75 will provide the "proof" of the drink.

Table 1.1 Alcohol Content of Various Drinks

Beverage	Alcohol Content (%)
Standard U.S. beer	4–7
Table wine	7–14
Sparkling wine	8–14
Fortified wine	14–24
Whiskey	40–75
Vodka	40–50
Gin	40–49
Rum	40–80
Tequila	45–50
Brandies	40–44

1.6.6 Alcohol Abuse Increases the Risk of Certain Cancers..........22
1.6.7 Fetal Alcohol Syndrome....................23
1.6.8 Alcohol Abuse and Reduced Life Span......24
1.6.9 Alcohol Abuse and Violent Behavior/ Homicide......................25
1.6.10 Alcohol Poisoning.....................26
1.7 Blood Alcohol Level 27
1.7.1 Alcohol Odor on Breath and Endogenous Alcohol Production......29
1.8 Conclusions..30
References 30

Currently in the United States, the alcohol content of a drink is measured by the percentage of alcohol by the volume. The *Code of Federal Regulations* requires that alcoholic beverage labels must state the alcohol content by volume. The regulation permits but does not require the "proof" of the drink to be printed. In the United Kingdom and in European countries, the alcohol content of a beverage is expressed also as the percentage of alcohol in the drink. Alcoholic drinks primarily consist of water, alcohol, and variable amounts of sugars and carbohydrates (residual sugar and starch left after fermentation); there are negligible amounts of other nutrients, such as proteins, vitamins, and minerals. However, distilled liquors such as cognac, vodka, whiskey, and rum contain no sugars. Red wine and dry white wines contain 2–10 g of sugar per liter, whereas sweet wines and port wines may contain up to120 g of sugar per liter of wine. Beer and dry sherry contain 30 g of sugar per liter [6].

1.4 GUIDELINES FOR ALCOHOL CONSUMPTION

The U.S. Department of Agriculture (USDA) and the Center for Nutrition Policy and Promotion publish "Dietary Guidelines for Americans." The latest 2010 guidelines define consumption of alcohol as follows:

- Recommended moderate consumption: Up to 1 standard drink per day for women and up to 2 standard drinks per day for men. However, individuals 65 years of age or older (both men and women) should consume only 1 standard drink per day.
- Heavy or high-risk drinking: Consumption of more than 3 standard drinks in any day or more than 7 standard drinks in 1 week for women. For men, consumption of 4 drinks a day or more exceeding 14 drinks per week is considered high-risk drinking.
- Binge drinking: This is defined as consumption of 4 or more drinks in a 2-hr period for women and 5 or more drinks in a 2-hr period for men.

In the United States, approximately 50% of adults consume alcohol on a regular basis, and another 14% are infrequent drinkers. An estimated 9% of men consume more than two drinks per day, and an estimated 4% of women consume more than one drink per day. Of those who consume alcohol, 29% report binge drinking within the past month, usually on multiple occasions. Excessive drinking is hazardous to health. Therefore, individuals who cannot restrict their drinking to moderate consumption should not consume alcohol at all. Other people who should not consume alcohol are listed in Box 1.1 [7].

Drinking more than the recommended amount can cause serious problems because the health benefits of drinking in moderation disappear quickly with

BOX 1.1 INDIVIDUALS WHO SHOULD NOT CONSUME ALCOHOL

- Individuals younger than 21 years old (the legal drinking age in the United States)
- Individuals who cannot restrict their drinking to a moderate level
- Pregnant women and women who want to become pregnant: No safe level of alcohol consumption during pregnancy has been established
- Individuals taking prescription or over-the-counter medicine that may interact with alcohol
- Individuals who plan to drive, operate machinery, or take part in an activity requiring attention, skill, or coordination or who are in situations in which impaired judgment may cause injury or death (e.g., swimming)

the consumption of more than three or four drinks per day. Alcohol abuse is a leading cause of mortality and morbidity internationally and is ranked by the World Health Organization (WHO) as one of the top five risk factors for disease burden. Without treatment, approximately 16% of all hazardous or heavy alcohol consumers will progress to become alcoholics [8]. Heavy consumption of alcohol leads to not only increased domestic violence, decreased productivity, and increased risk of motor vehicle as well as job-related accidents, but also to increased mortality from liver cirrhosis, stroke, and cancer. Alcohol overdose may also cause death.

Binge drinking is defined as heavy consumption of alcohol within a short period of time with the intention of becoming intoxicated. The National Institute of Alcohol Abuse and Alcoholism defines binge drinking as consuming an amount of alcohol needed to reach a blood alcohol level of 0.08%, which commonly represents 5 standard alcoholic beverages for men or 4 alcoholic beverages for women consumed within a 2-hr period. For younger drinkers, binge drinking increases the risk of short- and long-term blackouts, vehicle accidents, sexual assaults, homicide, and altered brain development. Despite the adoption by all states of a legal age for alcohol consumption of 21 years, binge drinking among U.S. high school seniors is a serious problem. Surveys indicate that 10.5% have consumed 10–14 drinks and 5.6% have consumed 15 or more drinks on one occasion [9]. Patrick *et al.* noted that between 2005 and 2011, 20.2% of high school seniors reported drinking 5 or more drinks on one occasion, 10.5% reported consuming 10 or more drinks, and 5.6% reported extreme binge drinking involving the consumption of 15 or more drinks at least once in the past 2 weeks. The authors concluded that binge drinking involving the consumption of 5 or more drinks on one occasion was common among high school seniors representative of all 12th graders throughout the United States [10]. Many college freshmen also consume alcohol at levels far beyond the binge drinking threshold. Approximately 1 in 5 males consumed 10 or more drinks and approximately 1 in 10 females consumed 8 or more drinks at least once during the previous 2-week period during a survey [11]. In another study, based on a survey of 14,150 binge drinkers, the authors found that 74.4% of binge drinkers consumed beer exclusively or predominantly, and at least 80.5% of binge drinkers consumed some beer. Wine accounted for only 10.9% of binge drinks consumed [12].

Alcohol is involved in many fatal car accidents. According to the U.S. Highway National Traffic Safety Administration, there were 12,998 fatalities related to alcohol use in 2007, which accounted for 31.7% of total traffic fatalities for the year. In addition, drivers between 16 and 24 years of age accounted for 23% of all alcohol-related fatal crashes. Alcohol-related fatal crashes involving female drivers have increased over time, but male drivers continue to surpass female drivers in the number of these traffic accidents. Restraint use during driving, such as the use of seat belts, decreased with increased blood alcohol level [13]. Excessive alcohol consumption is the cause of an average of 79,000 deaths annually, making alcohol abuse the third leading preventable cause of death in the United States. Moreover, the economic cost of excessive alcohol consumption is estimated to be $223.5 billion annually, as estimated in 2006 [14]. California is the largest alcohol market in the United States, and Californians consumed approximately 14 billion alcoholic drinks in 2005, resulting in an estimated 9439 deaths and 921,029 alcohol-related problems such as crime and injury. The economic burden was estimated to be $38.5 billion, of which $5.4 billion was for medical and mental health spending, $25.3 billion due to loss of work, and $7.8 billion for criminal justice spending [15]. In the United Kingdom, alcohol consumption was responsible for 31,000 deaths in 2005, and the National Health Services spent an estimated 3 billion pounds in 2005 and 2006 for treating alcohol-related illness and disability. Alcohol consumption was responsible for approximately 10% of disabilities [16].

1.5 BENEFITS OF DRINKING IN MODERATION

Consuming alcohol in moderation (up to two drinks per day for males, up to one drink per day for females, and up to one drink per day for both males and females 65 years of age or older) has many health benefits, including increased longevity. These health benefits are summarized in Box 1.2. Some of these benefits are attributable to alcohol, whereas many are due to the combined effect of alcohol and many beneficial phytochemicals present in beer and wine. More than 400 different phytochemicals are present in beer; some of these compounds originate from raw materials, whereas others are generated during the fermentation process. Melatonin is generated during the

BOX 1.2 BENEFITS OF CONSUMING ALCOHOL IN MODERATION

- Reduced risk of cardiovascular diseases including myocardial infarction
- Reduced risk of stroke
- Reduced risk of developing metabolic syndrome
- Reduced risk of developing type 2 diabetes
- Reduced risk of developing age-related dementia and Alzheimer's disease
- Reduced risk of certain types of cancer
- Reduced risk of forming gallstone
- Reduced risk of developing arthritis
- Increased longevity
- Less chance of getting common cold

brewing process. Beers with higher alcoholic content usually have higher amounts of melatonin [17]. More than 1600 phytochemicals are present in wine prepared from grapes [18].

1.5.1 Moderate Alcohol Consumption and Reduced Risk of Cardiovascular Disease

The beneficial effect of moderate alcohol consumption in reducing the risk of various cardiovascular diseases has been well characterized. The relationship between alcohol consumption and cardiovascular diseases was examined in the original Framingham Heart Study, which showed a U-shaped curve with a reduced risk of developing cardiovascular diseases with moderate drinking but a higher risk of developing such diseases with heavy drinking. Smoking is a risk factor for developing coronary heart disease, but moderate alcohol consumption may provide some protection against developing this disease among smokers [19]. An American Cancer Society prospective study of 276,802 American men found that during a period of 12 years, the relative risk (RR) of total mortality was 0.88 for occasional drinkers, 0.84 for those who drank one drink per day, and 1.38 for those who drank six or more drinks per day compared to nondrinkers. Interestingly, the risk of cardiovascular disease was mostly reduced in people who consumed one alcoholic beverage per day (RR = 0.79). This group also demonstrated the lowest all-cause mortality (RR = 0.84) among all groups studied [20].

Diabetic patients have a higher risk of developing cardiovascular disease. Moderate consumption of alcohol can help these patients to reduce this risk. In the Physician's Health Study, which involved 87,938 U.S. physicians (2970 diagnosed with diabetes mellitus) who were free of myocardial infarction, stroke, cancer, and liver disease on baseline and followed for an average of 5.5 years, it was observed that weekly consumption or daily consumption of alcohol reduced the risk of cardiovascular diseases in both diabetic and nondiabetic subjects [21]. Interestingly, women may receive a beneficial effect of alcohol from consuming lower amounts as well as consuming it less frequently than men. In a study of 28,448 women and 25,052 men between 50 and 65 years of age who were free of cardiovascular diseases at the time of enrollment in the study, during a 5.7-year follow-up, it was observed that women who consumed alcohol at least one day per week had a lower risk of coronary heart disease than those who drank alcohol less than one day a week. However, little difference was found between women who consumed at least one drink per week and women who consumed two to four drinks per week, five or six drinks per week, or even seven drinks per week. For men, lowest risk was found for those who consumed one drink per day. The authors concluded that for women, alcohol consumption can reduce the risk of heart disease and the frequency of drinking may not be an important factor, but for men drinking frequency is the determining factor in preventing heart disease [22]. Schroder *et al.* reported that consumption of up to 20 g alcohol per day through drinking wine, beer, and spirits significantly decrease the adjusted risk of myocardial infarction, but higher alcohol intake did not reduce the risk. The authors concluded that moderate alcohol consumption independent of the type of alcoholic beverages was associated with a reduced risk of nonfatal myocardial infarction [23].

Moderate alcohol consumption can not only reduce the risk of myocardial infarction but also provide protective effects against heart failure. In the Cardiovascular Health Study, which included 5595 subjects (age 65 years or older), it was observed that the risk of heart failure was reduced both in individuals who drank 1–6 drinks per week (hazard ratio (HR) = 0.82) and in individuals who drank 7–13 drinks per week (HR = 0.66). In addition, it was observed that moderate alcohol consumption reduced the risk of heart failure even in individuals who experienced myocardial infarction. The authors concluded that moderate alcohol use is associated with a lower incidence of congestive heart failure among older adults even after accounting for the incidence of myocardial infarction and other factors [24]. Another report based on a study of 1154 participants (580 men and 574 women) in Winnipeg, Manitoba, Canada, indicated that the well-established relationship between reduced risk of cardiovascular disease and moderate consumption of alcohol may not be evident until middle age (35–49 years) or older (50–64 years) in men. However, women may benefit from moderate consumption of alcohol at a much younger age (18–34 years). The beneficial effects of alcohol consumption are negated when alcohol is consumed in a heavy episodic drinking pattern (8 or more drinks per occasion), especially for middle-aged and older men [25].

There are several hypotheses on how moderate drinking can reduce the risk of developing heart disease (Box 1.3). Many studies have demonstrated increased high-density lipoprotein (HDL) cholesterol levels in drinkers compared to

BOX 1.3 HYPOTHESES ON HOW MODERATE ALCOHOL CONSUMPTION REDUCES THE RISK OF HEART DISEASE

- Increasing high-density lipoprotein cholesterol concentration
- Decreasing low-density lipoprotein cholesterol concentration
- Reduced plaque formation in coronary arteries
- Reduction in risk of blood clotting
- Reduced fibrinogen level
- Antioxidant effect of beer and wine

nondrinkers. The Honolulu Heart Study showed that men who consumed alcoholic beverages had a higher blood level of HDL cholesterol than that in nondrinkers. Gordon *et al.* reviewed data from 10 different studies, including the Honolulu Heart Study, and observed that there was a positive correlation between the amount of alcohol consumed and the serum level of HDL cholesterol. In the male population between ages 50 and 69 years, the average HDL cholesterol level was 41.9 mg/dL in those who consumed no alcohol, 47.6 mg/dL in those who consumed up to 16.9 g of alcohol per day (a single drink is 14 g of alcohol), 50.7 mg/dL in those who consumed 16.9–42.2 g of alcohol (one to three drinks) per day, and 55.3 mg/dL in those who consumed 42.3–84.5 g of alcohol (three to six drinks) per day [26]. In another study, the authors observed that the HDL cholesterol level in blood was increased by up to 33% in social drinkers compared to nondrinkers. A small experiment also revealed an average 15% reduction in HDL cholesterol levels among social drinkers who abstained from alcohol for a 2-week period [27]. In females, light drinking (one drink or fewer per day) was associated with a lower blood level of low-density lipoprotein (LDL) cholesterol and a higher level of HDL cholesterol [28]. Alcohol also diminishes thrombus formation on damaged walls of the coronary artery due to inhibition of platelet aggregation mediated through inhibiting phospholipase A_2 [29]. In 2013, Jones *et al.* commented that the effect of moderate alcohol consumption on lipid status or clotting does not fully explain the cardioprotective effect of alcohol. Exaggerated cardiovascular responses to mental stresses are detrimental to cardiovascular health. Using 88 healthy adults, Jones *et al.* demonstrated that alcohol consumption was inversely related to responses of heart rate, cardiac output, vascular resistance, and mean blood pressure provoked by stress (Montreal Imaging Stress Task). However, high alcohol consumers had larger cortisol (measured as salivary cortisol) stress responses than moderate alcohol consumers. The authors concluded that moderate alcohol consumption is associated with reduced cardiac responsiveness during mental stress, which has been linked to a lower risk of vascular diseases and hypertension among habitual drinkers. However, heavy alcohol consumption may negate such benefit due to greater cortisol stress response [30].

1.5.2 Is Red Wine More Effective than other Alcoholic Beverages for Protecting the Heart?

Studies indicate that the increased level in HDL cholesterol in blood may explain 50% of the protective effect of alcohol against cardiovascular disease and the other 50% may be partly related to inhibition of platelet aggregation and the antioxidant effect of various alcoholic beverages. Although alcohol is capable of increasing HDL cholesterol level and can inhibit platelet aggregation, polyphenolic antioxidant compounds found in abundance in red wine can further reduce platelet activity via other mechanisms. In addition, the polyphenolic compounds found in red wine can increase the level of vitamin E, thus providing further protection against various diseases. Therefore, it appears that red wine is more protective against cardiovascular diseases than are other alcoholic beverages [31]. Significant research to understand the epidemiological phenomenon known as the "French paradox" (a low incidence of cardiovascular diseases in the French population despite regular consumption of diet relatively rich in saturated fats, which is postulated to be a high-risk factor) indicates that the superiority of red wine in reducing the risk of cardiovascular diseases compared to other alcoholic beverages may be attributable to grape-derived polyphenols such as resveratrol in red wine [32]. It has been postulated that resveratrol (3,5,4′-trihydroxy-*trans*-stilbene), a polyphenol compound that is abundant in red wine in contrast to white wine, beer, or spirits, plays an important role as an antioxidant, and its inhibition of platelet aggregation may explain the increased cardioprotection from consuming red wine compared to other alcoholic beverages [33]. However, white wine can also provide cardioprotection due to the presence of tyrosol and hydroxytyrosol (both phenolic antioxidants) [34]. Using a pig model, Vilahur *et al.* observed that beer intake reduces oxidative stress and apoptosis and improves cardiac performance [35].

1.5.3 Moderate Consumption of Alcohol and Reduced Risk of Stroke

Another beneficial effect of consuming alcohol in moderation is a reduction in the risk of stroke among both men and women, regardless of age or ethnicity. The Copenhagen City Heart Study, which included 13,329 eligible men and women aged 45–84 years with a 16-year follow-up, indicated a U-shaped relationship between intake of alcohol and risk of stroke. People who consumed alcohol at low to moderate levels experienced a protective effect of alcohol against stroke, but heavy consumers of alcohol were at higher risk of suffering from a stroke compared to moderate drinkers or nondrinkers. However, there was no association between reduced risk of stroke and drinking beer or spirits, and only moderate consumption of wine was associated with a reduced risk of stroke [36]. In the second examination of the Copenhagen City Heart Study, involving 5373 men and 6723 women with a 16-year follow-up, it was observed that in individuals who experienced a high level of stress, weekly total consumption of 1–14 drinks was associated with lower risk of stroke in both men and women compared to individuals who also experienced a high level of stress but consumed no alcohol. However, no clear association was observed between risk of stroke and moderate consumption of alcohol in individuals who had a lower stress level. In addition, this study reported that drinking only beer or wine reduced the risk of stroke in individuals with a high stress level. It was suggested that alcohol may also alter psychological response to stress in addition to modifying physiological response [37]. Based on a study of 21,870 male physician participants (Physicians' Health Study) with an average follow-up of 12.2 years, Berger *et al.* concluded that light to moderate alcohol consumption reduced the overall risk of stroke and the risk of ischemic stroke in men. The benefit was apparent with as little as 1 drink per week, but greater consumption of alcohol did not increase the observed benefit [38]. Based on a prospective cohort study that included 45,449 Swedish women aged 30–50 years who were free of stroke and heart disease during enrollment and with an average follow-up of 11 years, Lu *et al.* observed that light (20 g of alcohol per week) to moderate alcohol consumption (20–69.9 g per week), regardless of alcoholic beverage type, reduced the risk of stroke in women younger than 60 years of age, especially in those who had never smoked. Smoking increased the risk of stroke, especially ischemic stroke, among women [39].

1.5.4 Moderate Consumption of Alcohol and Reduced Risk of Developing Metabolic Syndrome and Type 2 Diabetes

Metabolic syndrome or syndrome X was first described in 1988 by Gerald Reaven, who proposed the existence of a new syndrome characterized by insulin resistance, hyperinsulinemia, hyperglycemia, dyslipidemia, and arterial hypertension. Following this description of X syndrome, it became a major subject of research and public health concern. Individuals with this syndrome have a higher risk of coronary artery disease, stroke, and type 2 diabetes. The American Heart Association/National Heart, Lung, and Blood Institute criteria for metabolic syndrome include three or more of the following risk factors:

- Central obesity (waist circumference: 40 in. or more in men, 35 in. or more in women)
- Insulin resistance (fasting glucose >100 mg/dL)
- Elevated triglycerides (>150 mg/dL)
- Reduced HDL cholesterol (<40 mg/dL for men, <50 mg/dL for women)
- Elevated blood pressure (>130 mmHg for systolic or >85 mmHg for diastolic) or drug treatment for hypertension.

Other risk factors for metabolic syndrome include genetic makeup, advanced age, lack of exercise, and hormonal changes. Weight control, daily exercise, and healthy food habits are the primary goal of therapy. There is controversy regarding whether moderate alcohol consumption may reduce the risk of developing metabolic syndrome. In a study of 4505 Korean men without metabolic syndrome at baseline, Kim *et al.* observed that during 3 years of follow-up, the overall incidence of metabolic syndrome was 7.0% in nondrinkers, 10.3% in light drinkers, 13.8% in moderate drinkers, and 15.6% in heavy drinkers. The authors concluded that Korean men should restrict alcohol consumption to less than 15 g per day [40]. In contrast, in a prospective study of 7483 Caucasian men who were free of metabolic syndrome and cardiovascular diseases at baseline, Stoutenberg *et al.* observed that during follow-up clinical examination between 1979 and 2005, compared to no alcohol consumption, all levels of alcohol consumption (1–3 drinks per week to more than 14 drinks per week) provided a significant inverse association with the incidence of metabolic syndrome, although the multivariant hazard ratio was most favorable for moderate drinkers (0.68; 95% confidence interval: 0.57, 0.80) who consumed 4–7 drinks per week [41].

Moderate consumption of alcohol reduces the risk of developing type 2 diabetes. Based on 15 studies conducted in the United States, Finland, the Netherlands, Germany, the United Kingdom, and Japan involving 369,862 men and women with an average follow-up of 12 years, light drinkers and moderate drinkers had a lower risk of developing type 2 diabetes compared to nondrinkers. It made little difference whether an individual consumed beer, wine, or spirits, and it was better to consume alcohol frequently (e.g., daily or several times per week) rather than occasionally. The authors concluded that observational studies suggest an approximately 30% reduced risk of developing type 2 diabetes in moderate alcohol consumers, whereas no risk reduction was observed in heavy alcohol consumers (>48 g per day) [42]. In the Finnish twin study, in which twins with different drinking patterns (22,778 twins) were followed for 20 years, it was observed that moderate alcohol consumption (half a drink to two drinks (5–29.9 g) per day) for men and half to one and a half drinks (5–19.9 g) per day) for women) was associated with a lower risk of developing type 2 diabetes compared to light alcohol consumption (less than half a drink (<5 g) per day). Overweight subjects (body mass index ≥25.0 kg/m^2) received more beneficial effect from moderate alcohol consumption because the risk of developing diabetes was lower in overweight men and women who consumed alcohol than in nondrinkers. On the other hand, binge drinking and high alcohol consumption possibly increased the risk of type 2 diabetes in women, especially lean women, but it affected men to a lesser extent [43]. Based on a review of 20 studies, Baliunas *et al.* observed a U-shaped relationship between alcohol consumption and the risk of developing type 2 diabetes, where moderate alcohol consumption decreases the risk but heavy alcohol consumption increases the risk [44].

1.5.5 Moderate Alcohol Consumption and Reduced Risk of Dementia/Alzheimer's Disease

Moderate alcohol consumption can dramatically reduce the risk of age-related dementia and Alzheimer's disease. In a French study involving 3777 community residents aged 65 years or older, the authors observed that the subjects who consumed three or four alcoholic beverages (mostly wine) per day had an odds ratio (OR) of 0.18 for the incidence of dementia and an OR of 0.25 for Alzheimer's disease compared to nondrinkers. In mild drinkers (fewer than one or two drinks per day), there was a negative association only with Alzheimer's disease [45]. Mitchell *et al.* demonstrated that brain cultures preconditioned with moderate alcohol concentrations are resistant to neurotoxic Alzheimer's amyloid-β peptides. The mechanism of neuroprotection by moderate levels of alcohol consumption is probably related to early increases in NR1, NR2B, and NR2C subunits of *N*-methyl-D-aspartate (NMDA) receptors [46]. However, chronic abusers of alcohol are at higher risk of developing memory loss, dementia, and lack of appropriate motor control due to alcohol-related brain damage. Younger people, especially underage drinkers, are at higher risk of alcohol-related brain damage because alcohol has detrimental effects on the developing adolescent brain.

1.5.6 Association between Moderate Alcohol Consumption and Reduced Cancer Risk

Moderate consumption of alcohol may reduce the risk of certain types of cancer. It has been suggested that moderate drinking facilitates the elimination of *Helicobacter pylori*, which causes chronic atrophic gastritis (CAG) and gastric cancer. Gao *et al.*, in a study of 9444 subjects aged 50–74 years, observed that moderate drinkers (<60 g of alcohol per week or four drinks per week) had a significantly reduced chance of developing CAG compared to nondrinkers. Both beer and wine consumption provided protection against CAG. In addition to facilitating the elimination of *H. pylori*, other mechanisms may also have contributed to reducing the risk of CAG in moderate drinkers [47]. In the California Men's Health Study, which involved 84,170 men aged 45–69 years, consumption of one or more drinks of red wine per day was associated with an approximately 60% reduction in lung cancer risk, ever in smokers. In addition, even heavy smokers benefited from consuming red wine in moderation. No clear association was observed between moderate drinking in individuals who consumed white wine, beer, or other liquors [48]. In another study, it was observed that although moderate consumption of wine (one drink or fewer per day) was associated with a reduced risk of developing lung cancer (RR = 0.78), moderate consumption of beer (one or more drinks per day) increased the risk of developing lung cancer (RR = 1.23) in men but not in women [49]. Jiang *et al.* reported that people who consumed beer and wine but not spirits in moderation had a reduced risk of developing bladder cancer compared to nondrinkers [50]. Consumption of up to one drink per day reduced the risk of head and neck cancer in both men and women, but consuming more than three alcoholic beverages increased the risk of developing cancer [51]. In an Italian study, it was observed that moderate consumption of alcohol reduced the risk of developing renal cell carcinoma in both males and females [52]. However, based on a meta-analysis of 222 studies comprising approximately 92,000 light drinkers and 60,000 nondrinkers with cancer, Bagnardi *et al.* observed that light drinking was associated with the risk of oropharyngeal cancer (RR = 1.17), esophageal squamous cell carcinoma (RR = 1.30), and female breast cancer (RR = 1.05). The authors estimated that worldwide, 5000 deaths from oropharyngeal cancer, 24,000 deaths from esophageal squamous cell carcinoma, and 5000 deaths from breast cancer were attributable to light drinking in 2004 [53]. However, based on a systematic literature review, Menezes *et al.* commented that although moderate to heavy consumption of alcohol increased the risk of developing cancer of the oral cavity and pharynx, esophagus, stomach, larynx, colorectum, central nervous system, pancreas, breast, and prostate, no association was found between alcohol consumption and risk of developing lung, bladder, endometrium, and ovarian cancer. Alcohol consumption may be inversely related to the development of thyroid cancer [54].

1.5.7 Can Moderate Alcohol Consumption Prolong Life?

Because moderate consumption of alcohol can prevent many diseases, including the number one killer, the cardiovascular diseases, it is expected that moderate drinkers may live longer than lifetime abstainers from alcohol. In a study of 10,576 African American and 105,610 Caucasian postmenopausal women with an 8-year follow-up, Freiberg *et al.* demonstrated that moderate drinking (one to less than seven drinks per week) was associated with lower mortality among both hypertensive and nonhypertensive Caucasian women, but among African American women, only those who were hypertensive received benefit from moderate drinking. Consumption of only one drink or more per month was associated with increased longevity among Caucasian hypertensive and nonhypertensive women as well as among African American hypertensive women [55]. Klatsky *et al.* studied 10-year mortality in relation to alcohol in 8060 subjects and observed that persons who consumed two drinks or fewer daily fared best and had significant reductions in mortality rate compared to nondrinkers. The heaviest drinkers (six or more drinks per day) had a much higher mortality rate than moderate drinkers, whereas people who consumed three to five drinks per day had a similar mortality rate to that of nondrinkers. Therefore, consuming two drinks or fewer per day is the best practice [56]. In the Physicians' Health Study, which involved a 10-year follow-up of 22,071 male physicians in the United States between ages 40 and 84 years with no history of myocardial infarction, stroke, or cancer, the authors observed that men who consumed two to six drinks per week had more favorable results compared to those who consumed one drink per week. In contrast, men who consumed more than two drinks per day had higher mortality than those who consumed just one drink per week [57]. A study from the Netherlands reported that in the presence of stress, moderate drinkers were less likely to be absent from work than nondrinkers [58]. A 9-year prospective study indicated that moderate consumption of alcohol was associated with the most favorable health scores, indicating that these people in general enjoy a better overall health quality of life than abstainers [59].

1.5.8 Moderate Alcohol Consumption and Reduced Risk of Arthritis

Moderate alcohol consumption reduces the risk of developing rheumatoid arthritis. Results from two Scandinavian studies indicated that among moderate drinkers, the risk of rheumatoid arthritis was significantly reduced. Smokers had a higher risk of developing rheumatoid arthritis. The authors concluded that smokers should be advised to quit smoking in order to reduce the risk of developing arthritis, but moderate drinkers should not be discouraged from sensible alcohol consumption [60]. Moderate alcohol consumption not only reduces the risk of developing rheumatoid arthritis but also may slow the progression of disease. In a study of 2908 patients with rheumatoid arthritis, Nissen *et al.* reported that occasional or daily consumption of alcohol slowed the progression of the disease based on radiological studies (X-ray). The best results were observed in male patients [61].

1.5.9 Moderate Alcohol Consumption and Reduced Chance of Getting the Common Cold

The relationship between moderate alcohol consumption and reduced risk of the common cold has been studied. In one study, the authors observed that smokers are at greater risk of developing the common cold than nonsmokers. In addition, moderate alcohol consumption reduced the incidence of the common cold among nonsmokers but had no protective effect against the common cold in smokers [62]. In a large study of 4272 faculty and staff of five Spanish universities, the investigators observed that total alcohol intake from drinking beer and spirits had no protective effect against the common cold, whereas moderate wine consumption was associated with a reduced risk of the common cold. When individuals consumed 14 or more glasses of wine per week, the relative risk of developing the common cold was reduced (RR = 0.6) in these individuals compared to teetotalers. It was also observed that consumption of red wine provided superior protection against the common cold. The authors concluded that wine consumption, especially red wine, may have a protective effect against the common cold [63].

1.6 ADVERSE HEATH EFFECTS RELATED TO ALCOHOL DEPENDENCE

Many studies have demonstrated the harmful effects of alcohol on a variety of organ systems. Alcoholic liver disease and alcoholic liver cirrhosis are serious health hazards of alcohol abuse. Alcohol abuse affects multiple organ systems, including the brain, heart, bone, immune system, and endocrine system. Major adverse effects of chronic alcohol consumption include decreased life span; increased risk of violent behavior; alcoholic liver diseases, including cirrhosis of liver; mood disorder; and significantly increased risk of various cancers. Alcohol consumption during pregnancy may be associated with poor outcome of pregnancy, including fetal alcohol syndrome.

1.6.1 Liver Diseases and Cirrhosis of the Liver Associated with Alcohol Abuse

In the United States, approximately 60% of the general population admits to alcohol use, with 8–10% reporting heavy drinking (two or more drinks per day). Alcohol abuse is a leading cause of global morbidity and mortality, with the bulk of the alcohol-related disease burden resulting from alcoholic liver disease. Alcohol is a hepatotoxin if consumed in excess. In addition, alcohol consumption can potentiate other liver disease, such as viral hepatitis infection and non-alcoholic fatty liver disease. Alcoholic liver disease encompasses a wide range of diseases, including simple steatosis, inflammation, fibrosis, and eventually cirrhosis of the liver. Heavy drinking for only a few days may produce fatty change in the liver (steatosis) that is reversed after abstinence. Steatosis may be present in 90% of heavy drinkers. However, drinking heavily for a longer period may cause more severe alcohol-related liver injuries, such as alcoholic hepatitis. The prognosis for alcoholic hepatitis is variable, with a nearly 100% survival in mild cases, but mortality may occur in more severe cases. Individuals who continue to abuse alcohol may develop fibrosis and eventually cirrhosis. In general, 20–40% of patients with steatosis may eventually develop fibrosis; of these patients, 8–20% may develop liver cirrhosis. Women are at greater risk of developing alcoholic liver disease. Obesity as well as smoking may increase the risk of alcoholic liver disease [64].

The amount of alcohol consumed is one of the determining factors in the development of alcoholic hepatitis and liver cirrhosis. In one report, the authors commented that cirrhosis of the liver does not develop below a lifetime ingestion of 100 kg of alcohol (1 standard drink is approximately 14 g of alcohol, so this is the equivalent of a lifetime consumption of 7143 drinks). This amount corresponds to an average of five drinks a day for approximately 4 years. The authors also commented that consuming alcohol with food lowers the risk of developing cirrhosis of the liver compared to alcohol consumption on an empty stomach [65]. Although only a small percentage of alcohol-dependent individuals develop alcoholic hepatitis and liver cirrhosis, other alcohol-related liver damage occurs at a much lower intake of alcohol. In general, it is considered that the threshold of alcohol-induced liver toxicity is 40 g of alcohol per day (approximately 3 drinks per day) for men and 30 g (more than 2 drinks per day) for women for at least 5 years. However, based on a study of 6917 subjects, one report concluded that the alcohol consumption threshold for the risk of any alcohol-induced liver damage (non-cirrhotic liver damage) may be just 30 g (slightly more than 2 standard drinks) per day for both males and females with a lifetime drinking threshold of 100 kg of alcohol. In addition, the risk of liver cirrhosis, as expected, increases with increasing daily consumption, with the highest risk observed with alcohol consumption of more than 120 g per day. Drinking outside mealtimes and drinking multiple different alcoholic beverages increased the risk of alcohol-induced liver damage [66]. Walsh and Alexander commented that above a threshold of 7–13 drinks per week for women and 14–27 drinks per week for men, there is a risk of developing some alcohol-related liver problems. The greater sensitivity of women toward alcohol toxicity may be related to genetic predisposition of the pattern of metabolism of alcohol in women, in whom more oxidative by-products of alcohol are formed compared to men. Consumption of coffee may protect males against alcohol-induced liver damage, but no such data are available for females [67].

Hepatitis C is a liver disease caused by the hepatitis C virus. It is estimated that approximately 4 million Americans are infected with the hepatitis C virus and between 10,000 and 12,000 die annually. Hepatitis C infection is common among alcohol abusers, and this infection may even accelerate alcohol-related liver diseases, including cirrhosis of the liver and liver cancer. Therefore, whether or not a person infected with hepatitis C virus should consume alcohol is a controversial issue. In one study, the authors observed that moderate alcohol consumption of 31–50 g per day (2½–3½ drinks) for males and 21–50 g per day (1½–3½ drinks) for females could adversely affect the progression of liver damage [68].

The mechanism of alcohol-induced liver disease is complex. Whereas in moderate drinkers, alcohol is metabolized mostly by alcohol dehydrogenase in the liver, in alcoholics CYP2E1, a member of the cytochrome P450 drug-metabolizing family of enzymes in the liver, becomes activated. In this process, reactive oxygen species are generated. Hydroxyethyl radicals are probably involved in the alkylation of proteins found in hepatocytes. The formation of nitric oxide and elevated concentrations of stable metabolites, nitrites, and nitrates have been documented in alcoholics [69]. Singh *et al.* observed a significant positive correlation between γ-glutamyl transferase (GGT) and plasma malondialdehyde (a marker of lipid peroxidation) levels in alcoholics, indicating that alcoholics experienced increased oxidative stress. Moreover, reduced glutathione (an antioxidant enzyme) levels were depleted in alcoholics compared to controls [70]. In addition, acetaldehyde—a toxic product of alcohol metabolism if not removed quickly by further metabolism—may cause liver toxicity. In alcoholics, due to the tremendous burden of alcohol on the liver for metabolism, acetaldehyde and the reduced form of nicotinamide adenine dinucleotide (NADH) accumulation occurs, leading to oxidative stress to the liver and increased production of fatty acids. Metabolism of fatty acids is also impaired, causing their buildup in the liver, where they are eventually turned into fat (triglycerides) by the liver. With continued alcohol consumption, fatty liver may proceed to liver cirrhosis. Another mechanism of liver damage by alcohol is the excess cytokine production by Kupffer cells of the liver due to the release of bacterial endotoxin in the blood by the action of excess alcohol on bacteria present in the gut.

1.6.2 Alcohol Abuse and Neurological Damage

Although alcohol can cause relaxation and mild euphoria with moderate consumption, these pleasurable effects are reversed with blood alcohol levels greater than 100 mg/dL (0.1%). Alcohol has more damaging effects on the adolescent brain than on the adult brain. The commencement of alcohol consumption at an early age (13 years or younger) has devastating effects on the brain that may last a lifetime. There is also a link with a greater risk of alcohol dependence in adult life. Thiamine deficiency is one of the major factors involved in alcohol-related brain damage, and both alcohol and its toxic metabolite acetaldehyde exert toxic effects on neurons. Underage drinkers are also susceptible to immediate ill effects of alcohol use, such as blackouts, hangovers, and alcohol poisoning. These individuals are also at higher risk of neurodegeneration, impairment of functional brain activity, and neurocognitive deficits. Because adolescent drinking induces brain structure abnormalities, these changes lead to poor memory, impaired study habits, poor ability to learn, and poor academic performance [71]. Based on data from 8661 respondents to a survey in a 10-year study, Harford *et al.* concluded that education beyond high school has a protective effect against alcohol abuse and dependence. In contrast, people who do not attend college may also have a higher risk of alcohol abuse compared to people who do attend college [72]. Studies also show that children of alcoholics constitute a population at risk for skipping school days, poor performance, and dropping out of school. Children of alcoholics also have a higher incidence of repeating a grade [73].

Women are more susceptible to alcohol-related neurological damage than men. In particular, the female adolescent brain is more vulnerable to alcohol exposure that a male adolescent brain. Adolescents with alcohol abuse disorder have smaller prefrontal cortex volumes compared to those of healthy adolescents. The prefrontal cortex is located in the cortical region of the frontal lobe and is a crucial area of the brain responsible for planning complex cognitive behaviors such as learning, critical thinking, working with information held mentally, rational judgment, expression of personality, and appropriate social behavior. Consistent with the adult literature, alcohol use during adolescence is associated with prefrontal volume abnormalities, including differences in white matter; girls are affected more than boys by the adverse effects of alcohol [74].

Although the commencement of alcohol consumption at an early age carries a much higher risk of alcohol dependence and brain damage with long-lasting effects into adulthood, starting to drink at age 21 years followed by chronic abuse of alcohol can also cause significant damage to the human brain. The two major alcohol-related brain disorders are alcoholic Korsakoff's syndrome and alcoholic dementia. Korsakoff's syndrome is a brain disorder caused by a deficiency of thiamine, and major symptoms are severe memory loss, false memory, lack of insight, poor conversation skills, and apathy. Some heavy drinkers may also have a genetic predisposition to developing this syndrome. In Korsakoff's syndrome, loss of neurons is a common feature, including microbleeding in certain regions of gray matter [75]. When Wernicke's encephalopathy accompanies Korsakoff's syndrome in an alcoholic, the disorder is called Wernicke–Korsakoff syndrome. Wernicke's encephalopathy and Korsakoff syndrome are two related diseases; both are caused by thiamine deficiency, but clinical symptoms may be different. Alcoholics with Korsakoff's syndrome always have severe amnesic syndrome but may not have classical symptoms of Wernicke's encephalopathy, which include ophthalmoplegia, ataxia, and confusion. However, patients with Wernicke–Korsakoff syndrome show most of the symptoms found in both diseases. Damage to the anterior nucleus of the thalamus is commonly found in patients with Korsakoff's syndrome but may also be present in patients with Wernicke's encephalopathy. The anterior nucleus of the thalamus is involved in learning and memory as well as in the alertness of an individual. The Royal College of Physicians in London recommends that patients admitted to hospital who show evidence of chronic misuse of alcohol and poor diet should be treated with B vitamins [76]. Paparrigopoulos *et al.* reported a case in which a 52-year-old man with a 10-year history of heavy alcohol abuse was admitted to hospital and treated aggressively for Wernicke–Korsakoff syndrome with 600 mg per day oral thiamine supplement in addition to 300 mg of thiamine delivered intravenously every day; the patient fully recovered 2 months after therapy [77]. In addition to the development of Korsakoff's syndrome or Wernicke–Korsakoff syndrome, thiamine deficiency in chronic alcohol abusers is a major cause of alcohol-induced brain damage. Mild to moderate thiamine deficiency plays a role in the neurodegeneration observed in chronic alcoholics, and thiamine metabolism may also be altered in non-Wernicke's encephalopathy alcoholics. However, amnesic syndrome typical for Wernicke's encephalopathy is mainly due to damage in the diencephalic–hippocampal circuitry, including thalamic nuclei and mammillary bodies. The loss of cholinergic cells in the basal forebrain region may eventually cause decreased cholinergic input to the hippocampus and cortex, resulting in reduced choline acetyltransferase and acetylcholinesterase activity and function as well as acetylcholine downregulation within these brain regions [78].

Binge drinkers, both male and female, are at higher risk of developing alcohol-related brain damage. Chronic exposure to the high amounts of alcohol that are ingested during binge drinking leads to stimulation of NMDA receptors and calcium receptors that results in increased release of glucocorticoids (stress molecules such as cortisol that affect carbohydrate metabolism). NMDA-mediated mechanisms and glucocorticoid actions on the hippocampus are associated with brain damage. In addition, ethanol withdrawal becomes more difficult for binge drinkers [79]. Alcohol-related brain damage and loss of cognitive functions may be reversible, at least in part, if the brain damage is not permanent and the alcoholics can successfully complete a rehabilitation program and practice complete abstinence. Chronic alcoholism is often associated with brain shrinkage, but this may be reversed, at least in part, when abstinence is maintained, as demonstrated by Trabert *et al.* in a study of 28 male patients with severe alcohol dependence. Even with 3 weeks of abstinence, increased brain tissue densities were observed in these subjects [80].

CASE REPORT 1.1

A 42-year-old patient was admitted to the hospital for his slowly progressive impairment in recent memory; during the 6 months before admission, his supervisor had recognized that his memory disturbance was becoming worse. Neuropsychological tests on admission showed an immediate verbal memory decline. His laboratory tests were unremarkable except for slightly elevated levels of aspartate and alanine aminotransferase as well as triglycerides, suggestive of fatty liver, but thiamine levels were within the normal range. His memory improved during the hospital stay, and he was discharged with a low-dosage antidepressant (25 mg/day of maprotiline hydrochloride). One year after admission, he confessed that he had a history of alcohol abuse for several years (two drinks per day for several years but six drinks per day for 1 year prior to hospital admission). The diagnosis made by Asada *et al.* was that the patient was suffering from reversible alcohol-related dementia. However, he abstained from alcohol after discharge from the hospital. Initial studies using fluorodeoxyglucose-positron emission tomography (FDG-PET), an advanced imaging technique, indicated that glucose metabolism had slowed in the brain of the patient; glucose is the only fuel that brain cells can use. A 5-year follow-up study using PET imaging indicated that glucose metabolism in the brain had recovered to the normal level, and the patient showed dramatically improved cognitive functions [81].

1.6.3 Alcohol Abuse and Increased Risk of Cardiovascular Disease and Stroke

Although alcohol consumed in moderation can reduce the risk of both cardiovascular diseases and stroke, consuming more than three drinks per day (any type of beverage) may be harmful to the heart. Chronic alcohol abuse for several years may result in alcoholic cardiomyopathy and heart failure, systematic hypertension, heart rhythm disturbances, and hemorrhagic stroke [82]. Alcoholics who consume 90 g or more of alcohol per day (7 or 8 drinks) for 5 years are at high risk of developing alcoholic cardiomyopathy; if they continue to consume alcohol, cardiomyopathy may proceed to heart failure, a potentially fatal medical condition. Without complete abstinence, 50% of these patients will die within 4 years of developing heart failure [83]. Heavy drinking also increases the risk of stroke, particularly the risk of hemorrhagic stroke. In one study, it was observed that the risk of hemorrhagic stroke increased in an individual who drank 300 g or more of alcohol weekly (21 or more drinks) [84].

1.6.4 Alcohol Abuse and Damage to the Immune System

Alcohol abuse is associated with an increased risk of bacterial and viral infection due to impairment of the immune system by alcohol. Exposure to alcohol can result in reduced cytokine production. Mast cells are important immune cells that are widely distributed in tissues that are in contact with the external environment, such as skin, mucosa of lung, and the gastrointestinal tract. Mast cells produce a variety of compounds, including cytokines, histamine, eicosanoids, and tumor necrosis factor-α, which play important roles in defense against bacteria and parasites. Alcohol reduces the viability of mast cells and may cause cell death. Alcohol-induced reduction in the viability of mast cells may contribute to impaired immune system associated with alcohol abuse [85]. Alcohol also accelerates disease progression in patients with HIV infection because of immunosuppression. In a study of 231 patients with HIV infection who were undergoing antiretroviral therapy, it was observed that consumption of two or more drinks per day could cause a serious decline in $CD4^+$ cell count (higher $CD4^+$ counts indicate a good response to therapy) [86].

Adult respiratory distress syndrome (ARDS) is a severe form of lung injury. Approximately 200,000 individuals develop ARDS in the United States each year, and nearly 50% of these patients have a history of alcohol abuse. The mortality rate from ARDS is high (>40%), but it is especially high for alcohol abusers (~65%). In ARDS survivors, alcohol abuse was also associated with longer stays under ventilation in intensive care units. Alcohol impairs immune function and decreases pulmonary antioxidant capacity and thus may cause ARDS [87].

1.6.5 Alcohol Abuse and Damage to the Endocrine System and Bone

Alcohol abuse can have adverse effects on the human endocrine system. It may lead to a disease known as pseudo-Cushing's syndrome, which is indistinguishable from Cushing's syndrome and characterized by excess production of cortisol causing high blood pressure, muscle weakness, diabetes, obesity, and a variety of other physical disturbances. Diminished sexual function in alcoholic men has been described for many years. Administration of alcohol in healthy young male volunteers caused a diminished level of testosterone. Consuming three or more drinks per day may cause significant problems in women, including delayed ovulation or failure to ovulate and menstrual problems; however, such problems were not noticed in women who consumed two or fewer drinks per day. This may be related to alcohol-induced estrogen levels in women. Alcoholic women often experience reproductive problems. However, these problems may resolve when women practice abstinence from alcohol. To form healthy bone calcium, phosphorus and the active form of vitamin D are essential. Chronic consumption of alcohol may reduce bone mass through a complex process of inhibition of hormonal balance needed for bone growth, including testosterone in men, which is diminished in alcoholics. Alcohol abuse may also interfere with pancreatic secretion of insulin, causing diabetes [88].

1.6.6 Alcohol Abuse Increases the Risk of Certain Cancers

Epidemiological research has demonstrated a dose-dependent relationship between consumption of alcohol and certain types of cancers. The strongest link was found between alcohol abuse and cancer of the mouth, pharynx, larynx, and esophagus. An estimated 75% of all esophageal cancers are attributable to chronic alcohol abuse, whereas nearly 50% of cancers of the mouth, pharynx, and larynx are associated with chronic heavy consumption of alcohol. Prolonged drinking may result in alcoholic liver disease and cirrhosis of the liver, and such disease can progress to liver carcinoma. However, there are only weak links between alcohol abuse and cancer of the colon, stomach, lung, and pancreas. Disease of the pancreas (pancreatitis) and gallstones are common among alcohol abusers. In alcoholics, endotoxin may be released from gut bacteria by the action of excess alcohol, and such process may trigger progression of acute pancreatitis into chronic pancreatitis. Chronic pancreatitis may lead to pancreatic cancer [89]. The relationship between moderate alcohol consumption and the risk of breast cancer is controversial and there are conflicting reports in the medical literature. One Spanish study of 762 women between 18 and 75 years of age showed that even one drink per day may increase the risk of breast cancer, and women who consumed 20 g or more of alcohol per day (1½ drinks or more) had a 70% greater chance of developing breast cancer than nondrinkers [90]. In contrast, another study reported that women who consumed 10–12 g of wine per day (one glass of wine) had a lower risk of developing breast cancer compared to nondrinkers. However, the risk of breast cancer increases in women who have more than one drink per day [91]. Based on a review of 11 reports on the association between alcohol consumption and the risk of developing breast cancer, Nagata *et al.* concluded that epidemiological evidence of the link between the two remains insufficient [92].

1.6.7 Fetal Alcohol Syndrome

Fetal alcohol syndrome due to prenatal alcohol exposure was first reported by Jones and Smith in 1973 [93]. Since then, many publications have documented the teratogenic effects of alcohol in both human and animal studies. This syndrome is the most common noninherited cause of mental retardation in the United States. "Fetal alcohol spectrum disorders" was a term coined in 2004 to convey that exposure of the fetus to alcohol produces a continuum of effects and that many infants who do not fulfill all criteria for the diagnosis of fetal alcohol syndrome may nevertheless be profoundly negatively impacted throughout their lives due to exposure to alcohol. Therefore, fetal alcohol spectrum disorders include a wide range of permanent birth defects due to maternal consumption of alcohol during pregnancy, which also includes all serious complications in infants born with fetal alcohol syndrome. Other medical terminology used in relation to birth defects in infants whose mothers consumed alcohol during pregnancy include partial fetal alcohol syndrome, fetal alcohol effect, alcohol-related neurodevelopmental disorders, and alcohol-related birth defects.

Approximately 10–48 of every 10,000 children born in the United States have fetal alcohol syndrome, and as many as 91 out of 10,000 babies born have fetal alcohol spectrum disorder [94].

1.6.8 Alcohol Abuse and Reduced Life Span

Whereas moderate drinking is associated with increased longevity, alcohol abuse is associated with all-cause decreased longevity compared to that for abstainers. Even occasional heavy drinking may be detrimental to health. Dawson reported an increased risk of mortality among individuals who usually drank more than 5 drinks per occasion but consumed alcohol less than once per month [95]. Irregular heavy drinking even once a month (5 or more drinks per occasion) increases the risk of heart disease rather than protecting the heart as observed in moderate drinkers. The cardioprotective effect of moderate drinking also disappears when light to moderate drinking is mixed with occasional episodes of heavy drinking [96]. A British study of 5766 men aged 35–64 years with a 21-year follow-up observed that although compared to nondrinkers, individuals who consumed 8–14 standard alcohol drinks per week had slightly lower all-cause mortality, men who consumed more than 15 standard alcoholic beverages per week had a significantly higher risk of dying from all causes compared to nondrinkers. In addition, individuals who consumed more than 35 units per week had twice the risk of mortality compared to nondrinkers [97]. In a study of 20,765 drinkers with a 14-year follow-up during which 2547 people died, Breslow and Graubard observed that among men who consumed five or more drinks per drinking session, the adjusted RR of mortality from cardiovascular diseases was 1.30, that for cancer was 1.53, and that for other causes was 1.42 compared to those for men who consumed just one standard drink per day. The risk of mortality was also increased to some extent with the consumption of two drinks per day or more for males. Women drinkers who consumed two drinks or more in a session also showed all-cause higher mortality than moderate drinkers (one or fewer drinks per day). Among men, both quantity and frequency of drinking were significantly associated with mortality from cardiovascular disease, cancer, and other causes; among women, the quantity of alcohol was more important, and those who drank more than moderately showed a higher risk of mortality from cancer than did men [98]. The London-based Whitehall II cohort study, which involved 10,308 government employees aged 35–55 years with an 11-year follow-up, also concluded that optimal drinking is once or twice per week to daily consumption of one drink or fewer. People who consumed alcohol twice a day or more had an increased risk of mortality compared to those who consumed it once or twice per week [99]. Binge drinking is also hazardous. A study of a population of 1641 men who consumed beer found that the RR of all-cause death was 3.10 and that of fatal myocardial infarction was 6.50 for men who consumed six or more bottles of beer per session compared to men who consumed less than three bottles of beer per session [100]. Another study of 13,251 adults reported that individuals who consumed five or more drinks in a single session were significantly more likely to die from injuries than were persons who drank fewer than five drinks in a single session. Persons who consumed nine or more drinks in a single session had a much higher risk of dying from injuries compared to people who consumed less than five drinks [101].

In addition to increased mortality from various diseases, alcohol abuse is associated with increased risk of suicide, accidents, and violent crimes. Based on a survey of 31,953 school students, Schilling *et al.* observed that both drinking while depressed and episodic heavy drinking were associated with self-reported suicide attempts in adolescents [102]. Swahn *et al.* reported that in a high-risk school district in the United States, 35% of seventh graders reported alcohol abuse starting at age 13 years or younger. Preteen alcohol users were more involved in violent behavior than were nondrinkers. Early alcohol use was also associated with higher risk of suicide attempts among these adolescents [103].

1.6.9 Alcohol Abuse and Violent Behavior/Homicide

Many investigators have reported a close link between violent behavior, homicide, and alcohol intoxication. Studies conducted on convicted murderers suggest that approximately half of the murderers were under heavy influence of alcohol at the time of the murder [104]. When consumed in large quantities, alcohol may induce aggression and violent behavior by disrupting normal brain function. By impairing the normal information processing capability of the brain, a person can misjudge a perceived threat and may react more aggressively than warranted. Serotonin, a neurotransmitter, is considered to be a behavioral inhibitor. Alcohol abuse may lead to decreased serotonin activity, causing aggressive behavior. High testosterone concentrations in criminals have been associated with violent crimes. Adolescents and young adults with higher levels of testosterone compared to the general population are more often involved in heavy drinking and consequently violent behavior. Young men who exhibit antisocial behavior often "burn out" with older age due to a decreased level of testosterone and an increased level of serotonin. By modulating serotonin and testosterone concentration, alcohol may induce aggressive and violent behavior when consumed in excess [105]. Alcohol abuse by a husband may be related to husband-to-wife marital violence. Studies have shown a link between alcohol abuse by a husband before marriage and husband-to-wife aggression in the first year of marriage. The most violence abuse occurs in the first year of marriage in cases in which the husband was a heavy drinker before marriage and the wife was not [106].

1.6.10 Alcohol Poisoning

Drinking excessive alcohol on one occasion may cause alcohol poisoning, which if not treated promptly may be fatal. In general, drinking one or two standard alcoholic beverages produces a blood alcohol level appropriate for relaxation and mood elevation, but some impairment may occur at a level near or higher than the legal limit for driving while intoxicated (0.08% whole blood alcohol). Higher blood alcohol level leads to worsening of sensorimotor impairment, vomiting, respiratory depression, stupor, and coma (Table 1.2). In general, alcohol poisoning occurs at a blood alcohol level of 0.35% or higher, but a blood alcohol concentration greater than 0.25% (250 mg/dL) may place the patient at a higher risk of coma. Children and alcohol-naive individuals may experience toxicity at a blood alcohol concentration less than 100 mg/dL, whereas alcoholics may demonstrate significant impairment only at a blood alcohol level greater than 300 mg/dL [107]. Celik *et al.* reported that postmortem blood alcohol levels ranged from 136 to 608 mg/dL in 39 individuals who had died due to alcohol overdose. Most of these deceased were male [108]. The mechanism of death from alcohol poisoning is usually paralysis of respiratory and circulatory centers in the brain causing asphyxiation.

CASE REPORT 1.2

After consuming multiple mixed alcoholic beverages at an off-campus party, a 21-year-old college student suddenly became disoriented, developed slurred speech, vomited, and became unresponsive. He was transferred to the emergency department and on admission did not respond to voice commands and required repeated noxious stimuli for eye opening. On arrival at the emergency department, his Glasgow Coma Scale score was 8. Bedside cardiac monitoring was initiated, which showed sinus tachycardia. His blood chemistry results were unremarkable, but he showed a very high blood alcohol level of 350 mg/dL, which may be sufficient to cause coma, respiratory depression, and even death. However, other toxicology tests were negative. The serum osmolality was elevated to 338 mOsm/kg, which was consistent with alcohol poisoning. Moreover, the anion gap (30 mEq/L) was also elevated due to the high blood alcohol level. He was treated with supportive therapy in the intensive care unit and survived, but he did not recall the event. He was referred to an outpatient alcohol counseling program [109].

Table 1.2 Physiological Effects of Various Blood Alcohol Levels

Blood Alcohol Level	Physiological Effects
0.02–0.05% (20–50 mg/dL)	Relaxation and general positive mood; elevating effect of alcohol including increased social interaction
0.08% (80 mg/dL)	Legal limit of driving; minor impairment possible in people who drink rarely
0.1–0.15% (100–150 mg/dL)	Euphoria but sensory impairment and decreased cognitive ability; difficulty driving a motor vehicle
0.2% (200 mg/dL)	Worsening of sensorimotor impairment and inability to drive; decreased cognitive function and visual impairment
0.3% (300 mg/dL)	Vomiting; incontinence; symptoms of alcohol intoxication
0.4% (400 mg/dL)	Stupor; coma; respiratory depression; hypothermia
0.5% or higher (≥500 mg/dL)	Potentially lethal

1.7 BLOOD ALCOHOL LEVEL

Blood alcohol level depends on the number of alcoholic drinks consumed and the gender, body weight, age, as well as genetic makeup of the person. Currently, in all U.S. states, the legal limit for driving a vehicle is 0.08% alcohol in whole blood. The serum concentration of alcohol is higher than the whole blood concentration, and in order to calculate the whole blood concentration of alcohol, the measured serum concentration must be divided by a factor, most commonly 1.15. The legal limit for driving a vehicle in the United Kingdom and Canada is also 0.08%, but in other countries, lower levels of alcohol are mandated as the acceptable upper limit of driving under the influence of alcohol. In Switzerland, Denmark, Italy, the Netherlands, Austria, Australia, China, Thailand, and Turkey, the upper limit is 0.05% alcohol. In Japan, the upper acceptable limit is only 0.03%, and in other countries such as various Middle Eastern countries, Hungary, Romania, and Georgia, there is zero tolerance for blood alcohol in drivers.

Although the legal limit of blood alcohol in adult drivers in the United States is 0.08%, some driving impairment may occur even at lower blood alcohol levels. There is general agreement that some impairment of the ability to drive occurs at a blood alcohol level of 0.05%. Even a blood alcohol level of 0.03% affects some cognitive functions that rely on perception and processing of visual information [110]. Although a blood alcohol level of 0.05% usually produces more relaxation and more social interactions with other individuals, some intoxication can occur at a blood alcohol level of 0.1% and higher. The drunkest reported driver in Sweden had a blood alcohol level of 0.545% [111].

CASE REPORT 1.3

A citizen reported a person driving slowly and erratically, and the police found the car stopped with its lights on. When an officer approached the car, the driver tried to climb into the back seat, where there were two other passengers, and claimed that he was not the driver. He had a strong odor of alcohol and was unable to stand without support. The horizontal gaze nystagmus test showed a lack of smooth pursuit and nystagmus at maximum deviation. The subject was unable to recite the alphabet or follow instructions necessary to perform a breath alcohol test. The incident was reported at 12:15 AM, and his blood alcohol measured at 12:47 AM was 0.437% (437 mg/dL) [112].

In 1932, Swedish scientist Eric P. Widmark developed a formula that is still used today for the calculation of the amount of alcohol ingested and for assessing the concentration of blood alcohol prior to a blood alcohol analysis [113]. The Widmark formula estimates the blood alcohol level of a given amount of consumed alcohol taking into account the subject's body weight and gender:

$$A = C \times W \times r$$

where A is the total amount of alcohol consumed by the person in grams, C is the blood alcohol concentration in grams per liter, W is the body weight of the person expressed in kilograms, and r is a constant, which is assumed to be approximately 0.7 for men and 0.6 for women. The following form of the Widmark formula is commonly used to calculate blood alcohol concentration based on the amount of alcohol consumed by the individual, body weight, and gender:

$$C = (A/W \times r) - 0.015\,t$$

where t is the time that has passed since the beginning of the drinking session, and the 0.015 factor represents the average rate of elimination of alcohol (0.015%/h or 15 mg/dL/h).

In the United States, one standard drink of alcohol has 0.6 oz of alcohol, and the weight of a person is expressed in pounds. However, blood alcohol concentration is expressed as milligrams per 100 mL of whole blood. Taking into account all these factors, the previous formula can be modified for calculating blood alcohol concentration as follows:

$$C = (\text{total amount of alcohol consumed in ounces} \times 5.14/\text{weight in pounds} \times r) - 0.015\,t$$

where C is the blood alcohol in percent. Assuming each drink contains 0.6 oz of alcohol, this equation can be further modified as follows:

$$C = (\text{number of drinks} \times 3.1/\text{weight in pounds} \times r) - 0.015\,t$$

The blood alcohol level in women would be higher than that in men of the same weight. However, using the Widmark factor of 0.7 for men and 0.6 for women is more applicable to the Caucasian population. Tam *et al.* commented that for the Chinese population, a factor of 0.68 should be used for males and 0.59 should be used for females [114]. The Widmark formula provides a rough estimation of blood alcohol level based on gender, body weight, and the number of drinks consumed. However, blood alcohol level determined by the Widmark formula may differ significantly for some individuals compared to measured blood alcohol level or breath alcohol level. Thierauf *et al.* reported a case of a 75-year-old healthy man in which a discrepancy was observed between calculated blood alcohol level and measured blood alcohol level. The authors speculated that a reduction in total body water in the elderly may be the reason for such discrepancy [115].

1.7.1 Alcohol Odor on Breath and Endogenous Alcohol Production

Alcohol is almost odorless, and the alcohol smell perceived by people is due to the presence of many complex organic volatile compounds found in alcoholic beverages. Wine aroma is attributed to a large range of molecules from different chemical families, including esters, aldehydes, ketones, terpenes, tannins, and sulfur compounds. Some of these compounds originate from grapes, and others are formed during fermentation or aging. In general, more volatile substances are present in white wine compared to red wine [116]. Therefore, there is no correlation between blood alcohol level and alcohol odor. Such odor may also be present in an individual drinking nonalcoholic beer.

In general, the human body does not produce enough endogenous alcohol such that a measurable blood alcohol level is reached. There are reports of measurable endogenous ethanol production in patients with liver cirrhosis. In one report, after a meal eaten by such patients, negligible alcohol levels of 11.3 mg/dL (0.013%) and 8.2 mg/dL (0.0082%) were detected in two of eight patients. Small-intestinal bacterial overgrowth generates similarly small amounts of endogenous alcohol. Patients with liver cirrhosis often have small-intestinal bacterial overgrowth [117]. However, postmortem production of alcohol due to fermentation of sugar by bacteria is well documented. Toxicological analysis of a specimen from a deceased 14-year-old adolescent revealed high amounts of alcohol both in blood and in tissue, but ethyl glucuronide, a metabolite of alcohol, was not detected in the liver tissue. The authors concluded that postmortem alcohol in this adolescent was due to the action of the bacterial strain *Lactococcus garvieae* in the blood, which is capable of producing alcohol from glucose [118]. In another study, ethyl glucuronide was observed in the postmortem blood of 93 cases with antemortem blood alcohol, but it was not detected in 53 cases in which there was no indication of antemortem alcohol or use of alcohol by the deceased. The authors concluded that the presence of ethyl glucuronide in postmortem blood is a marker of antemortem ingestion of alcohol [119].

1.8 CONCLUSIONS

Although moderate alcohol consumption has many health benefits, heavy consumption of alcohol is detrimental to health. In moderation, alcohol use can increase longevity and reduce the risk of heart disease, stroke, and certain types of cancer. In addition, moderate alcohol consumption has a neuroprotective effect and reduces the risk of dementia, including Alzheimer's type. However, alcohol consumption in excess produces intoxication, withdrawal, brain trauma, central nervous system infection, hypoglycemia, hepatic failure, and Marchiafava–Bignami disease. Nutritional deficiency due to alcohol abuse also causes pellagra and Wernicke–Korsakoff disorder. In addition, alcohol is a neurotoxin, and in sufficient dosage it can cause lasting dementia [120]. It is generally accepted that alcohol consumption of no more than one drink per day for females and no more than two drinks per day for males younger than 65 years of age is safe. However, both males and females older than age 65 years should not consume more than one drink per day. There is no safe limit of alcohol consumption during pregnancy; to avoid fetal alcohol syndrome, pregnant women must practice total abstinence.

References

[1] Dudley R. Ethanol, fruit ripening and the historical origins of human alcoholism in primate frugivory. Integra Comp Biol 2004;44(4):315–23.

[2] Hanson D. History of alcohol and drinking around the world. Available from: <http://www2.potsdam.edu/alcohol/Controversies/1114796842.html>; [accessed 6.11.13].

[3] Vallee BL. Alcohol in the Western World. Sci Am 1998;278(6):80–5.

[4] Loyola Marymount University, Los Angeles Heads Up. History of alcohol use. Available from: <http://academics.lmu.edu/headsup/forstudents/historyofalcoholuse>; [accessed 6.11.13].

[5] Kerr WC, Greenfield TK, Tujague J, Brown SE. A drink is a drink? Variation in the amount of alcohol contained in beer, wine and spirits drinks in a U.S. methodological sample. Alcohol Clin Exp Res 2005;29:2015–21.

[6] Liber CS. Relationship between nutrition, alcohol use and liver disease. Alcohol Res Health 2003;27:220–31.

[7] U.S. Department of Agriculture/U.S. Department of Health and Human Services. Dietary Guidelines for Americans. 7th ed. Washington, DC: U.S. Government Printing Office; 2010. pp. 30–32. Available from: <http://www.cnpp.usda.gov/Publications/DietaryGuidelines/2010/PolicyDoc/Chapter3.pdf>; [accessed 7.11.13].

[8] Reid MC, Fiellin DA, O'Connor PG. Hazardous and harmful alcohol consumption in primary care. Arch Intern Med 2008;159:1681–9.

[9] Hingson RW, White A. Trends in extreme binge drinking among U.S. high school seniors. JAMA Pediatr 2013;167:996–8.

[10] Patrick ME, Schulenberg JE, Martz ME, Mags JL, et al. Extreme binge drinking among 12th grade students in the United States: prevalence and predictors. JAMA Pediatr 2013;167:1019–25.

[11] White AM, Kraus CL, Swartzwelder H. Many college freshmen drink at levels far beyond the binge threshold. Alcohol Clin Exp Res 2006;30:1006–10.

[12] Naimi TS, Brewer RD, Miller JW, Okoro C, et al. What do binge drinkers drink? Implications for alcohol control policy. Am J Prev Med 2007;33:188–93.

[13] Tsai VW, Anderson CL, Vaca FE. Alcohol involvement among young female drivers in U.S. fatal crashes: unfavorable trend. Inj Prev 2010;16:17–20.

[14] Bouchery EE, Harwood HJ, Sacks JJ, Simon CJ, et al. Economic costs of excessive alcohol consumption in the U.S., 2006. Am J Prev Med 2011;41:516–24.

[15] Rosen SM, Miller TR, Simon M. The cost of alcohol in California. Alcohol Clin Exp Res 2008;32:1925–36.

[16] Balakrishnan R, Allender S, Scarborough P, Webster P, et al. The burden of alcohol-related ill health in the United Kingdom. J Pub Health (Oxford) 2009;31:366–73.

[17] Garcia-Moreno H, Calvo JR, Maldonado MD. High levels of melatonin generated during the brewing process. J Pineal Res 2013;55:26–30.

[18] Murch SJ, Hall BA, Le CH, Saxena PK. Changes in the levels of indoleamine phytochemicals during veraison and ripening of wine grapes. J Pineal Res 2010;49:95–100.

[19] Friedman LA, Kimball AW. Coronary heart disease mortality and alcohol consumption in Framingham. Am J Epidemiol 1986;124:481–9.

[20] Boffetta P, Garfinkel L. Alcohol drinking and mortality among men enrolled in an American Cancer Society prospective study. Epidemiology 1990;1:342–8.

[21] Ajani UA, Gaziana JM, Lotufo PA, Liu S, et al. Alcohol consumption and risk of coronary heart disease by diabetes status. Circulation 2000;102:500–5.

[22] Tolstrup J, Jensen MK, Tjonneland A, Overvad K, et al. Prospective study of alcohol drinking patterns and coronary heart disease in women and men. BMJ 2006;332:1244–8.

[23] Schroder H, Masabeu A, Marti MJ, Cols M, et al. Myocardial infarction and alcohol consumption: a population based case–control study. Nutr Metab Cardiovasc Dis 2007;17:609–15.

[24] Bryson CL, Mukamal KJ, Mittleman MA, Fried LP, et al. The association of alcohol consumption and incident heart failure: the Cardiovascular Health Study. J Am Coll Cardiol 2006;48:305–11.

[25] Snow WM, Murray R, Ekuma O, Tyas SL, et al. Alcohol use and cardiovascular health outcomes: a comparison across age, gender in the Winnipeg Health and Drinking Survey Cohort. Age Aging 2009;38:206–12.

[26] Gordon T, Ernst N, Fisher M, Rifkind BM. Alcohol and high density lipoprotein cholesterol. Circulation 1981;64(3 Part 2, Suppl):III63–7.

[27] Hulley SB, Gordon S. Alcohol and high density lipoprotein cholesterol: casual inference from diverse study designs. Circulation 1981;64(3 Part 2, Suppl):III57–63.

[28] Wakabayashi I, Araki Y. Association of alcohol consumption with blood pressure and serum lipid in Japanese female smokers and nonsmokers. Gender Med 2009;6:290–9.

[29] Rubin R. Effect of ethanol on platelet function. Alcohol Clin Exp Res 1999;23:1114–18.

[30] Jones A, McMillan MR, Jones RW, Kowalik GT, et al. Habitual alcohol consumption is associated with lower cardiovascular stress responses—A novel explanation for the known cardiovascular benefits of alcohol. Stress 2013;16:369–76.

[31] Ruf JC. Alcohol, wine and platelet function. Biol Res 2004;37:209–15.

[32] Wu JM, Hiieh TC. Resveratrol: a cardioprotective substance. Ann N Y Acad Sci 2011;1215:16–21.

[33] Wu JM, Wang ZR, Hsieh TC, Bruder JL, et al. Mechanism of cardioprotection by resveratrol, a phenolic antioxidant present in red wine. Int J Mol Med 2001;8:3–17.

[34] Mukherjee S, Lekli I, Gurusamy N, Bertelli AA, et al. Expression of the longevity proteins by both red and white wines and their cardioprotective components, resveratrol, tyrosol and hydroxytyrosol. Free Rad Biol Med 2009;46:573–8.

[35] Vilahur G, Casani L, Guerra JM, Badimon L. Intake of fermented beverages protect against acute myocardial injury: target organ cardiac effects and vasculoprotective effects. Basic Res Cardiol 2012;107:291.

[36] Truelsen T, Gronbaek M, Schnohr P, Boyen G. Intake of beer, wine and spirits and risk of stroke: the Copenhagen City Heart Study. Stroke 1998;29:2467–72.

[37] Nielsen NR, Truelsen T, Barefoot JC, Johnsen SP, et al. Is the effect of alcohol on risk of stroke confined to highly stressed persons? Neuroepidemiology 2005;25(3):105–13.

[38] Berger K, Ajani UA, Kase CS, Gaziano JM, et al. Light to moderate alcohol consumption and risk of stroke among U.S. male physicians. N Engl J Med 1999;341:1557–64.

[39] Lu M, Ye W, Adami HO, Weiderpass E. Stroke incidence in woman under 60 years of age related to alcohol intake and smoking habit. Cerebrovasc Dis 2008;25:517–25.

[40] Kim BJ, Kim BS, Kang JH. Alcohol consumption and incidence of metabolic syndrome in Korean men: a 3-year follow up study. Circ J 2012;76:2363–71.

[41] Stoutenberg M, Lee DC, Sui X, Hooker S, et al. Prospective study of alcohol consumption and the incidence of the metabolic syndrome in U.S. men. Br J Nutr 2013;110:901–10.

[42] Koppes LL, Dekker JM, Hendriks HF, Bouter LM, et al. Moderate alcohol consumption lowers the risk of type 2 diabetes: a meta-analysis of prospective observational studies. Diabetes Care 2005;28(3):719–25.

[43] Carrison SW, Hammar N, Grill V, Kaprio J. Alcohol consumption and the incidence of type 2 diabetes: 20 years follow up of the Finnish twin cohort study. Diabetes Care 2003;26:2785–90.

[44] Baliunas DO, Taylor BJ, Irving H, Roereke M, et al. Alcohol as a risk factor for type 2 diabetes: a systematic review and meta-analysis. Diabetes Care 2009;32:2123–32.

[45] Orgogozo JM, Dartigues JF, Lafont S, Letenneur L, et al. Wine consumption and dementia in the elderly: a prospective study in Bordeaux area. Revue Neurologique (Paris) 1997;153:185–92.

[46] Mitchell RM, Neafsey EJ, Collins MA. Essential involvement of the NMDA receptor in ethanol preconditioning-dependent neuroprotection from amyloid-beta in vitro. J Neurochem 2009;111:580–8.

[47] Gao L, Weck MN, Stegmaier C, Rothenbacher D, et al. Alcohol consumption and chronic atrophic gastritis: population-based study among 9444 older adults from Germany. Int J Cancer 2009;125:2918–22.

[48] Chao C, Slezak JM, Caan BJ, Quinn VP. Alcoholic beverage intake and risk of lung cancer: the California Men's Health Study. Cancer Epidemiol Biomarkers Prev 2008;17:2692–9.

[49] Chao C. Association between beer, wine, and liquor consumption and lung cancer risk: a meta-analysis. Cancer Epidemiol Biomarkers Prev 2007;16:2436–47.

[50] Jiang X, Castelao KE, Cortessis VK, Ross RK, et al. Alcohol consumption and risk of bladder cancer in Los Angeles County. Int J Cancer 2007;121:839–45.

[51] Freedman ND, Schatzkin A, Leitzmann MF, Hollenbeck MF, et al. Alcohol and head and neck cancer risk in a prospective study. B J Cancer 2007;96:1469–74.

[52] Pelucchi C, Galeone C, Montella M, Polesel J, et al. Alcohol consumption and renal cell cancer risk in two Italian case controlled study. Ann Oncol 2008;19:1003–8.

[53] Bagnardi V, Rota M, Botteri E, Tramacere I, et al. Light alcohol drinking and cancer: a meta-analysis. Ann Oncol 2013;24:301–8.

[54] Menezes RF, Bergmann A, Thuler LC. Alcohol consumption and risk of cancer: a systematic literature review. Asian Pac J Cancer Prev 2013;14:4965–72.

[55] Freiberg MS, Chang YF, Kraemer KL, Robinson JG, et al. Alcohol consumption, hypertension, and total mortality among women. Am J Hypertens 2009;22:1212–18.

[56] Klatsky AL, Friedman GD, Siegekaub AB. Alcohol and mortality: a ten year Kaiser-Permanente experience. Ann Int Med 1981;95:139–45.

[57] Camargo CA, Hennekens CH, Gaziano JM, Glynn RJ, et al. Prospective study of moderate alcohol consumption and mortality in U.S. male physicians. Arch Intern Med 1997; 157:79–85.

[58] Vasse RM, Nijhuis FJ, Kok G. Association between work stress, alcohol consumption and sickness absence. Addiction 1998;93:231–41.

[59] Wiley JA, Camacho TC. Life-style and future health: evidence from Alameda County study. Prev Med 1980;9:1–21.

[60] Kallberg H, Jacobsen S, Bengtsson C, Pedersen M, et al. Alcohol consumption is associated with decreased risk of rheumatoid arthritis: results from two Scandinavian studies. Ann Rheumatoid Dis 2009;68:222–7.

[61] Nissen MJ, Gabay C, Scherer A, Finchk A. The effect of alcohol on radiographic progression in rheumatoid arthritis. Arthritis Rheum 2010;62:1265–72.

[62] Cohen S, Tyrell DA, Russell MA, Jarvis NJ, et al. Smoking, alcohol consumption and susceptibility to the common cold. Am J Public Health 1993;83:1277–83.

[63] Takkouch B, Regueira-Mendez C, Garcia-Closas R, Figueiras A, et al. Intake of wine, beer, and spirits and the risk of clinical common cold. Am J Epidemiol 2002;155:853–8.

[64] Orman ES, Odena G, Bataller R. Alcoholic liver disease: pathogenesis, management and novel targets for therapy. J Gastroenterol Hepatol 2013;28(Suppl. 1):77–84.

[65] Belentani S, Tribelli C. Spectrum of liver disease in general population: lessons from Dionysus study. J Hepatol 2001;35:531–7.

[66] Bellentani S, Saccoccio G, Costa G, Tribelli C, et al. Drinking habits as cofactors of risk of alcohol induced liver damage. Gut 1997;42:845–50.

[67] Walsh K, Alexander G. Alcoholic liver disease. Postgrad Med 2000;281:280–6.

[68] Hezode C, Lonjon I, Roudot-Thorval F, Pawlotsky JM, et al. Impact of moderate alcohol consumption on histological activity and fibrosis in patients with chronic hepatitis C, and specific influence of steatosis: a prospective study. Aliment Pharmacol Ther 2003;17:1031–7.

[69] Zima T, Fialova L, Mestek O, Janebova M, et al. Oxidative stress, metabolism of ethanol and alcohol related disease. J Biomed Sci 2001;8:59–70.

[70] Singh M, Gupta S, Singhal U, Pandey R, et al. Evaluation of the oxidative stress in chronic alcoholics. J Clin Diagn Res 2013;7:1568–71.

[71] Zeigler DW, Wang CC, Yoast RA, Dickinson BD, et al. The neurocognitive effects of alcohol on adolescents and college students. Prev Med 2005;40:23–32.

[72] Harford TC, Yi HY, Hilton ME. Alcohol abuse and dependence in college and non-college samples: a ten year prospective follow up in a national survey. J Stud Alcohol 2006;67:803–9.

[73] Casas-Gil MJ, Navarro-Guzman JL. School characteristics among children of alcohol parents. Psychol Rep 2002;90:341–8.

[74] Medina KL, McQueeny T, Nagel BJ, Hanson KL, et al. Prefrontal cortex volumes in adolescents with alcohol use disorders: unique gender effects. Alcohol Clin Exp Res 2008;32:386–94.

[75] Kopelman MD, Thomson AD, Guerrini I, Marshall EJ. The Korsakoff syndrome: clinical aspect, psychology and treatment. Alcohol Alcohol 2009;44:148–54.

[76] Harper C. The neurotoxicity of alcohol. Hum Exp Toxicol 2007;26(3):251–7.

[77] Paparrigopoulos T, Tzavellas E, Karaiskos D, Kouzoupis A, et al. Complete recovery from undertreated Wernicke–Korsakoff syndrome following aggressive thiamine treatment. In Vivo 2010;24:231–3.

[78] Nardone R, Holler Y, Storti M, Christova M, et al. Thiamine deficiency induced neurochemical, neuroanatomical and neuropsychological alterations: a reappraisal. ScientificWorldJournal 2013;2013:309143.

[79] Hunt WA. Are binge drinkers more at risk of developing brain damage? Alcohol 1993;10(6):559–61.

[80] Trabert W, Betz T, Niewald M, Huber G. Significant reversibility of alcohol brain shrinkage within 3 weeks of abstinence. Acta Psychiatr Scand 1995;92:87–90.

[81] Asada T, Takaya S, Takayama Y, Yamauchi H, et al. Reversible alcohol-related dementia: a five-year follow up using FDG-PET and neuropsychological tests. Intern Med 2010;49:283–7.

[82] Klatsky AL. Alcohol and cardiovascular health. Physiol Behavior 2010;100:76–81.

[83] Laonigro I, Correale M, Di Biase M, Altomare E. Alcohol abuse and heart failure. Eur J Heart Fail 2009;11:453–62.

[84] Ikehara S, Iso H, Yamagishi K, Yamamoto S. Alcohol consumption, social support and risk of stroke and coronary heart disease among Japanese men: the JPHC study. Alcohol Clin Exp Res 2009;33:1025–32.

[85] Numi K, Methuen T, Maki T, Lindstedt KA, et al. Ethanol induces apoptosis in human mast cells. Life Sci 2009;85:678–84.

[86] Baum MK, Rafie C, Lai C, Sales S, et al. Alcohol use accelerates HIV disease progression. AIDS Res Hum Retrovir 2010;26:511–18.

[87] Boe DM, Vandivier RW, Burnham EL, Moss M. Alcohol abuse and pulmonary disease. J Leukocyte Biol 2009;76:1097–104.

[88] Emanuele N, Emanuele MA. Alcohol alters critical hormonal balance. Alcohol Health Res World 1997;21:53–64.

[89] Apte M, Pirola R, Wilson J. New insights into alcoholic pancreatitis and pancreatic cancer. J Gastroenterol Hepatol 2009;24(Suppl. 3):S351–6.

[90] Martin-Moreno JM, Boyle P, Gorgojo L, Willett WC, et al. Alcoholic beverage consumption and risk of breast cancer in Spain. Cancer Causes Control 1993;4:345–53.

[91] Bessaoud F, Daures JP. Pattern of alcohol (especially wine) consumption and breast cancer risk: a case controlled study among population in southern France. Ann Epidemiol 2008;18:467–75.

[92] Nagata C, Mizoue T, Tanaka K, Tsuji I, et al. Alcohol drinking and breast cancer risk: an evaluation based on a systematic review of epidemiological evidence among Japanese population. Jpn J Clin Oncol 2007;37:568–74.

[93] Jones KL, Smith DW. Recognition of the fetal alcohol syndrome in early infancy. Lancet 1973;302:999–1001.

[94] Sampson PD, Streissguth AP, Bookstein FL. Incidence of fetal alcohol syndrome and prevalence of alcohol related neurodevelopmental disorder. Teratology 1997;56:317–26.

[95] Dawson DA. Alcohol and mortality from external causes. J Stu Alcohol Drug 2001;62 (6):790–7.

[96] Roerecke M, Rehm J. Irregular heavy drinking occasions and risk of ischemic heart disease: a systematic review and meta-analysis. Am J Epidemiol 2010;171(6):633–44.

[97] Hart CL, Smith GD, Hole DJ, Hawthorne VM. Alcohol consumption and mortality from all causes, coronary heart disease, and stroke: results from a prospective cohort study of Scottish men with 21 years of follow up. BMJ 1999;318(7200):1725–9.

[98] Breslow RA, Graubard BI. Prospective study of alcohol consumption in the United States: quantity, frequency and cause of cause specific mortality. Alcohol Clin Exp Res 2008;32:513–21.

[99] Britton A, Marmot M. Different measures of alcohol consumption and risk of coronary heart disease and all-cause mortality: an 11 year follow up of the Whitehall II Cohort Study. Addiction 2004;99:109–16.

[100] Kauhanen J, Kaplan GA, Goldberg DE, Salonen JT. Beer binging and mortality: results from Kuopio Ischemic Heart Disease Risk Factor Study, a prospective population based study. BMJ 1997;315:846–51.

[101] Anda RF, Williamson DF, Remington PL. Alcohol and fatal injuries among U.S. adults: finding from the NHANES I Epidemiologic Follow up Study. JAMA 1998;260 (17):2529 32.

[102] Schilling EA, Aseltine RH, Glanovsky JL, James A, et al. Adolescent alcohol use, suicidal indention and suicide attempts. J Adolesc Health 2009;44:335–41.

[103] Swahn MH, Bossarte RM, Sullivent III EE. Age of alcohol use initiation, suicidal behavior, and peer and dating violence victimization and perception among high risk seventh grade adolescents. Pediatrics 2008;121:297–305.

[104] Palijan TZ, Kovacevic D, Radeljak S, Kovac M, et al. Forensic aspects of alcohol abuse and homicide. Coll Antropol 2009;33:893–7.

[105] National Institute on Alcohol Abuse and Alcoholism. Alcohol violence and aggression. Alcohol Alert, No 38, October 1997. Available from: <http://pubs.niaaa.nih.gov/publications/aa38.htm>; [accessed 18.09.14].

[106] Quigley BM, Leonard KE. Alcohol and the continuation of early marital aggression. Alcohol Clin Exp Res 2000;24:1003–10.

[107] Adinoff N, Bone GH, Linnoila M. Acute ethanol poisoning and ethanol withdrawal syndrome. Med Toxicol Adverse Drug Exp 1988;3:172–96.

[108] Celik S, Karapirli M, Kandemir E, Ucar F, et al. Fatal ethyl and methyl alcohol related poisoning in Ankara: a retrospective analysis of 10,720 cases between 2001 and 2011. J Forensic Leg Med 2013;20:151–4.

[109] Vacca Jr VM, Correllus III DF. Alcohol poisoning. Nursing 2013;43:14–16.

[110] Breitmeier D, Seeland-Schulze I, Hecker H, Schneider U. The influence of blood alcohol concentrations around 0.03% on neuropsychological functions: a double-blind, placebo-controlled investigation. Addict Biol 2007;12(2):183–9.

[111] Jones AW. The drunkest driver in Sweden: blood alcohol concentration of 0.545%. J Stud Alcohol 1999;60(3):400–6.

[112] Jones AW, Harding P. Driving under the influence with blood alcohol concentration over 0.4g%. Forensic Sci Int 2013;231:349–53.

[113] Brouwer IG. The Widmark formula for alcohol quantification. SADJ 2004;59(10):427–8.

[114] Tam TW, Yang CT, Fung WK, Mok VK. Widmark factors for local Chinese in Hong Kong: a statistical determination on the effects of various physiological factors. Forensic Sci Int 2005;30:23–9.

[115] Thierauf A, Kempf J, Eschbach J, Auwarter V, et al. A case of a distinct difference between the measured blood ethanol concentration and the concentration estimated by Widmark's equation. Med Sci Law 2013;53:96–9.

[116] Torrens J, Riu-Aumatell M, Lopez-Tamames E, Buxaderas S. Volatile compounds of red and white wines by headspace–solid-phase microextraction using different fibers. J Chromatogr Sci 2004;42(6):310–16.

[117] Madrid AM, Hurtado C, Gatica S, Chacon I, et al. Endogenous ethanol production in patients with liver cirrhosis, motor alteration and bacterial overgrowth. Rev Med Chil 2002;130:1329–34.

[118] Appenzeller BM, Schuman M, Wennig R. Was a child poisoned by ethanol? Discrimination between ante-mortem and postmortem formation. Int J Legal Med 2008;122(5):429–34.

[119] Hoiseth G, Karinen R, Christophersen AS, Olsen L, et al. A study of ethyl glucuronide in postmortem blood as a marker of ante-mortem ingestion of alcohol. Forensic Sci Int 2007;165:41–5.

[120] Brust JC. Ethanol and cognition—Indirect effects, neurotoxicity and neuroprotection: a review. Int J Environ Res Public Health 2010;7:1540–57.

CHAPTER 2

Genetic Aspects of Alcohol Metabolism and Drinking Behavior

2.1 INTRODUCTION

Ethanol, commonly referred to as "alcohol," is a small water-soluble polar molecule with a molecular weight of 46. The ethanol molecule contains a hydroxyl (–OH) functional group. Alcohol (ethanol) is a nutrient with a caloric value of approximately 7 kcal/g, whereas protein has a caloric value of 4 kcal/g and fat produces 9 kcal/g. After ingestion, alcohol is readily absorbed, but a small amount also undergoes first-pass metabolism. Absorption of alcohol from the gastrointestinal tract depends on how fast the person is drinking as well as whether or not alcohol is consumed with food. After absorption, alcohol is distributed in various tissues and also undergoes extensive metabolism and finally elimination. Although the majority of alcohol is metabolized via the oxidative pathway mainly involving two enzymes—alcohol dehydrogenase and aldehyde dehydrogenase—a small amount is also oxidized by the liver cytochrome P450 enzyme system, most commonly CYP2E1, especially in the presence of a high blood alcohol level. Other enzymes, such as catalase, may also be capable of metabolizing alcohol, but they represent a minor pathway. Polymorphisms of genes coding both alcohol dehydrogenase and aldehyde dehydrogenase enzymes affect blood alcohol level, and some polymorphisms may protect an individual from alcohol abuse. Minor non-oxidative metabolic pathways for alcohol involve conjugation with glucuronic acid yielding ethyl glucuronide and conjugation with sulfate to produce ethyl sulfate.

2.2 ALCOHOL ABSORPTION: EFFECT OF FOOD

Alcohol is absorbed from both the stomach and small intestine. A small amount of alcohol that is not absorbed is found in the breath and is the basis of breath analysis of drivers suspected of driving while intoxicated. In general, it is assumed that approximately 1–5% of alcohol is excreted by the

CONTENTS

2.1 Introduction..37
2.2 Alcohol Absorption: Effect of Food..........37
2.3 First-Pass Metabolism of Alcohol..........39
2.4 Alcohol Metabolism..........40
2.4.1 Non-Oxidative Pathways of Alcohol Metabolism..........42
2.4.2 Factors Affecting Alcohol Metabolism....43
2.5 Genes Encoding Alcohol Dehydrogenase...45
2.5.1 Polymorphism of Alcohol Dehydrogenase Genes..........46
2.6 Genes Encoding Aldehyde Dehydrogenase...48
2.6.1 Polymorphisms of Alcohol Dehydrogenase and Aldehyde Dehydrogenase Genes that Protect Against the Development of Alcohol Use Disorder..........51

A. Dasgupta: Alcohol and Its Biomarkers. DOI: http://dx.doi.org/10.1016/B978-0-12-800339-8.00002-X

2.6.2 Polymorphism of Alcohol Dehydrogenase and Aldehyde Dehydrogenase Genes that may Increase the Risk of Developing Alcohol Use Disorder..56

2.7 Polymorphism of the *CYP2E1* Gene.....................59

2.8 Conclusions..59

References..........60

lungs, and 1–3% is excreted via other routes such as urine (0.5–2.0%) and sweat (up to 0.5%). A very small amount of alcohol is also metabolized by non-oxidative pathways, and products of such reactions are often used as alcohol biomarkers, such as ethyl glucuronide and ethyl sulfate. The overall elimination process of alcohol can be described by a capacity-limited model similar to the Michaelis–Menten model of enzyme kinetics [1]. When alcohol is consumed, approximately 20% is absorbed by the stomach and the rest is absorbed by the small intestine by passive diffusion. A peak concentration is usually achieved 30–60 min after consumption. Food substantially slows the absorption of alcohol, and sipping of alcohol instead of drinking also slows the absorption of alcohol from the gastrointestinal tract. The presence of food in the stomach before alcohol consumption delays gastric emptying and reduces the rate of delivery of alcohol in the duodenum, thus reducing the alcohol absorption rate.

The effect of food on absorption and metabolism of alcohol has been widely studied and reported in the medical literature. In one study, 10 healthy men took in, via drinking, a moderate dosage of alcohol (0.80 g of alcohol per kilogram of body weight) in the morning after an overnight fast or immediately after breakfast (two cheese sandwiches, one boiled egg, orange juice, and fruit yogurt). Subjects who consumed alcohol on an empty stomach felt more intoxicated than those who consumed the same amount of alcohol after eating breakfast. The blood alcohol analysis revealed that the average peak blood alcohol in subjects who consumed alcohol on an empty stomach was 104 mg/dL. In contrast, the average peak blood alcohol in subjects who consumed alcohol after eating breakfast was 67 mg/dL. The time required to metabolize the total amount of alcohol was on average 2 hr shorter in subjects who consumed alcohol after eating breakfast compared to subjects who consumed alcohol on an empty stomach. The authors concluded that food in the stomach before alcohol consumption not only reduces the peak blood alcohol concentration but also increases elimination of alcohol from the body [2]. The effect of the nature of the food, such as high fat versus high protein or high carbohydrate, on the magnitude of the reduction of absorption of alcohol has also been studied. Jones *et al.* reported that the average peak blood alcohol level was 30.8 mg/dL in volunteers when alcohol was consumed on an empty stomach, but levels were respectively 16.6, 17.77, and 13.3 mg/dL when alcohol was consumed after eating a fatty meal, a meal rich in carbohydrates, or a meal rich in protein. The peak blood concentration was reached between 30 and 60 min after alcohol was consumed on an empty stomach or after eating a meal rich in protein. However, peak alcohol level was reached between 30 and 90 min when alcohol was consumed after eating a meal rich in fat or rich in carbohydrates. As expected, intravenous infusion of alcohol resulted in a higher average blood alcohol level of 54.3 mg/mL, and peak blood alcohol was observed within 30 min. The authors concluded that regardless of the composition of the meal, food in the stomach decreases the systemic availability of alcohol, probably due to slower gastric emptying time. Moreover, with food in the stomach, a portion of alcohol may be trapped by the constituents of the meal [3].

2.3 FIRST-PASS METABOLISM OF ALCOHOL

After alcohol is ingested, a small portion of alcohol enters the hepatic portal system and undergoes first-pass metabolism (FPM) by the liver, but gut alcohol dehydrogenase enzymes also play an important role in FPM of alcohol. Lim *et al.* commented that FPM of alcohol is predominantly gastric in nature [4]. In relation to alcohol, many factors affect FPM of alcohol, including gastrointestinal motility, nutritional status, liver function, gender, and age, as well as the genetic makeup of the person. After drinking the same amount of alcohol, men have a lower peak blood alcohol level compared to women with the same body weight. This gender difference in the blood alcohol level is partly related to the different body water content in men and women: women have a lower amount of body water (52% body water content is the average in women) compared to men (61% average). Therefore, less body water is available to dissolve the same amount of alcohol, which is water soluble, in women compared to men. However, gender difference in the rate of metabolism of alcohol by the alcohol dehydrogenase enzyme present in gastric mucosa is also responsible for higher blood alcohol in women compared to men with same body weight after consumption of the same amount of alcohol. Ammon *et al.* reported that total FPM on average accounted for 9.1% of alcohol metabolism in men and 8.4% in women. Dose-corrected values for the area under the blood alcohol concentration–time curves (AUC) over time were on average 28% higher in women than in men [5]. Marshall *et al.*, who studied 9 normal women and 10 normal men, observed that, after oral ethanol administration (0.5 g/kg body weight), women showed higher peak blood alcohol levels than did men (mean: 88 mg/dL in women and 75 mg/dL in men). The mean apparent volume of distribution of alcohol was 0.59 L/kg in females and 0.73 L/kg in men. Both apparent volume of distribution of alcohol and AUC were significantly correlated with total body water (measured by ^{3}H water dilution), suggesting that gender difference in ethanol pharmacokinetics may be related to gender difference relating to body water content [6].

First-pass metabolism of alcohol by gastric alcohol dehydrogenase (ADH) is slightly higher in men than in women, and this may be related to higher ADH activity in men than in women, especially women younger than age

50 years. Commonly used drugs, such as acetaminophen, aspirin, and H2 blockers (ranitidine, cimetidine, etc.), may decrease the activity of gastric ADH, thus reducing FPM of alcohol. As a result, elevated blood alcohol levels may be observed. Fasting strikingly decreases FPM of alcohol probably due to accelerated gastric emptying resulting in a higher blood alcohol level [7]. However, older men show lower alcohol dehydrogenase activities than do younger men. Therefore, gender difference in gastric ADH activities equalize in the elderly or are even reversed due to gastric mucosal atrophy, which occurs more often in males than in females [8]. The gender difference in alcohol level may be related only to lower gastric FPM of alcohol in women compared to men rather than to differences in gastric emptying or hepatic oxidation of alcohol. Higher blood alcohol levels in women compared to men after consuming the same amount of alcohol may be responsible for the increased vulnerability of women to the toxic effects of alcohol [9]. In a study of 20 men and 23 women (6 in each group were alcoholics), Frezza *et al.* observed that, in non-alcoholic subjects, FPM and gastric mucosal ADH activity of women were 23 and 59%, respectively, of those in men. In addition, gastric mucosal ADH activities (measured in endoscopic gastric biopsies) correlated with FPM of alcohol. In alcoholic men, FPM and gastric mucosal ADH activities were approximately 50% lower than those in non-alcoholic men. Moreover, FPM of alcohol was virtually abolished in alcoholic women [10].

When alcohol enters the circulation, it is distributed in various tissues, and a portion of alcohol is also metabolized. The tissue content of alcohol depends on the water content of the tissue, the rate of blood flow, and tissue mass, as well as blood alcohol concentration. The same dosage of alcohol per unit of body weight may produce different blood alcohol levels in different individuals due to wide variations in the proportions of fats and water in their bodies and the low lipid-to-water partition coefficient of alcohol. Compared to men, women have a lower volume of distribution of alcohol due to a higher percentage of body fat in women. Blood alcohol concentration also depends on race and ethnicity. American Indians absorb alcohol at a higher rate than Caucasians, and they may show higher blood alcohol levels than do Caucasians for similar amounts of alcohol consumption [11].

2.4 ALCOHOL METABOLISM

Metabolism of alcohol can generally be divided into major pathways and several minor pathways. The majority of alcohol is metabolized in the liver by ADH present in the cytosol. ADH has many different variants (isoenzymes) with various capacities to transform ethanol to acetaldehyde, a toxic metabolite. Acetaldehyde is then rapidly transformed into acetate by aldehyde dehydrogenase (ALDH), which also exists in several isoforms. ALDH2, found mostly in mitochondria, is mainly responsible for metabolism of acetaldehyde to acetate, although ALDH1 present in liver cytosol can also metabolize acetaldehyde into acetate. Nicotinamide adenine dinucleotide (NAD^+) is a required cofactor for both enzymatic reactions, in which NAD^+ is converted into NADH (the reduced form of NAD^+) in the cytosol (mostly during conversion of alcohol into acetaldehyde by ADH); as a result, the $NADH/NAD^+$ ratio in the cytosol is increased, causing a shift in the redox potential):

$$CH_3CH_2OH \xrightarrow{\text{Alcohol dehydrogenase}} CH_3CHO$$

$$NAD^+ \rightarrow NADH$$

Then acetaldehyde is further metabolized by ALDH into acetate CH_3COO^-:

$$CH_3CHO \xrightarrow{\text{Aldehyde dehydrogenase 2}} CH_3COO^-$$

$$NAD \rightarrow NADH$$

Acetate produced as the end product of alcohol metabolism is oxidized to carbon dioxide. Acetate is also converted into acetyl coenzyme A, which is involved in lipid and cholesterol biosynthesis in the mitochondria of peripheral tissue and brain. In chronic alcoholics, the brain may use acetate instead of glucose for energy [12].

The microsomal ethanol oxidizing pathway is a minor secondary pathway for ethanol metabolism. However, chronic alcohol consumption can stimulate this pathway, and especially CYP2E1, which is the ethanol inducible form of the cytochrome P450 family of enzymes; in this process, ethanol is also oxidized to acetaldehyde. However, the high redox potential of CYP2E1 for nicotinamide adenine dinucleotide phosphate (NADP) as a cofactor leads to the formation of reactive oxygen species, especially superoxide anion, hydroxyl radicals, and hydroxyethyl radicals; as a result, this pathway of ethanol metabolism induces oxidative stress and lipid peroxidation [13]:

$$CH_3CH_2OH \xrightarrow{\text{CYP2E1}} CH_3CHO$$

$$NADPH \rightarrow NADP^+$$

Alcohol-induced oxidative stress plays an important role in the pathogenesis of much alcohol-related damage, including liver damage. Individuals who abuse alcohol chronically also experience elevated oxidative stress compared to individuals who do not drink or who drink in moderation. In a study of 60 male alcoholics with a history of alcohol abuse for more than 5 years and 20 healthy male volunteers who served as a control population, Singh *et al.* demonstrated that serum malondialdehyde (a marker of *in vivo* lipid peroxidation) levels were significantly elevated in alcoholics compared to healthy volunteers, indicating that alcoholics experienced elevated oxidative stress compared to controls. Glutathione levels were also decreased in alcoholics compared to controls [14].

Another very minor pathway of ethanol metabolism involves the enzyme catalase, which is present in peroxisomes of the liver. Catalase is considered as an antioxidant enzyme because it is capable of removing hydrogen peroxide, a reactive oxygen species, from the body. However, catalase is also capable of oxidizing ethanol into acetaldehyde in the presence of hydrogen peroxide [13]. Acetaldehyde produced in the brain (ADH is inactive in the brain) due to metabolism of alcohol by catalase has been suggested to play a role in the development of alcohol tolerance as well as positive reinforcing actions of alcohol possibly via interaction of acetaldehyde with catecholamines to produce various condensation products [15]:

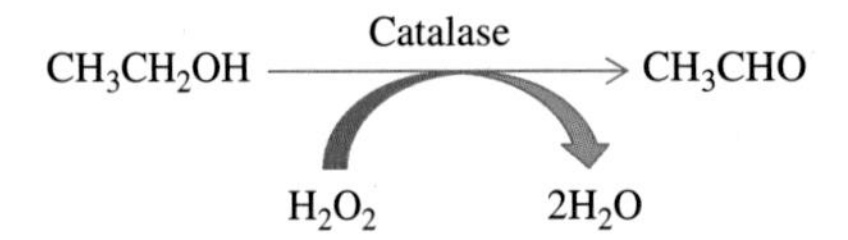

2.4.1 Non-Oxidative Pathways of Alcohol Metabolism

Other minor pathways of alcohol metabolism do not involve oxidative metabolism of ethanol into acetaldehyde. Conjugation of ethanol with glucuronic acid yielding ethyl glucuronide and conjugation with sulfate to produce ethyl sulfate are considered minor pathways for ethanol metabolism. These metabolites are used as alcohol biomarkers. In addition, fatty acid ethyl ester synthases, which are present in most tissues but predominantly in liver and pancreas, catalyze the reaction of ethanol and fatty acids, yielding fatty acid ethyl esters. These esters are synthesized in the endoplasmic reticulum and transported to the plasma membrane. Then fatty acid ethyl esters are removed from the cell by binding to lipoproteins and albumin. Fatty acid ethyl esters are detected in blood when alcohol is no longer detectable. Therefore, fatty acid ethyl ester can also be used as an alcohol biomarker [16]. Phospholipase D can convert ethanol into phosphatidylethanol; this is also a minor metabolic pathway of ethanol [17]. Phosphatidylethanol can also be used as a biomarker of alcohol abuse. Major and minor metabolic pathways of ethanol metabolism are summarized in Table 2.1.

Table 2.1 Metabolic Pathways of Ethanol

Pathway type	Enzyme	Comments
Major	Alcohol dehydrogenase (ADH) and aldehyde dehydrogenase 2 (ALDH2)	ADH present in the liver oxidizes ethanol into acetaldehyde, and NAD^+ is used as a cofactor in this reaction. This reaction takes place in cytosol. Then acetaldehyde is further oxidized into acetate by mitochondrial ALDH2, and again NAD^+ is used as a cofactor.
Minor but may be activated in alcoholics	CYP2E1	CYP2E1 present in the liver also oxidizes ethanol into acetaldehyde, and this is a minor metabolic pathway. However, this pathway is activated due to chronic ingestion of alcohol, especially in alcoholics. This pathway of ethanol metabolism also produces harmful reactive oxygen species.
Minor	Catalase	Catalase present in peroxisomes can also convert ethanol into acetaldehyde in the presence of hydrogen peroxide. In this enzymatic reaction, hydrogen peroxide is converted into water.
Minor	Uridine diphosphate glucuronosyltransferase	Ethanol is conjugated with glucuronic acid, producing ethyl glucuronide, a minor metabolite. This conjugation occurs in the liver, and ethyl glucuronide is an alcohol biomarker.
Minor	Sulfotransferase	Conjugation of ethanol with sulfate produces ethyl sulfate, another minor metabolite. Ethyl sulfate is also an alcohol biomarker.
Minor	Fatty acid ethyl ester synthase	Esterification of fatty acids with ethanol produces fatty acid ethyl esters. This is also a minor metabolic pathway, but fatty acid ethyl ester (combination of several esters) is an alcohol biomarker.
Minor	Phospholipase D	Ethanol is converted into phosphatidyl ethanol, a minor metabolic pathway for ethanol. Phosphatidyl ethanol is an alcohol biomarker.

2.4.2 Factors Affecting Alcohol Metabolism

Conversion of ethanol to acetaldehyde by ADH is the rate-limiting step in alcohol metabolism. Although many drugs are metabolized following first-order kinetics, alcohol metabolism usually follows zero-order kinetics, indicating that regardless of blood alcohol level, alcohol is metabolized by the liver ADH enzyme at a constant rate. This is due to the fact that the Michaelis constant (K_m) of most isoenzymes of ADH for ethanol is low ($\sim$2–10 mg/100 mL). Thus, the ADH enzyme is saturated at a relatively low blood ethanol concentration, resulting in the elimination of ethanol at a constant rate independent of blood alcohol level (zero-order kinetics). However, at a very low blood alcohol level, ADH may not be saturated, and alcohol metabolism may follow first-order kinetics. Moreover, in alcoholics with a high blood alcohol level, CYP2E1 plays an important role in the metabolism of alcohol, and elimination follows first-order kinetics (elimination is proportional to blood alcohol concentration) because the Michaelis constant is higher (60–80 mg/100 mL) compared to that for ADH and enzyme saturation does not occur even with a higher blood alcohol level. Therefore, it is usually assumed that ethanol elimination may follow first-order kinetics at very low blood alcohol level (<20 mg/dL; 0.02%) or high blood alcohol level in alcoholics (due to metabolism by CYP2E1). After consuming alcohol on an empty stomach, elimination of alcohol from the body is usually 10–15 mg/dL per hour, but the rate of elimination is usually higher if alcohol is consumed after eating a meal (15–20 mg/dL/h). Alcoholics may metabolize alcohol at a much higher rate, with an elimination rate of 25–35 mg/dL per hour due to activation of the CYP2E1 pathway. In general, it can be assumed that in moderate drinkers, the average elimination rate of alcohol is 15 mg/dL per hour [18].

There is controversy regarding the elimination kinetics of alcohol in men and women. Some studies have observed no significant difference in alcohol elimination kinetics between men and women, whereas others have documented differences. Faster elimination rates of alcohol are observed in women compared to men when rates are corrected for lean body mass. Dettling *et al.* reported that the elimination rate of alcohol was slightly higher in women (mean: 17.9 mg/dL per hour) than in men (15.5 mg/dL per hour). However, when corrected for the calculated liver weight, no statistically significant difference was found between elimination rates of alcohol observed in women and those observed in men [19]. Hormonal changes also play a role in the metabolism of alcohol in women, although this finding has been disputed in the medical literature. Some publications have indicated that women metabolize alcohol at a higher rate during the luteal phase of the menstrual cycle (19–22 days of the cycle), but a few days before menstruation, women's alcohol metabolism may be slow [20]. Alcohol may interfere with hormonal balance and reproductive function of women by modulating the neurohormonal axis as well as by altering hepatic metabolism of hormones. The menstrual cycle disturbances in alcoholic women are most prominent during the middle of the cycle [21].

Older people may metabolize alcohol more slowly than young people. Recommended alcohol intake for older individuals (>65 years old), regardless of gender, is one standard drink per day (see Chapter 1). Lean body mass decreases and adipose tissue increases with advancing age, resulting in a corresponding decrease in volume of total body water. Therefore, an older person may experience a higher blood alcohol level than a younger person with the same body weight after taking in the same amount of alcohol [22]. Metabolism of alcohol also changes with advanced age because the activity of enzymes involved in alcohol metabolism, including ADH, ALDH, and CYP2E1, diminishes with age. Moreover, older people may take prescription drugs and may experience adverse drug–alcohol interactions. In addition, some elderly people may suffer from liver diseases, and alcohol may exacerbate such diseases [23]. In Table 2.2 are listed various factors that affect alcohol metabolism.

Table 2.2 Factors that Affect Alcohol Metabolism

Factor	Comments
Alcohol consumed with food/empty stomach	Blood alcohol level is higher when alcohol is consumed on an empty stomach. Alcohol metabolism is higher in fed nutritional state compared to fasted state. Food increases liver blood flow. Moreover, sucrose present in food may increase alcohol metabolism by providing substrates that help to convert NADH to NAD because NAD is required as a cofactor for alcohol dehydrogenase.
How quickly alcohol is consumed	Sipping alcohol rather than drinking produces a lower blood alcohol level. If alcohol is consumed quickly, a higher blood alcohol level is observed rather than if alcohol is consumed slowly because metabolism of alcohol is initiated as soon as alcohol enters the circulation.
Body weight	Blood alcohol level is inversely proportional to body weight. However, the blood alcohol level may be higher in a woman with the same body weight as a man after drinking the same amount of alcohol due to different body water content between men and women.
Gender	For the same amount of alcohol consumed based on body weight, women may experience a higher blood alcohol level than men due to slower first-pass metabolism by the gut enzyme alcohol dehydrogenase.
Age	The activity of alcohol dehydrogenase may decrease with advanced age, and older individuals may metabolize alcohol slowly. Older people (>65 years) should not consume more than one drink per day.
Ethnicity	Polymorphism of genes encoding alcohol dehydrogenase and aldehyde dehydrogenase may significantly impact alcohol metabolism. For example, East Asians with the *ALDH2*2* allele cannot metabolize acetaldehyde properly due to defective aldehyde dehydrogenase enzyme; as a result, acetaldehyde may build up in blood, causing unpleasant reactions and thus deterring these people from drinking alcohol. Therefore, this genetic polymorphism has a protective effect on the habit of drinking and alcoholism.
History of alcohol abuse	Although metabolism of ethanol into acetaldehyde by CYP2E1 is a minor pathway in moderate drinkers, chronic abuse of alcohol usually induces this alternative pathway of alcohol metabolism; as a result, an alcoholic may metabolize alcohol at a much faster rate than a moderate drinker.

In addition, genetic polymorphism of both ADH and acetaldehyde dehydrogenase has been reported, and certain polymorphisms can significantly impact the metabolism of alcohol.

2.5 GENES ENCODING ALCOHOL DEHYDROGENASE

ADH is a zinc-containing enzyme that consists of two protein subunits (dimeric enzyme). ADH can oxidize endogenous ethanol produced by microorganisms in the gut as well as consumed ethanol. ADH has broad substrate specificity and can metabolize primary as well as secondary alcohol, such as methanol, propanol, and ethylene glycol. ADH is found mostly in cytosol and is present in highest amounts in the liver, followed by the gastrointestinal tract, kidneys, nasal mucosa, testes, and uterus [16].

In humans, seven genes encode ADH enzymes, and all these genes are clustered (*ADH7–ADH1C–ADH1B–ADH1A–ADH6–ADH4–ADH5*) in a small region of chromosome 4 in a head-to-tail array approximately 370 kb long (4q21–24). Based on similarities in their amino acid sequence and kinetic properties, ADH can be broadly categorized into five different classes with a total of seven ADH isoforms (usually enzyme terms are printed in regular font and gene terms encoding such enzymes are written in *italics*). The three class I genes—namely *ADH1A*, *ADH1B*, and *ADH1C*—are closely related to each other. These genes respectively encode alpha (α), beta (β), and gamma (γ) subunits, resulting in class I ADH enzymes that may be homodimer or heterodimer. These class I ADH isoforms are responsible for the majority of ethanol metabolized by the liver (almost 70% of total ethanol metabolizing capacity). The *ADH4* gene codes for the pi (π) subunit that produces a homodimer class II ADH enzyme in the liver and, to a lesser extent, the kidney. This enzyme has a higher Michaelis constant (in contrast to most ADH) and plays an important role in ethanol metabolism by the liver, especially at high blood alcohol concentration, accounting for approximately 30% of ethanol metabolism. The *ADH5* gene encodes the chi (χ) subunit, producing class III ADH, which is usually a homodimer. This enzyme is a ubiquitously expressed formaldehyde dehydrogenase with a very low affinity for ethanol. The major function of this enzyme is to oxidize formaldehyde to formic acid and to terminate nitric oxide signaling. The class III ADH is the only enzyme detected in the brain and plays a minor role in overall metabolism of ethanol. The class IV ADH is encoded by the *ADH7* gene, which produces the sigma (σ) subunit. This ADH enzyme is capable of converting retinol to retinal as well as ethanol to acetaldehyde. The *ADH6* gene produces mRNA, which can be detected in both fetal and adult liver, but the enzyme has not been isolated [24]. Various classes of ADH enzymes and the genes that encode them are listed in Table 2.3.

2.5.1 Polymorphism of Alcohol Dehydrogenase Genes

Most of the variants of *ADH* genes involve a single nucleotide polymorphism (SNP). Some of these variations occur in the part of the gene that is responsible for coding the protein; such SNPs are called "coding variations" that may result in the generation of ADH enzymes with altered activities. However, some variations may also occur in noncoding areas of the gene; such alterations are called "noncoding variations," but they may also affect the expression of the gene.

As mentioned previously, class I ADH enzymes encoded by *ADH1A*, *ADH1B*, and *ADH1C* genes play the most important role in metabolism of ethanol [25]. Although there are no reports of coding variation in the

Table 2.3 Isoforms of ADH Enzymes and Genes that Code for Such Enzymes

Class of ADH Enzyme	Gene Name	Enzyme (Dimeric Structure)	Comments
Class I	*ADH1A*	αα	Class I alcohol dehydrogenase enzymes are responsible for almost 70% of total ethanol oxidation capability.
	ADH1B	ββ	
	ADH1C	γγ	
Class II	*ADH4*	ππ	Class II ADH plays an important role in ethanol metabolism at high blood alcohol concentration and may contribute up to 30% of total ethanol oxidation capability.
Class III	*ADH5*	χχ	Class III ADH has a low affinity for ethanol and is the only enzyme detected in the brain.
Class IV	*ADH7*	σσ	Class IV ADH contributes to both ethanol and retinol oxidation. It is expressed in the upper digestive tract, and it oxidizes ethanol at higher concentration.
Class V	*ADH6*	Enzyme not characterized	The mRNA product is present in fetal and adult liver. The catalytic activity of this enzyme is unknown due to its labile nature.

ADH1A gene, SNPs in the *ADH1B* and *ADH1C* genes may encode ADH enzymes with different enzymatic activities. There are three different *ADH1B* alleles (including wild type) that alter the amino acid sequence of the encoded β subunit. There are also polymorphisms of the *ADH1C* gene that result in altered amino acid sequence in the γ subunit.

The reference allele for the *ADH1B* gene is *ADH1B*1* (wild type), which encodes the β_1 subunit of the ADH enzyme with an arginine at amino acid positions 48 (Arg48) and 370 (Arg370). This allele is the predominant allele observed worldwide among Caucasians, Native Americans, and people of African descent, but it is less common among East Asians. However, the common polymorphism *ADH1B*2*, which encodes the β_2 subunit, has histidine at position 48 (Arg48His, rs1229984) instead of arginine. This polymorphism is found commonly among East Asians (Han Chinese, Japanese, Koreans, Filipinos, Malays, etc. and aborigines of Australia and New Zealand) and also among approximately 25% of people of Jewish origin. This allele is encountered with a low frequency among Caucasians. The polymorphism *ADH1B*3*, which encodes the β_3 subunit, has cysteine at position 370 instead of arginine (Arg370Cys, rs2066702). This allele is found primarily in people of African descent and Native Americans. ADH isoenzymes encoded by *ADH1B*2* or *ADH1B*3* alleles are superactive enzymes (due to the significantly higher turnover rate of these enzymes), resulting in a 30- to 40-fold increase in metabolism of ethanol compared to normally functioning enzymes encoded by the wild-type *ADH1B*1* gene. As a result, acetaldehyde may build up in the blood,

Table 2.4 Major Alleles of *ADH1B* and *ADH1C* Genes and Their Association with Various Ethnic Groups

Allelic Variant	Enzyme (Dimeric Structure)	Association with Ethnic Groups
*ADH1B*1* (wild type)	$\beta_1\beta_1$	This is the most commonly observed allele, found in varying frequencies throughout the world population. It is common among Caucasians and people of Africa and African decent; it is less common among East Asians.
*ADH1B*2*	$\beta_2\beta_2$	This is most prevalent among East Asians, including Han Chinese, Japanese, Koreans, Filipinos, Malays, Mongolians, and aborigines of Australia and New Zealand. This allele is also found among approximately 25% of people of Jewish origin. There is an approximately 20% frequency among Middle Eastern people. It is also observed in smaller frequency among Caucasians and people of African origin.
*ADH1B*3*	$\beta_3\beta_3$	This is primarily found in people of African descent (up to 25%) and Native Americans.
*ADH1C*1* (wild type)	$\gamma_1\gamma_1$	This is a very common allele among Asians and people of Africa or African descent. It is also common ($\sim$50%) among Caucasians.
*ADH1C*2*	$\gamma_2\gamma_2$	This is commonly observed in Caucasians and is less common among Asians and people of African descent.

causing facial flushing and adverse side effects after alcohol use, thus discouraging people who carry such alleles from drinking [24].

The reference allele of the *ADH1C* gene is *ADH1C*1* (wild type), which encodes the γ_1 subunit with arginine at position 272 and isoleucine at position 350 (Arg272 Ile350). However, the enzyme encoded by two *ADH1C*2*-linked SNPs (Arg272Gln, rs1693482, and Ile350Val, rs698; these two SNPs occur together in almost all cases due to very high linkage disequilibrium) shows two amino acid exchanges. Moreover, enzymatic activity is approximately 2.5 times greater when the enzyme is encoded by the *ADH1C*1* reference allele than when it is encoded by the *ADH1C*2* haplotype (Gln272Val350). *ADH1C*1* is also in strong linkage disequilibrium with *ADH1B*2*. The *ADH1C*1* allele has been associated with alcohol-related cancer, especially in heavy drinkers [26]. Major alleles of *ADH1B* and *ADH1C* genes and their associations with various ethnic groups are summarized in Table 2.4. In addition to polymorphisms of *ADH1B* and *ADH1C* genes, polymorphisms of other *ADH* genes have also been described. Studies indicate a link between polymorphism of the *ADH4* gene and alcohol dependence. Polymorphisms of *ADH5* and *ADH7* genes have also been described.

2.6 GENES ENCODING ALDEHYDE DEHYDROGENASE

ALDH represents a group of NAD-dependent enzymes that metabolize a wide variety of aliphatic and aromatic aldehydes generated from both endogenous and exogenous precursors [27]. Unlike *ADH* genes, which are located on a single chromosome (chromosome 4), *ALDH* genes are not localized on a single chromosome. Humans have 19 genes and three pseudogenes in the *ALDH* gene superfamily encoding the ALDH superfamily of isoenzymes; mutation of these genes leading to defective aldehyde metabolism is the molecular basis of several diseases, including γ-hydroxybutyric aciduria, Sjögren–Larsson syndrome, and type II hyperprolinemia [28]. However, three of these genes—*ALDH1A1*, *ALDH1B1*, and *ALDH2*—are relevant to acetaldehyde metabolism. The ALDH1A enzyme is usually found in cytosol, whereas the ALDH1B1 and ALDH2 enzymes are produced in the nucleus but have leader sequences that direct them to mitochondria, where they exert their function [29].

The ALDH1A enzyme is encoded by the *ALDH1A1* gene, which is located on chromosome 9 (9q21.13) and extends over a 52-kb region. The ALDH2 enzyme is encoded by the *ALDH2* gene, which is located on chromosome 12 (12q24.2) and extends over 43 kb. Both genes have a similar structure with 13 exons, and the proteins they encode are 70% similar in sequence. The ALDH2 enzyme seems to play the most important role in the metabolism of acetaldehyde into acetate; the medication disulfiram (Antabuse), which is used in the treatment of alcohol detoxification, inhibits this key enzyme. In the ALDH2 enzyme encoded by the wild-type gene (*ALDH2*1*), glutamic acid is present at position 504 of the amino acid sequence. Polymorphism of the *ALDH2* gene has been described, and the *ALDH2*2* allele is relatively common in East Asians. This allele (Glu504Lys, rs671 G>A) encodes a defective ALDH2 enzyme in which lysine replaces glutamic acid at position 504 (this substitution corresponds to substitution at position 487 in the mature ALDH2 enzyme), and as a result, the enzyme is virtually inactive. Acetaldehyde levels increase in individuals carrying this allele, resulting in an unpleasant reaction from drinking alcohol. Therefore, this polymorphism may deter these individuals from drinking, thus protecting these individuals from alcohol abuse. This polymorphism may also increase the risk of both coronary heart disease and myocardial infarction among the Asian population [30].

When the ALDH2 enzyme is not functional due to the presence of the *ALDH2*2* allele, the ALDH1A enzyme present in liver cytosol plays an important role in the metabolism of acetaldehyde [31]. The effectiveness of the cytosolic ALDH1A enzyme encoded by the *ALDH1A1* gene in the metabolism of acetaldehyde has been demonstrated in a study in which decreased red blood cell ALDH1A enzyme activities in a small group of Caucasians and Asians were related to alcohol flushing reaction probably due to acetaldehyde buildup in blood. In addition, acetaldehyde clearance was considerably slower in Japanese individuals with extremely low red blood cell levels of the

ALDH1A enzyme and the *ALDH2*1/*2* genotype. The wild-type *ALDH1A1* gene is *ALDH1A1*1*, which is observed among the worldwide population. Two polymorphisms in the promoter region of the *ALDH1A1* gene have been described. The *ALDH1A1*2* allele is a 17-bp deletion (from position −416 to position −432), and the *ALDH1A1*3* allele is a 3-bp insertion (−524).

The ALDH1B enzyme, which is mitochondrial in nature, is encoded by the *ALDH1B1* gene located on chromosome 9. This gene does not contain introns in the coding sequence. There are four common missense *ALDH1B1* variants (rs2073478, rs2228093, rs4878199, and rs113083991). Analyzing the linkage disequilibrium between these mutations shows haplotypes encoding differing ALDH1B isoenzymes [32]. However, the effects of these variants on alcohol-related issues have been poorly investigated. Important allelic variations of the ALDH gene that are involved in the metabolism of acetaldehyde in humans are summarized in Table 2.5.

Table 2.5 Aldehyde Dehydrogenase Genes and Their Allelic Variations That Are Involved in Human Acetaldehyde Metabolism

Class of Aldehyde Dehydrogenase Enzyme	Gene	Allele	Enzyme Composition	Comments
I	*ALDH1A1*	*ALDH1A1*1* (wild type) *ALDH1A1*2* *ALDH1A1*3*	Tetramer	Cytosolic aldehyde dehydrogenase encoded by the *ALDH1A1*2* allele; it is present in low frequency among Asians, Caucasians, the Jewish population, and African Americans. However, the *ALDH1A1*3* allele is present only in African Americans.
II	*ALDH2*	*ALDH2*1* (wild type)	Tetramer	Mitochondrial aldehyde dehydrogenase. It is a major enzyme responsible for acetaldehyde metabolism; enzyme encoded by wild-type gene is found in various frequencies among the world population.
		*ALDH2*2*	Tetramer	Mitochondrial aldehyde dehydrogenase enzyme that is inactive. The allelic frequency of the *ALDH2*2* variant is approximately 50% among the Han Chinese, Taiwanese, and Japanese populations. It is also observed significantly among Koreans and Vietnamese and at a frequency of 1–10% in Tibetans, Mongolians, Thais, Malays, Filipinos, and Taiwanese aborigines. However, this allele is not found among the non-East Asian population.

2.6.1 Polymorphisms of Alcohol Dehydrogenase and Aldehyde Dehydrogenase Genes that Protect Against the Development of Alcohol Use Disorder

Although alcoholism is a complex behavioral disorder involving a myriad of gene–gene and gene–environment interactions, polymorphisms of genes, including the ALDH gene (*ALDH2*) and the ADH gene (*ADH1B*), have been proposed as particularly compelling genes associated with etiology of alcohol use disorders, especially protection from alcohol abuse [33]. In addition, other genes may play a significant role in precipitating alcoholism, including genes encoding GABA, glutamate receptor, dopamine receptors and transporters, serotonin receptors, and cholinergic receptors [34,35]. See Chapter 10 for more details.

Acetaldehyde buildup in the blood may deter a person from consuming alcohol because many unpleasant reactions from doing so, such as flushing (alcohol flush reaction or Asian flush), severe nausea, asthma attack, rapid heartbeat, and psychological distress, are related to increased blood levels of acetaldehyde. Therefore, genetic polymorphisms of ADH genes (*ADH1B*2* and *ADH1B*3*), as well as the wild-type *ADH1C*1* allele, that lead to highly active ADH enzymes result in rapid conversion of ethanol to acetaldehyde. As a result of acetaldehyde buildup in blood, individuals carrying such alleles may be deterred from drinking alcohol because acetaldehyde in blood is responsible for the adverse effects of alcohol, including facial flushing [26]. Similarly, genetic polymorphism of the *ALDH2* gene, such as the *ALDH2*2* allele that results in inactive ALDH2 enzyme, should also cause acetaldehyde buildup in blood due to slow removal of acetaldehyde and should deter a person from drinking alcohol. Therefore, it can be assumed that alleles that encode highly active ADH enzymes should have a protective effect against alcohol use disorders:

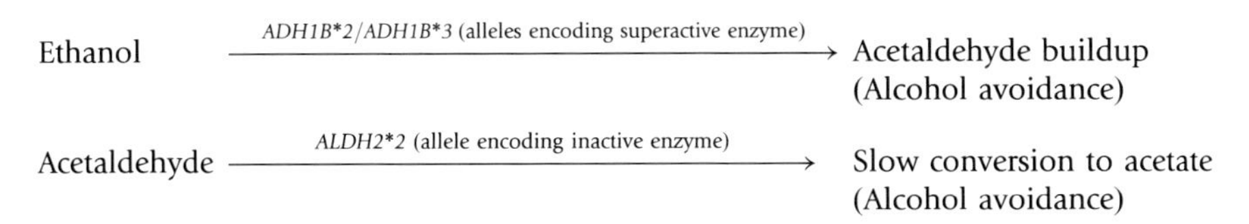

If ethanol is metabolized normally due to the presence of the wild-type allele (*ADH1B*1*) producing the ADH enzyme with normal activity, it may result in alcohol tolerance, thus increasing the risk of alcohol abuse. However, the *ADH1C*2* allele (Gln272Val350) that encodes the enzyme also has lower enzymatic activity than the enzyme coded by wild-type *ADH1C*1* (Arg272Ile350). Therefore, individuals carrying *ADH1B*1* and *ADH1C*2* may metabolize ethanol slowly with no acetaldehyde buildup in blood, and these individuals may be at risk of consuming high amounts of alcohol.

Similarly, if the ALDH enzyme has normal activity, it may also increase the risk of alcohol-related disorders because a person may not experience any unpleasant effects of alcohol due to no acetaldehyde buildup in the blood:

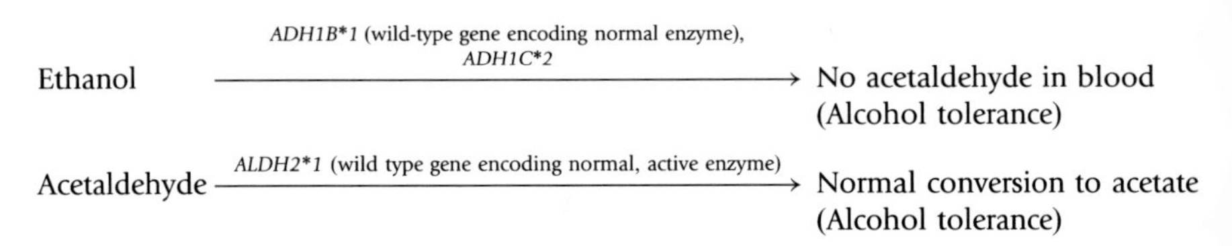

There is strong evidence that the variant allele *ALDH2*2* (rs671 G>A) resulting from a single nucleotide exchange causing substitution of lysine for glutamate at position 487 of the mature ALDH2 enzyme protects individuals from alcohol abuse. *ALDH2*2* is a common variant among 45% of East Asians, including Han Chinese, Japanese, and Koreans, but it is rare in other ethnic groups. It has been estimated that 540 million people worldwide (8% of the world population) carry this allele. *ALDH2*2* homozygotes (*ALDH2*2/*2*) exhibit essentially no ALDH2 enzymatic activity, but heterozygotes (*ALDH2*1/*2*) may show partial ALDH2 activity. East Asians deficient in ALDH2 enzyme activity exhibit the accumulation of acetaldehyde in blood, even after consumption of small amounts of alcohol, and may experience acetaldehyde-induced alcohol sensitivity. Genetic and epidemiological studies have shown that *ALDH2*2* homozygous individuals are almost completely protected from developing alcohol use disorders, but heterozygous individuals have partial protection (approximately 60%). Analysis of liver samples collected from surgical biopsies indicated that ALDH2 activity in liver was too low to estimate in individuals with genotype *ALDH2*2/*2* (homozygous individuals), but ALDH2 activity in heterozygous individuals (*ALDH2*1/*2*) was 17% of individuals with normal ALDH2 activity (*ALDH2*1/*1* genotype) [36]. In a study of 93 Asian subjects, Yamamoto *et al.* found that none of the subjects homozygous for the *ALDH2*1* allele showed a detectable blood acetaldehyde level or facial flushing, but all subjects homozygous and heterozygous for the *ALDH2*2* allele exhibited facial flushing and detectable blood acetaldehyde levels. As expected, blood acetaldehyde levels were significantly more elevated in *ALDH2*2* homozygous (*ALDH2*2/*2*) individuals than in heterozygous (*ALDH2*1/*2*) individuals [37]. Peng and Yin reported that in individuals with the homozygous *ALDH2*2/*2* genotype, peak blood acetaldehyde levels were 1.6- to 3.3-fold higher than those of individuals carrying the *ALDH2*1/*2* heterozygous after a low to moderate intake of alcohol. Interestingly, 130 min after consuming a small amount of alcohol, individuals carrying *ALDH2*2/*2* still showed detectable blood acetaldehyde levels due to almost complete loss of ALDH2 enzyme activity. The authors concluded that *ALDH2* polymorphism, which most commonly occurs in East Asians, is the strongest genetic modifier of drinking behavior and protects against alcoholism.

CASE REPORT 2.1

A 25-year-old man was found dead in his bedroom by his brother at approximately 6:30 PM on November 30. Empty snap-out sheets of flunitrazepam corresponding to 52 tablets were found in the trash at his bedside. The deceased had been beaten up at approximately 2:20 AM by approximately 10 boys and girls, all of whom were junior high school students. They also stole approximately 50,000 yen from him. He reported the incident to the police and returned home at approximately 9:10 AM. He had been attending a mental hospital every week and took flunitrazepam for depression. At medicolegal autopsy, only slight subdural hemorrhage was noticed, and none of the bruises observed was considered to be the cause of death. Using gas chromatography/mass spectrometry, no flunitrazepam or its metabolite was detected in body fluid. His blood alcohol concentration from the femoral vein was 2.0 mg/mL (200 mg/dL); usually, blood alcohol levels in fatalities range from 2.25 to 6.23 mg/mL. However, genotype analysis showed that the deceased was *ALDH2*1/*2* heterozygous, and such individuals are known to metabolize acetaldehyde slowly, thus showing a high concentration of acetaldehyde in blood despite a relatively low level of alcohol. Consequently, the cause of death was established as acute alcohol intoxication including acetaldehyde poisoning related to this genotype [39].

These individuals do not consume alcohol due to the negative symptoms that occur after drinking, which are attributable to acetaldehyde buildup secondary to nonfunctional ALDH2 enzyme [38].

It has been speculated that homozygotes of the *ALDH2*2* allele in the Asian population probably have almost 100% protection from alcohol abuse. Peng *et al.* reported that homozygotes (*ALDH2*2/*2*) were found to be strikingly responsive to small amounts of alcohol, as evidenced by pronounced cardiovascular hemodynamic effect as well as subjective perception of general discomfort as long as 2 hr following ingestion of alcohol [40]. In addition to the *ALDH2*2* allele, polymorphism in the ADH gene may also protect from alcoholism, especially if the *ADH1B*2* allele is present. In Asians, *ADH1B* has two common allelic forms—the wild-type *ADH1B*1* and the variant *ADH1B*2*. Interestingly, the *ADH1B*2* allele is found in almost 90% of the Asian population but in lower frequency in Caucasians (5–10%, but it may be found in higher percentages among eastern European Jews and Russians). This allele is associated with lower rates of alcohol dependence because the ADH enzyme encoded by this allele is superactive, resulting in rapid conversion of ethanol into acetaldehyde. This relationship was observed after taking into account the *ALDH2*2* allele in Asians. However, the combination of *ALDH2*2* and *ADH1B*2* alleles may provide additional protection from the development of alcohol use disorder. Luczak *et al.* reported that men and women with *ALDH2*1/*2* had greater pulse rate increases, greater observed flushing responses, and greater subjective feeling of being dizzy, drunk, and high compared with *ALDH2*1/*1* (wild-type) individuals, despite having an equivalent breath alcohol concentration. The authors concluded that the low risk of alcoholism based on possession of the *ALDH2*2* allele relates to greater response to alcohol-related adverse reaction in both men and women [41]. Whitfield reported that the protective effect of *ADH1B*2* in

Caucasians was approximately half that observed in Asians, suggesting that *ALDH2*2* and *ADH1B*2* each contribute unique protective effects against developing alcohol dependence [42].

The mechanism by which the *ADH1B*2* allele protects from alcohol abuse is related to faster conversion of alcohol into acetaldehyde by the ADH superactive enzyme encoded by this allele compared to normal activity of the ADH enzyme encoded by the wild-type gene. Therefore, individuals who possess the *ADH1B*2* allele may experience flushing, headache, and other alcohol-related adverse reactions due to acetaldehyde buildup. Moreover, an individual who possesses both *ADH1B*2* and *ALDH2*2* alleles may experience further buildup of acetaldehyde in blood due to faster conversion of alcohol into acetaldehyde by *ADH1B*2* but slower removal of acetaldehyde from blood due to the *ALDH2*2* allele. Based on a study of 784 Asian American students (Chinese and Korean descent), Luczak *et al.* reported that the effect of the *ADH1B*2* allele on alcohol-related hypersensitivity was experienced more strongly in Asians who already had heightened sensitivity to alcohol from possessing one *ALDH2*2* allele. As expected, individuals who possessed one *ALDH2*2* allele reported heightened initial response to alcohol even after consuming a low dose (one or two drinks) [43]. Although *ADH1B*2* is more prevalent in Asians and protects carriers of this allele from alcohol abuse, in Caucasians the *ADH1B*2* allele also provides protection from alcohol use disorders [44]. Toth *et al.* reported that Hungarians who carried the *ADH1B*2* allele showed reduced risk of developing alcoholism [45]. The *ADH1B*2* allele is also observed in higher frequency in the Jewish population compared to other Caucasian populations, and as expected, this allele is also protective against alcoholism in the Jewish population. Carr *et al.* noted that men in the *ADH1B*2* allele group reported more unpleasant reactions to alcohol compared to men who carried the wild-type *ADH1B*1* gene. The authors concluded that *ADH1B* gene polymorphism was associated with unpleasant reactions to alcohol, and carriers of the *ADH1B*2* allele consumed alcohol less frequently than men who were homozygous for *ADH1B*1* [46].

The presence of the *ADH1B*3* allele may lead to higher oxidative capacity of ethanol (i.e., more rapid ethanol oxidation to acetaldehyde). *ADH1B*3* is found almost exclusively in populations of African ancestry, and this allele has been shown to be associated with lower rates of alcohol dependence [47]. Studies have shown that up to one-third of participants in African American samples may have the *ADH1B*3* allele. One hypothesis to explain the protective effect of this allele from the development of alcohol use disorder is the faster conversion of alcohol into acetaldehyde in these individuals and the fact that the higher level of acetaldehyde may trigger stronger alcohol-related adverse symptoms. In a study of 91 African American young adults, McCarthy *et al.* observed that one-third of the participants had the *ADH1B*3* allele, and even individuals with one *ADH1B*3* allele reported experiencing greater sedation and increased pulse rate after consumption of alcohol compared to those homozygous for the *ADH1B*1* allele. The increased pulse rate was probably related to acetaldehyde burst. The authors concluded that lower rates of alcohol dependence in individuals carrying the *ADH1B*3* allele may be related to sedation and the adverse response to alcohol [48].

The population of Trinidad is composed mainly of people from East India (Indo-Trinidadians) and Africa (Afro-Trinidadians). Moore *et al.* reported that the *ADH1B*3* allele was associated with a reduced risk of alcoholism in Afro-Trinidadians, and a variant of the gene encoding *ADH1C* (i.e., *ADH1C*1*) provided protection from alcoholism in Indo-Trinidadians [49]. Interestingly, studies involving African American mothers have shown that the *ADH1B*3* allele in the mother may reduce the risk of alcohol-induced fetal/infant impairment. The higher alcohol elimination rate of ADH encoded by this allele probably has a protective effect against alcohol-induced developmental changes [50]. The *ADH1C*1* allele appears to have some protective effects against alcoholism in Asian populations. However, this allele is usually co-inherited with the *ADH1B*2* allele and may not have an independent effect on protecting an individual from alcohol abuse because the *ADH1C*1* allele is in linkage disequilibrium with the *ADH1B*2* allele [51]. Polymorphisms of ADH and ALDH genes that have a protective effect against the development of alcohol use disorder are listed in Table 2.6.

Table 2.6 Polymorphisms of Alcohol Dehydrogenase and Aldehyde Dehydrogenase Genes that Have a Protective Effect against the Development of Alcohol Use Disorder

Allele	Comments
*ALDH2*2*	This gene encodes almost inactive ALDH2 enzyme in homozygotes (*ALDH2*2/*2*) and significantly reduced enzyme activity in heterozygotes (*ALDH2*1/*2*). Asians carrying two alleles (homozygotes) have almost complete protection from developing alcoholism. Carriers of one allele also have significant protection from developing alcoholism.
*ADH1B*2*	This gene encodes ADH enzyme with much higher activity than normal ADH enzyme. Therefore, due to the accelerated metabolism of alcohol, individuals may experience a negative reaction after drinking due to acetaldehyde buildup. However, both *ALDH2*2* and *ADH1B*2* alleles may be present simultaneously in Asians, and such combination provides the highest protection from alcoholism.
*ADH1B*3*	This allele is found in people of African American descent, and it also provides some protection from developing alcoholism. This allele may also be present in Native Americans.
*ADH1C*1*	This appears to have a protective effect against alcoholism.

Although many studies have shown that individuals with *ALDH2*2*, *ADH1B*2*, and *ADH1C*3* alleles are protected from the development of alcohol abuse, a variant of the *ALDH1B1* gene (rs2228093; Ala69Val) was also associated with nondrinking and decreased total alcohol intake in a study based on a Caucasian population (1216 Danish men and women). However, more studies are needed to establish this initial observation [52]. Linneberg *et al.* reported an association of a variant of the *ALDH1B1* gene (rs2228093) with alcohol-induced hypersensitivity in a study based on a population residing in Copenhagen. The alcohol-induced hypersensitivity in this Caucasian population was genetically determined and was probably related to the histamine-releasing effect of acetaldehyde [53].

2.6.2 Polymorphism of Alcohol Dehydrogenase and Aldehyde Dehydrogenase Genes that may Increase the Risk of Developing Alcohol Use Disorder

Although the *ALDH2*2* allele has a protective effect against alcoholism (due to acetaldehyde buildup and negative reaction after consuming alcohol), the functional wild-type gene *ALDH2*1* may increase the risk of alcohol use disorder because no acetaldehyde buildup is associated with a functional allele. Similarly, the wild-type *ADH1B*1* allele may also increase the risk of alcohol use disorder because ethanol is metabolized normally to acetaldehyde with no acetaldehyde buildup in blood. Several studies have shown an association of *ADH1B*1* and *ALDH2*1* with alcoholism, especially in the Asian population. Kim *et al.* studied 1032 Korean subjects and concluded that alcohol dependence in 86.5% of alcoholic subjects can be attributed to *ADH1B*1* and/or *ALDH2*1* [54]. In a large study involving genotyping of 9080 Caucasian men and women from the general population, Tolstrup *et al.* observed that weekly alcohol intake on average was 9.8 drinks in men with the *ADH1B*1/*1* genotype compared to 7.5 drinks per week for men with the *ADH1B*1/*2* genotype. In addition, the odds of any daily heavy and excessive drinking were two to four times greater among men and women with *ADH1B*1* homozygotes compared to both men and women with *ADH1B*2* homozygotes or heterozygotes. Using the Brief Michigan Alcoholism Screening Test (Brief MAST), the authors observed that men with the *ADH1B*1/*1* genotype had a two- to fourfold increased risk of developing alcoholism compared to men with the *ADH1B*2/*1* or *ADH1B*2/*2* genotype. For *ADH1C*, the risk of heavy and excessive alcohol intake was 40–70% greater in men who were hetero- or homozygotes for the *ADH1C*2* allele (slow ethanol metabolism) compared to men who were homozygotes for the *ADHC*1* allele (fast ethanol metabolism). The authors further found that because the *ADH1B*1* allele is found in more than 90% of Caucasians and less than 10% of East Asians, the population attributable risk of heavy drinking and alcoholism by the *ADHB*1/*1* genotype was 67 and 62%, respectively, among Caucasians compared to 9 and 24%, respectively, among East Asians [55].

Associations between polymorphism of *ADH1B* and *ADH1C* genes and protection from alcohol use disorder or increased risk of alcohol use disorder have been extensively studied. In contrast, the potential association between *ADH1A* and alcohol dependence has been poorly studied because there is no report of coding variation in the *ADH1A* gene. However, synonymous variation (silent variation) may affect transcription level and translation efficacy of density of αα ADH. Zuo *et al.* reported that *ADH1A* variation may contribute to the genetic component of variation in personality trait and substance abuse disorder [56].

Polymorphism of the *ADH4* gene may be associated with alcohol and drug dependence in European Americans. Luo *et al.* genotyped seven SNPs spanning the *ADH4* gene in 365 healthy controls and 561 alcohol- and/or drug-dependent subjects and observed that SNP2 (rs1042363) at axon 9 or SNP6 (rs1800759) at the promoter region showed the greatest degree of Hardy–Weinberg disequilibrium with either alcohol dependence or drug dependence [57]. In addition, other rarely encountered polymorphisms of alcohol dehydrogenase genes may lead to alcohol dependence. In a study of 49,353 subjects, Zuo *et al.* observed that a rare variant constellation across the entire *ADH* gene cluster (*AD7–ADH1C–ADH1B–ADH1A–ADH6–ADH4–ADH5*) was associated with alcohol dependence in European Americans, European Australians, and African Americans. Association signal in this region came principally from *ADH6, ADH7, ADH1B*, and *ADH1C*; in particular, a rare *ADH6* variant constellation showed a replicable association with alcohol dependence across these three independent cohorts [58]. Van Beek *et al.* reported an association between age of onset of regular alcohol use and an SNP just upstream of *ADH7* (rs2654849) in the Dutch population. A significant association was also found between reactions to alcohol and polymorphism in *ADH5* (rs6827292) [59].

Liver cytosolic aldehyde dehydrogenase (ALDH1A) can also convert acetaldehyde into acetate, especially when mitochondrial enzyme (ALDH2) is inactive. Moreover, no individual has been identified with a total absence of catalytic activity of ALDH1A. In addition, ALDH1A is involved in the degradation of dopamine in the ventral tegmental area as well as the synthesis of retinoic acid. Sequence analysis of the promoter region of the gene encoding the ALDH1A enzyme indicated the presence of two polymorphisms: *ALDH1A1*2* (17-bp deletion from position −416 to −432) and *ALDH1A1*3* (3-bp insertion at −524). These polymorphisms may influence *ALDH1A1* gene expression; however, both polymorphisms are rarely encountered. The *ALDH1A1*2* allele is observed in low frequencies among Asians (0.035), Caucasians (0.023), the Jewish population (0.023), and the African American population (0.012). The *ALDH1A1*3* allele is found in low frequency among African Americans (0.029). Moore *et al.* reported that frequencies of *ALDH1A1*1* (wild type), *ALDH1A1*2*, and *ALDH1A1*3* in Afro-Trinidadians were 0.941, 0.035, and 0.024, respectively. However, the frequencies were 0.926, 0.074, and 0.000 among Indo-Trinidadians, indicating that the *ALDHA1*3* allele is found only among people of African descent. The authors also described a new polymorphism, *ALDH1A1*4* [60]. Based on a study of 463 southwest Californian Indians, Ehlers *et al.* reported that individuals with the *ALDH1A1*2* allele had lower rates of alcohol dependence [61]. In contrast, other studies have indicated that this allele may be associated with alcohol dependence. Moore *et al.* observed that *ALDH1A1*2* was associated with alcohol dependence in Indo-Trinidadians [49]. Spencer *et al.* commented that both *ALDH1A1*2* and *ALDH1A1*3* alleles produced a trend in an African American population that may be indicative of association with alcoholism, but more observations are required to validate this observation [62]. Liu *et al.* observed a modest association between alcohol dependence and the *ALDH1A1* gene (yin–yang haplotypes; i.e., haplotypes that have opposite allelic configuration) as well as the *ADH4* gene (SNP rs3762894) in Finnish Caucasians, African Americans, plains American Indians, and southwest American Indians [26]. Polymorphisms of ADH and ALDH genes that may increase the risk of developing alcohol use disorder are listed in Table 2.7.

Table 2.7 Polymorphisms of Alcohol Dehydrogenase and Aldehyde Dehydrogenase Genes that may Increase the Risk of Developing Alcohol Use Disorder[a]

Allele	Comments
*ADH1B*1*	Homozygote men with wild-type *ADH1B*1* allele have a two- to fourfold higher risk of alcoholism compared to men with *ADH1B*2/*1* or *ADH1B*2/*2* genotype.
*ADH1C*2*	Men with the *ADH1C*2* allele have a higher risk of heavy drinking than men with *ADH1C*1*, but the *ADHIC* genotype in general is not associated with alcoholism.
*ALDH2*1*	This wild-type gene leads to normal acetaldehyde metabolism and is associated with an increased risk of alcohol abuse because there is a lack of acetaldehyde buildup so that individuals do not experience the negative effects of drinking, such as alcohol flushing. Kim *et al.* studied 1032 Korean subjects and concluded that alcoholism in 86.5% of the alcoholic subjects could be attributed to *ADH1B*1* and/or the *ALDH2*1* allele [54].

[a] *There are polymorphisms of* ADH4, ADH6, *and* ADH7 *that may be associated with an increased risk of developing alcohol use disorder, but more studies are needed to establish their role in the development of this disorder.*

2.7 POLYMORPHISM OF THE *CYP2E1* GENE

With higher consumption of alcohol, especially in individuals with alcohol use disorder, the liver CYP2E1 enzyme is activated and contributes significantly to the oxidation of ethanol into acetaldehyde in chronic alcohol abusers. The CYP2E1 enzyme is encoded by the *CYP2E1* gene located on chromosome 10 (10q26.3). The effect of polymorphism of the *CYP2E1* gene on ethanol metabolism has also been studied. Although several *CYP2E1* polymorphisms have been reported, four polymorphisms—*CYP2E1**5B* (c2 allele), *CYP2E1**6* (C allele), *CYP2E1**1B* (A1 allele), and *CYP2E1**1D* (IC allele)—may have an association with alcoholism and related disorders. Carriers of the c2 allele (*CYP2E1**5B*) have often been found to have an increased risk for alcoholic liver disease, possibly due to an increased tendency to consume excessive amounts of alcohol. The C allele (*CYP2E1**6*) has been associated with predisposition of alcoholism in Japanese men. The A1 allele was found more commonly in alcoholics than in nonalcoholic individuals in a Mexican Indian population. Webb *et al.* observed an association of SNP rs10776687 that is in complete linkage disequilibrium with the c1 allele of *CYP2E1**5B* (rs2031920), indicating that this marker is associated with the level of response to alcohol [63]. However, some studies have found no association between *CYP2E1* gene polymorphism and alcohol use disorder. Cichoz-Lach *et al.* determined allele and genotype frequency of *ADH1B*, *ADH1C*, and *CYP2E1* in 204 alcohol-dependent Polish men and 172 healthy volunteers who did not drink alcohol (control). The authors observed no statistical difference between the distribution of the *CYP2E1* allele and genotype between alcohol-dependent subjects and controls [64].

2.8 CONCLUSIONS

Alcohol oxidation occurs mainly in the liver, catalyzed by cytosolic ADH and mitochondrial ALDH2 enzymes. Genetic polymorphism of the *ADH1B* gene results in genes that encode the more active form of the ADH enzyme compared to enzyme activity encoded by the wild type. As a result, some ADH isoenzymes may cause acetaldehyde buildup in the blood due to their superactivity, and individuals with these genetic makeups tend to avoid alcohol because many unpleasant reactions of alcohol are related to acetaldehyde accumulation in blood. However, Kang *et al.* reported that blood ethanol concentrations of the *ADH1B**2/*2* group were higher than those of the *ADH1B**1/*2* group, regardless of *ALDH2* genotype, and blood acetaldehyde concentrations of the *ADH1B**2/*2* group were also higher than those of *ADH1B**1/*2* only in the *ALDH2**1/*2* group. It is known that the *ADH1B**2* allele encodes the ADH enzyme, which is much more active than the normal

ADH enzyme encoded by the wild-type gene. Therefore, the finding that individuals with the *ADH1B*2/*2* genotype showed higher blood alcohol levels than individuals with the *ADH1B*1/*2* genotype is surprising and difficult to explain because a lower level of blood ethanol is expected in individuals carrying the *ADH1B*2* allele. Other investigators reported that *ADH1B* polymorphism did not affect peak blood ethanol level or AUC of ethanol, but *ALDH2* polymorphism had an effect on both blood ethanol and acetaldehyde level. Because this study was based on 24 subjects, the authors commented that larger follow-up studies are needed to validate their initial findings. Nevertheless, they concluded that higher blood ethanol and acetaldehyde levels in the *ADH1B*2/*2* genotype-carrying individuals may constitute the mechanism of protection against alcoholism by this genotype [65].

In general, it is assumed that carriers of *ADH1B*2* and *ADH1B*3* (which encodes superactive ADH isoenzymes) alleles are protected from alcohol use disorder because these individuals may refrain from drinking due to acetaldehyde buildup in blood following imbibition of alcoholic beverages. Acetaldehyde buildup in blood is associated with many unpleasant reactions following the consumption of alcohol. Similarly, the *ALDH2*2* allele results in an inactive ALDH2 enzyme, and individuals who are homozygotes (*ALDH2*2/*2*) (observed mostly in the East Asian population) are almost 100% protected from alcohol use disorders. Carriers of both *ALDH2*2* and *ADH1B*2* alleles are most protected from developing alcohol use disorders. However, alcohol use disorder is a complex disorder involving extensive interactions between genes and environment.

References

[1] Holford NH. Clinical pharmacokinetics of ethanol. Clin Pharmacokinet 1987;13:273–92.

[2] Jones AW, Jonsson KA. Food-induced lowering of blood ethanol profiles and increased rate of elimination immediately after a meal. J Forensic Sci 1994;39:1084–93.

[3] Jones AW, Jonsson KA, Kechagias S. Effect of high-fat, high-protein and high-carbohydrate meals on the pharmacokinetics of a small dose of alcohol. Br J Clin Pharmacol 1997;44:521–6.

[4] Lim Jr RT, Gentry RT, Ito D, Yokoyama H, et al. First-pass metabolism of ethanol is predominantly gastric. Alcohol Clin Exp Res 1993;17:1337–44.

[5] Ammon E, Schafer C, Hofmann U, Klotz U. Disposition and first-pass metabolism of ethanol in humans: is it gastric or hepatic and does it depends on gender? Clin Pharmacol Ther 1996;59:503–13.

[6] Marshall AW, Kingstone D, Boss M, Morgan MY. Ethanol elimination in males and females: relationship to menstrual cycle and body composition. Hepatology 1983;3:701–6.

[7] Lieber CS, Gentry RT, Baraona E. First-pass metabolism of ethanol. Alcohol Alcohol Suppl 1994;2:163–9.

[8] Pozzato G, Moretti M, Franzin F, Croce LS, et al. Ethanol metabolism and aging: the role of "first-pass metabolism" and gastric alcohol dehydrogenase activity. J Gerontol A Biol Sci Med Sci 1995;50:B135–41.

[9] Baraona E, Abittan CS, Dohmen K, Moretti M, et al. Gender difference in pharmacokinetics of alcohol. Alcohol Clin Exp Res 2001;25:502–7.

[10] Frezza M, di Padova C, Pozzato G, Terpin M, et al. High blood alcohol levels in women: the role of decreased gastric alcohol dehydrogenase activity and first-pass metabolism. N Engl J Med 1990;322:95–9.

[11] Farris JJ, Jones BM. Ethanol metabolism in male American Indians and whites. Alcohol Clin Exp Res 1978;2:77–81.

[12] Zakhari S. Overview: how is alcohol metabolized by the body? Alcohol Res Health 2006;29:245–54.

[13] Pisa PT, Loots D, Nienaber C. Alcohol metabolism and health hazards associated with alcohol abuse in a South African contest: a review. S Afr J Clin Nutr 2010;23(Suppl. 1):S4–10.

[14] Singh M, Gupta S, Singhal U, Pandey R, et al. Evaluation of the oxidative stress in chronic alcoholics. J Clin Diagn Res 2013;7:1568–71.

[15] Deng XS, Deitrich RA. Putative role of brain acetaldehyde in ethanol addiction. Curr Drug Abuse Rev 2008;1:3–8.

[16] Cederbaum AI. Alcohol metabolism. Clin Liver Dis 2012;16:667–85.

[17] Kobayashi M, Kanfer JN. Phosphatidyl ethanol formation via transphosphatidylation by rat brain synaptosomal phospholipase D. J Neurochem 1987;48:1597–603.

[18] Jones AW. Evidence-based survey of the literature of the elimination rates of ethanol from blood with applications in forensic casework. Forensic Sci Int 2010;200:1–20.

[19] Dettling A, Fischer F, Bohler S, Ulrichs F, et al. Ethanol elimination in men and women in consideration of the calculated liver weight. Alcohol 2007;41:415–20.

[20] Gill J. Women, alcohol and the menstrual cycle. Alcohol Alcohol 1977;32:435–41.

[21] Augustynska B, Ziolkowski M, Odrowaz-Sypmiewska G, Kielpinski A, et al. Menstrual cycle in women addicted to alcohol during the first week following drinking cessation: changes in sex hormone levels in relation to detected clinical features. Alcohol Alcohol 2007;42:80–3.

[22] Dufour MC, Archer L, Gordis E. Alcohol and elderly. Clin Geriatr Med 1992;8:127–41.

[23] Meier P, Seitz HK. Age, alcohol metabolism and liver disease. Curr Opin Clin Nutr Metab Care 2008;11:21–6.

[24] Edenberg HJ. The genetics of alcohol metabolism: role of alcohol dehydrogenase and aldehyde dehydrogenase variants. Alcohol Res Health 2007;30:5–13.

[25] Hurley TD, Edenberg HJ, Li TK. The pharmacogenomics of alcoholism. In: Licinio J, Wong ML, editors. Pharmacogenomics: The search for individualized therapeutics. Weinheim, Germany: Wiley–VCH; 2002. pp. 417–41.

[26] Liu J, Zhou Z, Hodgkinson CA, Yuan Q, et al. Haplotypes-based study of the association of alcohol metabolizing genes with alcohol dependence in four independent populations. Alcohol Clin Exp Res 2011;35:304–16.

[27] Vasiliou V, Pappa A. Polymorphisms of human aldehyde dehydrogenase. Pharmacology 2000;61:192–8.

[28] Jackson B, Brocker C, Thompson DC, Black W, et al. Update on the aldehyde dehydrogenase gene (ALDH) superfamily. Hum Genomics 2011;5:283–303.

[29] Rodriguez-Zavala JS, Weiner H. Structural aspects of aldehyde dehydrogenase that influence dimer–tetramer formation. Biochemistry 2002;41:8229–37.

[30] Xiao Q, Weiner H, Crabb D. Studies on the dominant negative effect of ALDH2*2 allele. Adv Exp Med Biol 1997;414:187–94.

[31] Hurley TD, Edenberg HJ. Genes coding enzymes involved in ethanol metabolism. Alcohol Res 2012;34:339–44.

[32] Way MJ. Computational modelling of ALDH1B1 tetramer formation and the effect of coding variants. Chem Biol Interact 2014;207:23.

[33] Li TK. Pharmacogenetics of responses to alcohol and genes that influence alcohol drinking. J Stud Alcohol 2000;61:5–12.

[34] Morozova TV, Goldman D, Mackay TF, Anholt RR. The genetic basis of alcoholism: multiple phenotypes, many genes, complex network. Genome Biol 2012;13:239.

[35] Rietschel M, Treutlein J. The genetics of alcohol dependence. Ann N Y Acad Sci 2013;1282:39–70.

[36] Lai CL, Yao CT, Chau GY, Yang LF, et al. Dominance of the inactive Asian variant over activity and protein content of mitochondrial aldehyde dehydrogenase 2 in human livers. Alcohol Clin Exp Res 2014;38:44–50.

[37] Yamamoto K, Ueno Y, Mizoi Y, Tatsuno Y. Genetic polymorphism of alcohol and aldehyde dehydrogenase and the effects on alcohol metabolism. Arukoru Kenkyuto Yakubutsu Ison 1993;28:13–25.

[38] Peng GS, Yin SJ. Effect of the allelic variants of aldehyde dehydrogenase ALDH2*2 and alcohol dehydrogenase ADH1B*2 on blood acetaldehyde concentrations. Hum Genomics 2009;3:121–7.

[39] Yamamoto H, Tanegashima A, Hosoe H, Fukunaga T. Fatal acute alcohol intoxication in an ALDH2 heterozygote: a case report. Forensic Sci Int 2000;112:201–7.

[40] Peng GS, Wang MF, Chen CY, Luu SU, et al. Involvement of acetaldehyde full protection against alcoholism by homozygosity of the variant allele of mitochondrial aldehyde dehydrogenase gene in Asians. Pharmacogenomics 1999;9:463–76.

[41] Luczak SE, Elvine-Kreis B, Shea SH, Carr LG, et al. Genetic risk of alcoholism related to level of response to alcohol in Asian-American men and women. J Stud Alcohol 2002;63:74–82.

[42] Whitfield B. Alcohol dehydrogenase and alcohol dependence: variation in genotype-associated risk between populations. Am J Hum Genet 2002;71:1247–50.

[43] Luczak SE, Pandika D, Shea SH, Eng MY, et al. ALDH2 and ADH1B interactions in retrospective reports of low-dose reactions and initial sensitivity to alcohol in Asian American college students. Alcohol Clin Exp Res 2011;35:1238–45.

[44] Wall TL, Shea SH, Luczak SE, Cook TA, et al. Genetic association of alcohol dehydrogenase with alcohol use disorders and endophenotypes in white college students. J Abnorm Psychol 2005;114:456–65.

[45] Toth R, Pocsai Z, Fiatal S, Szeles G, et al. ADH1B*2 allele is protective against alcoholism but not chronic liver disease in Hungarian population. Addiction 2010;105:891–6.

[46] Carr LG, Foroud T, Stewart T, Castelluccio P, et al. Influence of ADH1B polymorphism on alcohol use and its subjective effects in Jewish population. Am J Med Genet 2002;112:138–43.

[47] Wall TL, Carr LG, Ehlers CL. Protective association of genetic variation in alcohol dehydrogenase with alcohol dependence in Native American mission Indians. Am J Psychiatry 2003;160:41–6.

[48] McCarthy DM, Pedersen SL, Lobos EA, Todd RD, et al. ADH1B*3 and response to alcohol in African Americans. Alcohol Clin Exp Res 2010;34:1274–81.

[49] Moore S, Montane-Jaime LK, Carr LG, Ehlers CL. Variation in alcohol metabolizing enzymes in people of East Indian and African descent from Trinidad and Tobago. Alcohol Res Health 2007;30:28–30.

[50] Scott DM, Taylor RE. Health related effects of genetic variations of alcohol metabolizing enzymes in African Americans. Alcohol Res Health 2007;30:18–21.

[51] Choi OG, Son HG, Yang BH, Kim SH, et al. Scanning of genetic effects of alcohol metabolism gene (ADH1B and ADH1C) polymorphism on the risk of alcoholism. Hum Mutat 2005;26:224–34.

[52] Husemoen LL, Fenger M, Friedrich N, Tolstrup JS, et al. The association of ADH and ALDH gene variants with alcohol drinking habits and cardiovascular disease risk factors. Alcohol Clin Exp Res 2008;32:1984–91.

[53] Linneberg A, Gonzalez-Quintela A, Vidal C, Jorgensen T, et al. Genetic determinants of both ethanol and acetaldehyde metabolism influence hypersensitivity and drinking behavior among Scandinavians. Clin Exp Allergy 2010;40:123–30.

[54] Kim DJ, Choi IG, Park BL, Lee BC, et al. Major genetic components underlying alcoholism in Korean population. Hum Mol Genet 2008;17:854–8.

[55] Tolstrup JS, Nordestgaard BG, Rasmussen S, Tybjaerg-Hansen A, et al. Alcoholism and alcohol drinking habits predicted from alcohol dehydrogenase genes. Pharmacogenomics J 2008;8:220–7.

[56] Zuo L, Gelernter J, Kranzler HR, Stein MB, et al. ADH1A variation predisposes to personality traits and substance dependence. Am J Med Genet B Neuropsychiatr Genet 2010;153B:376–86.

[57] Luo X, Kranzler HR, Zuo L, Lappalainen J, et al. ADH4 gene variation is associated with alcohol dependence and drug dependence in European Americans: results from HWD tests and case–control association study. Neuropsychopharmacology 2006;31:1085–95.

[58] Zuo L, Zhang H, Malison RT, Li CS, et al. Rare ADH variant constellations are specific for alcohol dependence. Alcohol Alcohol 2013;48:9–14.

[59] Van Beek JH, Willemsen G, de Moor MH, Hottenga JJ, Boomsma DI. Association between ADH gene variants and alcohol phenotypes in Dutch adults. Twin Res Hum Genet 2010;13:30–42.

[60] Moore AM, Liang T, Graves TJ, McCall KM, et al. Identification of a novel cytosolic aldehyde dehydrogenase allele, ALDH1A1*4. Hum Genomics 2008;3:24–35.

[61] Ehlers CL, Spence JP, Wall TL, Gilder DA, et al. Association of ALDH1 promoter polymorphisms with alcohol-related phenotypes in southwest California Indians. Alcohol Clin Exp Res 2004;28:1481–6.

[62] Spencer JP, Liang T, Eriksson CJ, Taylor RE, et al. Evaluation of aldehyde dehydrogenase 1 promoter polymorphisms identified in human populations. Alcohol Clin Exp Res 2003;27:1389–94.

[63] Webb A, Lind PA, Kalmijn J, Feiler HS, et al. The investigation of CYP2E1 in relation to the level of response to alcohol through a combination of linkage and association analysis. Alcohol Clin Exp Res 2011;35:10–18.

[64] Cichoz-Lach H, Celinski K, Wojcierowski J, Slomka M, et al. Genetic polymorphism of alcohol metabolizing enzyme and alcohol dependence in Polish men. Braz J Med Biol Res 2010;43:257–61.

[65] Kang G, Bae KY, Kim SW, Kim J, et al. Effect of the allelic variant of alcohol dehydrogenase ADH1B*2 on ethanol metabolism. Alcohol Clin Exp Res 2014;38:1502–9.

CHAPTER 3

Measurement of Alcohol Levels in Body Fluids and Transdermal Alcohol Sensors

CONTENTS

3.1 Introduction .. 65
3.2 Breath Alcohol Determination 68
3.2.1 Chemical Principle of Breath Alcohol Analyzers 69
3.2.2 Effect of Breathing Pattern on Breath Alcohol Test Results ... 71
3.2.3 Interference in Various Breath Alcohol Analyzers 71
3.3 Blood Alcohol Determination 74
3.3.1 Enzymatic Alcohol Assays and Limitations 75
3.3.2 Gas Chromatography in Blood Alcohol Determination 78
3.3.3 Stability of Alcohol in Blood During Storage 79
3.3.4 Correlation between Blood and Breath Alcohol 80
3.4 Endogenous Production of Alcohol 81

3.1 INTRODUCTION

Alcohol levels are measured in various blood fluids, including blood, breath, urine, and saliva. In addition, transdermal alcohol sensors are used in the criminal justice system. Measurement of alcohol in various specimens is summarized in Table 3.1. Although blood alcohol measurement can be considered the most reliable test in legal situations, especially for prosecuting individuals who were impaired due to alcohol consumption at the time of the incident, breath alcohol is more commonly measured in such situations and a blood alcohol determination may or may not follow. Breath alcohol results can be converted into blood alcohol results using the known partition of alcohol between blood and alveolar air. If an evidentiary breath analyzer is used by a police officer, such evidence is admissible in a court of law. Emergency room physicians may use breath analyzers to determine alcohol levels in patients who appear intoxicated after motor vehicle accidents, or they may order a blood alcohol test, which could be determined in the clinical laboratory.

DWI stands for "driving while intoxicated" or "driving while impaired." A similar term is DUI, which stands for "driving under the influence." Both offenses may be related to alcohol and/or drug use (illegal drugs as well as some prescription medications may impair driving) and the terms may be used interchangeably depending on the state. However, in some states, the drunk driving laws are different for DUI and DWI: A DUI charge may indicate a lesser degree of intoxication and may carry a lesser penalty than DWI. Although impairment may also be drug related, alcohol is the major cause of DWI not only in the United States but also worldwide. Alcohol-related motor vehicle accidents kill approximately 17,000 Americans annually and are associated with more than $51 billion in total costs annually. There is a strong correlation between binge drinkers and alcohol-impaired drivers in the United States. In one study, it was found that overall, 84% of all alcohol-impaired

A. Dasgupta: Alcohol and Its Biomarkers. DOI: http://dx.doi.org/10.1016/B978-0-12-800339-8.00003-1

3.5 Urine Alcohol Determination 82
3.6 Saliva Alcohol Determination 84
3.7 Transdermal Alcohol Sensors .. 85
3.8 Conclusions .. 87
References 87

Table 3.1 Alcohol Testing of Various Specimens

Specimen	Analytical Method	Limitations/Comments
Blood (whole blood or serum)	Enzymatic assays are usually applicable for determination of serum/plasma ethanol level, whereas more specific gas chromatography or gas chromatography/mass spectrometric methods can be applied for analysis of either serum or whole blood ethanol level.	■ Blood specimen is the preferred specimen for legal alcohol determination. ■ Enzymatic assays suffer from interference if high amounts of lactate and lactate dehydrogenase are present in serum/plasma. ■ Propyl alcohol may also interfere. ■ Chromatographic methods are very specific and relatively free from interference.
Breath	Breath analyzers may use colorimetry, infrared spectroscopy, fuel cell technology, or mixed technology (infrared and fuel cell). Ignition lock devices also measure alcohol in exhaled breath and use similar techniques.	■ The breath alcohol test is the most commonly used alcohol test, but it is subject to interference (see Box 3.1). ■ In general, for the majority of individuals, breath analyzers underestimate blood alcohol level by approximately 15%. However, in smaller people, breath alcohol tests may overestimate blood alcohol due to small lung volume. ■ Hyperventilation prior to testing may lower the breath alcohol test result, whereas holding breath before testing may increase the reading.
Urine	Enzymatic assay used for serum ethanol determination may be modified to determine urine ethanol concentration, but gas chromatography or gas chromatography/mass spectrometric methods are more specific for urine alcohol determination.	■ Urine alcohol is determined less often than blood or breath alcohol. ■ Urine must be collected with a preservative such as sodium fluoride to prevent post collection ethanol production if sugar and yeast such as *Candida* are present in urine.
Saliva	Enzymatic assay can be used to measure ethanol levels in saliva. However, chromatographic methods are more specific.	■ Ethanol is less commonly measured in saliva. ■ QED device for measuring ethanol in saliva suffers from interference from propyl alcohol. Isopropyl alcohol may also cause interference, but this is much lower in magnitude than that due to propyl alcohol. ■ Alco-Screen is a saliva dipstick that can be used as a screening device, but for definitive diagnosis, a confirmatory test such as blood or breath alcohol must be conducted.
Perspiration/sweat	Transdermal alcohol sensors such as SCRAM are based on electrochemical determination of alcohol (fuel cell technology) that is excreted transdermally (~1% of total alcohol consumed).	■ Transdermal alcohol monitoring is a continuous alcohol monitoring process usually used to ensure abstinence from alcohol in the criminal justice system. ■ There is a lag time (33–53 min, depending on the amount of alcohol consumed) between the appearance of peak blood alcohol and peak transdermal alcohol.

Table 3.2 Legal Limit for Driving in Various Countries

Legal Limit	Countries
0.08% BAC	United States, Canada, Brazil, Chile, Ecuador, Mexico, New Zealand, Ireland, Malta, Singapore, Uganda, Zimbabwe
0.05% BAC	Austria, Belgium, Bulgaria, Costa Rica, Denmark, Finland, Greece, Hong Kong, Israel, Peru, Portugal, Serbia, Spain, Switzerland, Thailand, Turkey, Italy, Korea, Taiwan, Russia, Yugoslavia
0.04% BAC	Belarus, Lithuania
0.03% BAC	India, Japan, China, Moldova, Turkmenistan
0.02% BAC	China, Poland, Norway, Sweden, Estonia
0.01% BAC	Albania
Zero	Saudi Arabia, United Arab Emirates, Brazil, Bangladesh, Hungary, Czech Republic, Romania, Jordan, Bahrain, Mali, Pakistan, Saudi Arabia

BAC, blood alcohol content.

drivers were binge drinkers. Non-heavy drinkers are also involved in alcohol-related motor vehicle accidents [1].

Currently, in all states in the United States, the legal limit for driving is 0.08% alcohol in whole blood. Alcohol concentration is higher in serum than in whole blood; to calculate the whole blood concentration of alcohol, the measured serum concentration must be multiplied by a factor that is generally taken as 0.87. In Switzerland, Denmark, Italy, The Netherlands, Austria, Australia, China, Thailand, and Turkey, the upper acceptable limit for driving is an alcohol level in blood of 0.05%. In Japan, the upper acceptable limit is only 0.03%, and in countries such as various Middle Eastern countries, Hungary, Romania, and Georgia, there is a zero tolerance for blood alcohol in drivers. The legal limits for driving in some countries are listed in Table 3.2.

In general, alcohol is eliminated from blood following zero-order kinetics, and the average rate of elimination is 15 mg/dL per hour. Usually, ethanol in blood can be detected 6–8 hr after consumption, depending on the amount of alcohol consumed. For example, if initial blood alcohol is 0.1% (100 mg/dL), it will take approximately 6 hr for the blood alcohol level to be less than 20 mg/dL—the limit of quantification in many enzymatic alcohol assays. With very high blood alcohol, such as 0.25% (250 mg/dL), it will take 15 hr for the blood concentration to be reduced to 25 mg/dL, assuming the average elimination rate of 15 mg/dL per hr. However, such a high alcohol level is expected only in alcoholics, who usually have much higher rates of elimination due to activation of the CYP2E1 pathway (see Chapter 2 for details). Assuming a typical elimination rate of 25 mg/dL per hour in alcoholics, ethanol will be detected in blood approximately 9 hr after consumption. In general, the disappearance of ethanol from blood and breath follows a similar pattern. Jones *et al.* reported that the average disappearance rate of alcohol from venous blood was 15.2 mg/dL per hour, whereas the average rate of disappearance of alcohol from breath was equivalent to 16.3 mg/dL per hour of blood alcohol [2]. Pavlic *et al.* reported that the average rate of elimination of blood alcohol was 16.9 mg/dL per hour, whereas the average elimination rate of breath alcohol was 0.082 mg/L per hour measured by an Alcotest 7110MK III breath analyzer [3]. This elimination rate of breath alcohol is equivalent to 17.2 mg/dL per hour blood alcohol (breath alcohol value in milligrams per liter is multiplied by 210 to produce the equivalent blood alcohol value expressed as milligrams per deciliter). In general, alcohol can be detected for a slightly longer time in urine compared to blood.

3.2 BREATH ALCOHOL DETERMINATION

Breath analyzers have been reliably used to estimate blood alcohol concentration since the 1970s. Depending on the state, for prosecution of DWI, alcohol determined by an evidentiary breath analyzer may be admissible in the court of law. In general, blood alcohol determined by gas chromatography (GC) is subjected to very little interference, but breath alcohol may be affected by certain interfering substances. Nevertheless, evidentiary breath analyzers that are approved by the National Highway Traffic Safety Administration and used by law enforcement are reliable, and evidence is admissible in court. Moreover, interlock devices in vehicles also work on the principle of breath alcohol analysis.

A very small amount of alcohol is found in human breath. Only air in the deepest portion of the lung known as the alveolar sacs comes into contact with alcohol if present in blood, and there is equilibrium between alcohol in the exhaled air and alcohol in blood. The estimated ratio between breath alcohol and blood alcohol is 1:2100. This ratio is used in various breath alcohol analyzers to calculate blood alcohol level based on the concentration of ethanol in exhaled air by multiplying breath alcohol concentration expressed in milligrams per liter by 2100 to obtain the ethanol level per liter of blood (or by multiplying by 210 to obtain blood alcohol expressed as milligrams per deciliter). Some breath analyzer software may automatically calculate blood alcohol values from the observed breath alcohol level. This process of equilibrium of alcohol between alveolar air and blood is based on Henry's law, which states that the ratio between alcohol in blood and alcohol in deep lung air is constant. However, blood alcohol measurement is a direct measurement, and there are established guidelines for assessing the degree of impairment of individuals based on blood alcohol level and drinking history.

3.2.1 Chemical Principle of Breath Alcohol Analyzers

There are four types of breath analyzer:

- Analyzers that use color change due to a chemical reaction to determine alcohol level
- Analyzers based on infrared spectroscopy
- Analyzers based on fuel cell technology
- Analyzers based on mixed technology (infrared and fuel cell) and other techniques.

The earliest developed breath analyzer was based on a chemical principle in which exhaled air passed through a cocktail of chemicals containing sulfuric acid, potassium dichromate, silver nitrate, and water. Silver nitrate catalyzes the reaction in which alcohol, in the presence of sulfuric acid, turns orange potassium dichromate solution green due to conversion of potassium dichromate into chromium sulfate. The intensity of the green color can be used to estimate the amount of alcohol in the exhaled air. Captain Robert Borkenstein of the Indiana State Police used this chemical principle to develop breath analyzers in 1954, and some breath alcohol analyzers still use this principle. The Breathalyzer is the oldest breath alcohol analyzer; it is based on the principle of color change of potassium dichromate solution in the presence of alcohol and then analysis using spectroscopy after a specified time to ensure complete reaction. The analyzer contains two vials of chemical cocktail. After a subject exhales into the device, the air is passed through one vial; if alcohol is present in the exhale, a color change occurs. A system of photocells is connected to a meter to measure color change associated with the chemical reaction by comparing the response from the second vial (through which no air is passed), thus producing an electrical signal proportional to color change in the reaction vial. This electrical signal can cause the meter indicator to move (more alcohol, more signal and higher reading), and the alcohol level in subjects can be thus determined. Breathalyzer was the brand name originally developed and marketed by Smith & Wesson, which sold the brand to the German company Draeger. The Breathalyzer 900 model was replaced by newer versions, such as model 1100, but this technology is subject to interference from a variety of substances; as such, other companies have focused on developing more robust technology for breath alcohol analysis.

There are many evidentiary breath alcohol analyzers that are based on the principle of infrared (IR) spectroscopy for quantitative determination of alcohol in exhaled air. The Intoxilyzer was originally developed by Omicron (Palo Alto, CA) and later sold to CMI (Owensboro, KY). The earlier models were 4011 A, 4011 S, and the Intoxilyzer 5000, which is used as an evidentiary breath alcohol analyzer; an Intoxilyzer 8000 model is now commercially available. Many states use this analyzer as the evidentiary breath analyzer. In addition to the Intoxilyzer, DataMaster cdm (National Patent Analytical System, Mansfield, OH), which is also used in many states as the evidentiary breath alcohol analyzer, is based on IR spectroscopy technology. The Intoxilyzer 5000 uses a five-wavelength filter at 3.36, 3.4, 3.47, 3.52, and 3.8 μm and thus can differentiate between ethanol and common sources of interference in exhaled air, such as acetone, acetaldehyde, and toluene. The 3.4-μm wavelength is used to detect alcohol, the 3.47-μm wavelength identifies interfering substances, and the 3.9-μm wavelength is used as the reference wavelength. The Intoxilyzer 8000 uses a pulsed IR source instead of moving wavelength filter and also uses dual wavelength for measuring alcohol in breath (3.4 and 9.36 μm). It also has more advanced computer technology to provide more accurate alcohol level results. DataMaster cdm, an evidentiary breath analyzer widely used by police officers in many states, is also based on the principle of IR spectra, where alcohol is detected using two different wavelengths (3.37 and 3.44 μm).

Several different brands of evidentiary breath alcohol analyzers are based on the principle of fuel cell technology, including Alcotest models 6510, 6810, and 7410 (National Drager, Durango, CO) and Alco-Sensor III and IV (Intoximeters, St. Louis, MO), which are evidentiary breath analyzers. The fuel cell is a porous disk coated on both sides with platinum oxide (also called platinum black). The porous layer is impregnated with acidic solution containing various electrolytes so that charged particles such as hydrogen ions can travel through the medium. In addition, both sides of the disk containing platinum oxide are connected with a platinum wire. The fuel cell is mounted in a case along with the entire assembly so that when a person blows into the disposable mouthpiece, the air travels through the fuel cell. If any alcohol is present in the exhaled air, the alcohol is converted into acetic acid, hydrogen ions, and electrons on the top surface by the platinum oxide. Then the hydrogen ions travel to the bottom surface (which also contains platinum oxide) and water is formed by their combining with oxygen present in the air. In this process, electrons are removed from the platinum oxide. Because there is an electron excess on the top surface and an electron deficit on the bottom surface, electrons flow from one surface to another, generating an electric current that flows through the platinum wire. The intensity of the current is proportional to the amount of alcohol present in the exhaled air. The instrument's microprocessor then converts the current to equivalent blood alcohol.

Some evidentiary breath analyzers are based on both fuel cell and IR spectroscopy technology, which gives them good sensitivity and specificity for analyzing alcohol on breath. Various models of the Intox EC/IR I desktop evidentiary alcohol breath analyzers (Intoximeters) combine reliable fuel cell analysis with the real-time analytical advantages of IR technology. Semiconductor alcohol sensors are used in inexpensive breath analyzers marketed to the general public. However, sensor response is nonspecific to alcohol and nonlinear in response. For example, semiconductor sensors will respond to particles and gases present in cigarette smoke. The gas chromatography technique (discussed later) applied for analysis of blood alcohol may also be used for breath alcohol analysis [4].

3.2.2 Effect of Breathing Pattern on Breath Alcohol Test Results

The breath alcohol test is a single exhalation maneuver in which a subject is asked to inhale air (preferably a full inhalation to total lung capacity) and then exhale (preferably a full exhalation to residual volume) into the breath analyzer instrument. The assumption is that alcohol concentration in exhaled breath is equal to that in alveolar air. Very few restrictions, such as exhaled volume, exhaled flow rate, inhaled volume, and pretest breathing pattern, are placed on the breathing maneuver. In general, the ethanol level in end-exhaled air is always lower than that in alveolar air. When performing a breath alcohol test, the subject is asked to inhale ambient air and exhale into the breath analyzer (usually 1.1–1.5 L of exhaled air is required for the test). Therefore, smaller subjects with smaller lung capacity must exhale a greater fraction of air in their lungs to fulfill the minimum volume requirement of the analyzer; as a result, the alcohol breath test could overestimate the blood alcohol level in smaller subjects compared to larger subjects with larger lung capacity. In addition, in normal circumstances, a single exhalation alcohol breath test shows a gradual and continually increasing breath alcohol level [5]. Breathing pattern may also affect breath alcohol test results because values may decrease by 11% in the case of pretest hyperventilation and may increase by 15% in the case of pretest breath hold. However, other investigators have reported an average decrease of 4.4% due to hyperventilation and an average increase of 6.7% due to breath hold (relative hypoventilation). George *et al.* used a mathematical model and reported that hyperventilation may cause an average 4.4% decrease in the breath alcohol test result, whereas hypoventilation may increase the value by 3.7%. Inhaling hot humid air may decrease the value by 2.9%, whereas inhaling hot dry air, cold humid air, or cold dry air has minimal effects [6].

3.2.3 Interference in Various Breath Alcohol Analyzers

Kechagias *et al.* compared blood alcohol values with values obtained by breath alcohol analyzer (DataMaster) in patients with gastroesophageal reflux disease and concluded that it is highly improbable that breath alcohol analyzers overestimate true blood alcohol values due to eruption of alcohol from the stomach to the mouth caused by gastric reflux [7]. Sometimes a driver stopped by police may use mouthwash to hide alcoholic breath. Because some mouthwashes contain alcohol, use of a mouthwash prior to taking a breath alcohol analysis may cause falsely elevated breath alcohol results. However, residual alcohol evaporates from the mouth rapidly; this is the reason why there is a 15-min waiting period in police stations that is supervised so that suspects cannot take anything by mouth during this period. Fessler *et al.* studied the effect of alcohol-based substances, such as mouthwash, cough mixture, and breath spray, just prior to breath alcohol measurement using the Drager evidentiary portable breath alcohol analyzer on 25 volunteers. The authors concluded that a 15-min waiting period was necessary to ensure that there was no residual alcohol in the mouth after using mouthwash and other alcohol-containing products. Otherwise, alcohol from mouthwash may interfere with breath alcohol analysis, causing falsely elevated values [8]. Harding *et al.* studied the effect of dentures and denture adhesives on mouth alcohol retention using the Intoxilyzer 5000 and concluded that dentures had no significant effect on breath alcohol test results as long as a waiting period of 20 min was observed prior to testing [9]. Logan *et al.* evaluated the effect of asthma inhalers and nasal decongestant sprays on breath alcohol tests and observed that the only product that had any effect on breath alcohol tests was Primatene Mist containing 34% ethyl alcohol, but alcohol was eliminated from the breath within 5 min. The authors concluded that inclusion of a 15-min deprivation period during which no food or drink could be consumed prior to an evidential breath test was an adequate safeguard against interference in the test caused by alcohol-containing inhalers [10].

Consuming energy drinks while driving is legal, but some energy drinks contain very low levels of alcohol. When volunteers drank various energy drinks, 11 of 27 drinks gave positive results using evidentiary breath analyzers when testing was done just after consumption. However, after a 15-min waiting period, all breath alcohol analysis reports were negative. The authors concluded that a 15-min waiting period eliminates the possibility of testing false positive after consuming an energy drink with low alcohol content [11]. Laakso *et al.* studied the effect of various volatile solvents for potential interference with breath alcohol analysis using the Drager 7110 evidentiary breath analyzer. They concluded that acetone, methyl ethyl ketone, methyl isobutyl ketone, ethyl acetate, and diethyl ether did not interfere with breath alcohol measurement significantly, but propyl alcohol and isopropyl alcohol had a significant effect on breath alcohol measurement [12]. Jones and Rossner

BOX 3.1 INTERFERENCE IN BREATH ALCOHOL TEST

- Mouthwash containing alcohol: Wait for at least 15 min prior to taking breath analyzer test.
- Some energy drinks may contain alcohol: Wait for at least 15 min prior to taking breath analyzer test.
- Propyl alcohol.
- Isopropyl alcohol.
- Ketogenic diet leads to a stage called ketonemia, in which concentrations of acetone, acetoacetic acid, and β-hydroxybutyric acid are high. Although acetone does not interfere with breath analyzer tests, it is known to be converted into isopropyl alcohol by the action of liver alcohol dehydrogenase, and isopropyl alcohol can interfere with breath analyzer tests.
- Glue sniffing may cause false-positive test results because glue contains hydrocarbons, ethyl acetate, and toluene.

described a case in which a 59-year-old man undergoing a weight loss program using a ketogenic diet attempted to drive a car that was fitted with an alcohol ignition interlock device, the vehicle proving impossible to start. Because he had completely stopped drinking, he was surprised and upset. The ketogenic diet used for treating obesity and controlling seizures in some epileptic children is high in fat, very low in carbohydrates, and also has adequate protein. The goal is to burn fat to get energy rather than getting it from glucose, which is formed by carbohydrate metabolism. However, consuming the ketogenic diet led to a stage called ketonemia, in which concentrations of acetone, acetoacetic acid, and β-hydroxybutyric acid are high. The high levels of acetone lead to its presence in exhaled air. The interlock device in the car determines alcohol by an electrochemical oxidation method, and acetone does not interfere with the process. However, acetone is known to be converted into isopropyl alcohol by the action of liver alcohol dehydrogenase, and isopropyl alcohol can be falsely identified as ethanol by the ignition interlock device. In addition, methanol and propanol can also be falsely identified as alcohol. The authors concluded that the side effects of ketogenic diets need further evaluation by authorities, especially for people involved in safety-sensitive positions such as airline pilots and bus drivers who are subjected to much tougher alcohol tolerance policies [13]. Glue sniffing may cause false-positive breath alcohol test results because glue contains aliphatic hydrocarbons, ethyl acetate, and toluene [14]. Methanol poisoning is dangerous because it may cause death or blindness. Methanol poisoning may cause false-positive test results with breath analyzers. In one report, the authors observed that toluene, xylene, methanol, and isopropyl alcohol in exhaled air were mistakenly identified as breath alcohol by the Intoxilyzer 5000 evidentiary breath alcohol analyzer [15]. Common sources of interference in breath alcohol analyzers are summarized in Box 3.1.

CASE REPORT 3.1

A 47-year-old man who was found at a public park acting in an intoxicated manner was given a breath analyzer test at a police station using the Intoxilyzer 5000 EN, which showed a concentration corresponding to 288 mg/dL blood alcohol. The subject admitted that he was suicidal and was transported to a hospital. In the emergency room, the patient admitted drinking gas line antifreeze, which contains 99% methanol. The serum drug screen for alcohol showed a negative result, indicating that the positive ethanol level recorded by the breath analyzer was false positive due to interference of methanol. As expected, the patient's serum methanol level was 589 mg/dL, but the serum ethylene glycol level was less than 5 mg/dL. The patient was initially treated with fomepizole and then with ethanol infusion to reduce metabolism of methanol to toxic formaldehyde metabolite. Hemodialysis was initiated 9 hr after ingestion, and 44 hr after ingestion his serum methanol level was reduced to 22 mg/dL. The visual activity of the patient was not affected by methanol intoxication. Finally, he was transferred to a psychiatric facility [16].

3.3 BLOOD ALCOHOL DETERMINATION

In general, whole blood alcohol level is determined using GC in a forensic laboratory, whereas serum or plasma alcohol is determined in a hospital laboratory either by an enzymatic method or by GC. Because the legal limit for driving a vehicle in the United States is 0.08% (80 mg/dL) of whole blood, if serum alcohol readings are used as evidence, the results must be converted into those for whole blood alcohol. In a study of 212 consecutive patients admitted to a hospital trauma center, serum was analyzed for ethanol using an enzymatic method, whereas whole blood was analyzed using GC in a forensic toxicology laboratory. The authors observed that the serum to whole blood alcohol ratio was dependent on alcohol concentration, but the values ranged from 1.12 to 1.18. However, using a linear regression model, adequate predictions of whole blood alcohol can be made based on the serum alcohol level. For example, at a 95% confidence level, a serum alcohol concentration of 103 mg/dL corresponds to a whole blood alcohol of 80 mg/dL [17]. Rainey reported that the ratio between serum and whole blood alcohol ranged from 0.88 to 1.59, but the median value was 1.15. Therefore, dividing the serum alcohol value by 1.15 provides the whole blood alcohol concentration, and multiplying the serum alcohol reading by 0.87 should also provide the whole blood alcohol concentration [18]. The molecular weight of ethanol is 46. Therefore, if ethanol concentration is expressed as millimoles per liter, then the value should be multiplied by 4.6 to obtain the ethanol concentration in milligrams per deciliter, a unit more commonly used in the United States. For example, the ethanol level of 17.4 mmol/L is equivalent to 80 mg/dL.

3.3.1 Enzymatic Alcohol Assays and Limitations

In hospital laboratories, ethyl alcohol is analyzed using enzymatic methods and automated analyzers. Several types of automated analyzer are available from various diagnostics companies that are capable of analyzing alcohol in serum or plasma. Enzyme-based automated methods are generally not applicable for analysis of whole blood, although modified methods are available for analysis of alcohol in urine specimens. Enzymatic assay of alcohol is based on the principle of conversion of alcohol to acetaldehyde by alcohol dehydrogenase, and in this process NAD is converted into NADH:

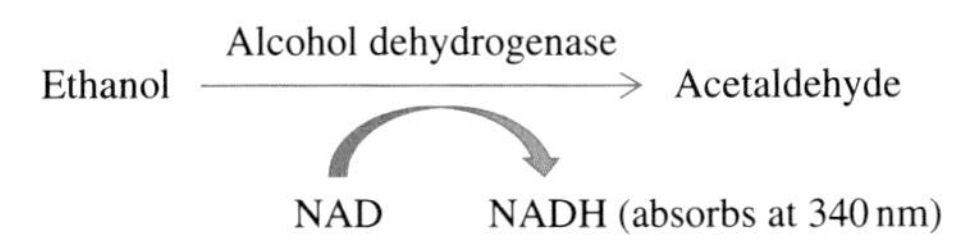

Whereas NAD has no absorption of ultraviolet light at 340-nm wavelength, NADH absorbs at 340 nm. Therefore, an absorption peak is observed when alcohol is converted into acetaldehyde due to simultaneous conversion of NAD into NADH. The intensity of the peak is proportional to the amount of alcohol present in the specimen. If no alcohol is present, no peak is observed. Usually, methanol, isopropyl alcohol, ethylene glycol, and acetone have negligible effects on alcohol determination using enzymatic methods, but propanol, if present, may elevate the true alcohol value by 15–20%. Whereas isopropyl alcohol, which is used as rubbing alcohol, is commonly used in households, propanol is used much less frequently in household products. Instead of ultraviolet detection, other detection methods may be used in enzymatic assays of alcohol based on alcohol dehydrogenase. For example, in the assay design, in addition to alcohol dehydrogenase, the enzyme diaphorase can be used to generate a color change in a dye. The method was originally designed for the Abbott TDx/FLx platforms and is currently available on the Abbott AxSYM. Radiative energy attenuation measures the degree of inhibition of the fluorescence of fluorescein dye resulting from the production of a colored product.

In the enzymatic alcohol method, major interference is caused by lactate dehydrogenase (LDH) and lactate, which may result in false-positive alcohol levels in patients with lactic acidosis. In addition, enzymatic alcohol assay is unsuitable for determination of alcohol in postmortem blood because high concentrations of LDH and lactate are present in postmortem blood. Therefore, only GC can be used to measure alcohol in postmortem specimens. In one report, the authors observed 690 mg/dL (0.69%) of alcohol in serum using an enzymatic method for alcohol in a patient, but GC did not show any alcohol in the same serum specimen [19]. This patient had end-stage renal disease and received a kidney transplant; at the time blood was drawn, she had severe metabolic acidosis and was admitted to hospital. The LDH concentration was 27,000 U/L, and the lactate concentration, 15.0 mmol/L. However, the authors observed no apparent alcohol level in any specimen containing normal levels of LDH and lactate [19]. Badcock and O'Reilly reported false-positive ethanol levels in postmortem plasma of infants who had died from sudden infant death syndrome, using the enzyme multiplied immunoassay technique (EMIT; Syva, San Jose, CA) for alcohol assay, which utilizes alcohol dehydrogenase. The authors suspected that elevated LDH and lactate were causing interference and established that an LDH concentration of 2800 U/L or greater and a lactate concentration of 20 mmol/L or greater were needed to obtain a false-positive alcohol result. The mean plasma LDH value in sera of infants with falsely elevated alcohol by the EMIT assay was 6430 U/L, whereas the mean plasma lactate level was 91 mmol/L, thus explaining the observed interference [20]. Nine *et al.* noted that elevated serum LDH and lactate can cause varying degrees of false-positive ethyl alcohol results in three enzymatic immunoassays (Syva EMIT, Abbott, and Roche alcohol assay) [21].

Lactate concentrations also tend to increase in trauma patients. Dunne *et al.* reported that 27% (3536) of 13,102 patients had positive alcohol screen (mean alcohol, 141 mg/dL; range, 10–508 mg/dL) [22]. In contrast, Winek *et al.* compared the alcohol concentration obtained by an enzyme assay (Dimension Analyzer, Siemens Diagnostics, Deerfield, IL) and by GC in trauma patients and observed no false-positive results using the enzyme assay. Alcohol concentrations obtained by the enzyme assay correlated well with GC values, and only in 6 specimens (out of 27) did the differences exceed 10%, with the highest difference being 22%. The authors concluded that the enzyme method can be used in hospital laboratories for the determination of alcohol concentrations in trauma patients [23]. Contrasting results may be related to differences in lactate concentrations of patients in two different studies. Powers and Dean described a case in which the result of a hospital alcohol dehydrogenase-based ethanol determination was challenged in court on the basis of potential LDH and lactate interference. Hepatic trauma was suggested as the cause of elevated LDH and lactate based on elevated serum liver enzyme levels. The authors evaluated the clinical laboratory test results—including alanine aminotransferase (ALT; 144 U/L), aspartate aminotransferase (AST; 229 U/L), sodium, potassium, chloride, and carbon dioxide levels—and observed an anion gap of 8 mEq/L (normal range, 8–15 mEq/L). Elevated serum lactate contributes to anion gap causing elevated levels. Therefore, serum lactate was not significantly elevated in this case due to the normal anion gap observed in this individual. Based on slightly elevated ALT and AST values, the authors concluded that the LDH concentration could not be elevated more than a maximum of 2000 U/L. Therefore, the contribution of lactate oxidation to a falsely elevated ethanol level should be negligible, and the observed serum ethanol level of 200 mg/dL could be in reality only 199 mg/dL [24].

Although elevated lactate and LDH are the major cause of interference in the enzymatic alcohol method, other interference has also been reported. Gharapetian *et al.* observed false positive ethanol levels in three patients with acetaminophen-induced hepatocellular necrosis. The authors used the ADVIA 1650 analyzer and the ethanol assay system manufactured by the same company (previously Bayer Diagnostics, now Siemens Diagnostics). However, when ethanol was measured by GC or the Dade Behring Dimension Flex alcohol cartridge, no ethanol was detected in these specimens. The false-positive ethanol results in three patients, using the ADVIA 1650 analyzer, were 151.8, 151.8, and 156.5 mg/dL (33, 33, and 34 mmol/L, respectively). The lactate levels were 22.5, 1.2, and 5.1 mmol/L, and LDH values were 8075, 1871, and 10,147 U/L, respectively. However, according to the manufacturer, the LDH value should be greater than 100,000 U/L and the lactate value should be greater than the normal reference range in order to observe a spurious ethanol level greater than 80 mg/dL. The authors speculated that endogenous dehydrogenases and substrates other than LDH and lactate may be implicated in NADH production causing false-positive test results in these three patients with the ADVIA 1650 analyzer [25]. Nonspecific interference in enzymatic alcohol assay has also been reported.

CASE REPORT 3.2

An 85-year-old woman presented in the emergency room with a history of unconscious collapse late on Christmas Day. Her Glasgow Come Scale score was 3, and she was intubated and transferred to the intensive care unit. Using the Olympus analyzer, her blood ethanol level was 200 mg/dL at the emergency room and 1 hr later had increased to 400 mg/dL, but it declined to 100 mg/dL after 8 hr. The patient improved over 48 hr and was extubated followed by transfer to a ward. She went home on day 18. The family was deeply offended when told that ethanol was the cause of her collapse and insisted that the woman did not consume any alcohol. Therefore, on day 11, three positive ethanol specimens were retrieved from the laboratory (all specimens were stored refrigerated) and sent to a different laboratory for testing using GC. No ethanol was detected in any specimen. Although lactate level was not measured, her serum LDH concentration was 294 U/L (normal, 110–230 U/L), which was slightly elevated and could not explain the false-positive test result using the enzymatic assay. The authors also ruled out propofol as the cause of interference. The specific cause of interference was not identified [26].

3.3.2 Gas Chromatography in Blood Alcohol Determination

The gold standard for determination of blood alcohol is GC or gas chromatography–mass spectrometry (GC–MS). Usually, headspace GC is used for the determination of alcohol in whole blood. However, direct injection of sample has also been described. Jain described a GC method in which blood can be injected directly into a column after addition of isobutanol as the internal standard. Ethanol (retention time, 3.5 min) eluted from the column before isobutanol (retention time, 7.25 min). Baseline separation was also observed for acetone, isopropyl alcohol toluene, and xylene [27]. Smith described determination of ethanol along with methanol, isopropanol, and acetone using capillary GC with direct sample injection. Smith used 1-propanol (*n*-propanol) as the internal standard. Serum (200 μL) was mixed with the 200-μL internal standard solution (prepared in deionized water) along with 200 μL of 200 mmol/L of sodium tungstate and 200 μL of 200 mmol/L cupric sulfate. After vortex mixing and centrifugation to sediment the precipitate, 1 μL of aqueous supernatant was directly injected into the gas chromatograph [28]. Baseline separation was obtained between ethanol, other analytes, and the internal standard (Figure 3.1). Maleki *et al.* described analysis of methanol and ethanol in human whole blood, urine, and saliva samples based on headspace solid phase microextraction using silver sulfide and polyvinyl chloride-coated silver wire. Unlike commercial fibers that are coated with fused silica, due to the metallic base of this fiber as described by the authors, this fiber is very durable and thermally stable up to 250°C [29].

Several authors have also used GC–MS for analysis of ethanol or ethanol along with other volatiles in biological specimens. Tiscione *et al.* described ethanol analysis by headspace GC with simultaneous flame ionization and MS detection [30]. Wasfi *et al.* developed static headspace capillary column GC–MS for analysis of ethanol along with acetone, methanol, acetaldehyde, and acetic acid in blood. The authors used *n*-propanol as the internal standard and used *m*/*z* 31 for quantification of ethanol as well as the internal standard. The qualifier ion for ethanol was *m*/*z* 46, whereas that for the internal standard was *m*/*z* 60. The limit of detection was between 0.02 and 0.02 mg/dL, and linearity was between 5 and 200 mg/dL [31]. Cordell *et al.* described a static headspace GC–MS for quantitative analysis of acetaldehyde, methanol, ethanol, and acetic acid in headspace microvolume of blood using *n*-propanol as the internal standard [32]. Xiao *et al.* described a GC–MS method for analysis of whole blood ethanol using *n*-propanol as the internal standard. The mass spectrometer was operated in selected ion monitoring mode (ionization source: electron ionization), monitoring *m*/*z* 31 and 45 for ethanol and *m*/*z* 31 and 59 for the internal standard.

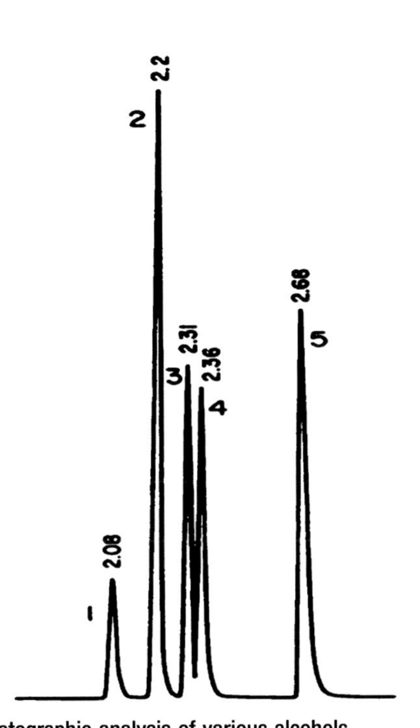

FIGURE 3.1 Gas chromatographic analysis of various alcohols.
Peak 1, methanol; peak 2, ethanol (alcohol); peak 3, acetone; peak 4, isopropyl alcohol; peak 5, propyl alcohol (internal standard). *Source: Norman B Smith* [28]*; Reproduced with permission from the American Association for Clinical Chemistry.*

The method showed linearity for whole blood ethanol concentrations between 4 and 126.3 mg/dL [33].

3.3.3 Stability of Alcohol in Blood During Storage

Although volatile, alcohol is stable in whole blood or serum if stored properly. Penetar *et al.* collected blood from five adult volunteers who consumed standard alcoholic drinks during a 15-min period and observed that ethanol at levels of 60–90 mg/dL in plasma or whole blood was not significantly altered by storage conditions (storage at room temperature vs. refrigerated storage) or choice of collection tube (tubes containing an anticoagulant vs. those containing an anticoagulant and a preservative) for a period of 10 days [34]. In another study, the authors did not observe any significant differences between whole blood alcohol levels determined initially and after 14 days of storage at 0–3° or 22–29°C, with or without preservative. All values were within ±5% difference, the experimental error of the method [35].

Sutlovic *et al.* assessed 79 postmortem blood samples for stability of alcohol after storage (range, 191–468 days) at 4°C, each sample being reanalyzed after storage and the concentration compared to the initial concentration. Approximately 90% of the results were within the 95% limit, and only 10% were outside the limit; these results indicate that when specimens are refrigerated, alcohol is relatively stable in most specimens, even for a long period. However, some specimens showed higher variation (up to 10%), which is not acceptable as a precise forensic standard [36]. Shan *et al.* observed that if blood collection tubes were opened during storage, some loss of alcohol could occur [37].

3.3.4 Correlation between Blood and Breath Alcohol

Breath alcohol results are subject to interference, and the forensic standard is analysis of blood alcohol using GC. Breath alcohol measurement assumes partition of alcohol between breath and blood, where the breath alcohol to blood alcohol ratio is 1:2100 in the postabsorptive phase (30–60 min after consumption); however, during alcohol absorption from the gut, this ratio is highly variable. Therefore, if the ratio of breath to blood alcohol in an individual is more than 1:2100, in the postabsorptive phase the blood alcohol level calculated from the breath alcohol value may be falsely lowered. In general, breath analyzers rely on a breath to blood alcohol ratio of 1:2100 and tend to underestimate true blood alcohol concentration due to the commonly accepted postabsorption ratio of 1:2300 [38]. Simpson found that in 68% of cases, the breath analyzer underestimated blood alcohol content (BAC), in 16% the values were acceptably close to the actual BAC, and in 16% the breath analyzer overestimated the true BAC [39]. Jones and Anderson compared 130 specimens for blood and breath alcohol results and observed that in 18 specimens, breath alcohol results were essentially zero, whereas actual blood alcohol was between 0.04 and 0.15 mg/g (4–15 mg/dL). The average breath to blood alcohol ratio was 1:2448. No significant gender difference was found. The authors concluded that breath test results obtained by the Intoxilyzer 5000S were generally lower than the coexisting concentration of ethanol in venous blood (mean bias, 6.8 mg/dL), which gives an advantage to suspects who provide breath samples compared to blood samples in cases where BAC readings are close to the threshold alcohol limit [40]. Based on a roadside study involving 1875 tests, Kriikku *et al.* observed that the mean venous blood alcohol level in traffic offenders was higher (1.82 g/kg; approximately 182 mg/dL) than predicted from breath alcohol tests (mean, 1.72 g/kg; approximately 172 mg/dL). In general, blood alcohol levels calculated from breath alcohol values were approximately 15% lower than corresponding blood alcohol values determined using blood specimens [41].

3.4 ENDOGENOUS PRODUCTION OF ALCOHOL

Endogenous production of alcohol (auto-brewery syndrome) is a common defense strategy adopted by some individuals charged with driving under the influence of alcohol. A small amount of alcohol is produced endogenously, but it should not alter blood alcohol level. Jones *et al.* reported that endogenous alcohol levels varied from none detected to up to 1.6 μg/mL (0.16 mg/dL), which is negligible [42]. Madrid *et al.* studied eight patients with liver cirrhosis and observed that during fasting, none of the patients showed any endogenous ethanol levels, but after a meal, two patients showed serum alcohol levels of 11.3 and 8.2 mg/dL, respectively, and another four patients showed negligible values [43]. In general, healthy individuals as well as patients with metabolic disease such as diabetes, hepatitis, or cirrhosis showed blood alcohol levels ranging from 0 to 0.08 mg/dL due to endogenous production of alcohol, which is negligible. However, there was an isolated report of a Japanese individual with an endogenous ethanol level exceeding 80 mg/dL who suffered from a very serious yeast infection, and endogenous ethanol was produced after he consumed a carbohydrate-rich meal [44].

In general, postmortem production of alcohol due to fermentation of sugar by bacteria is well documented. Toxicological analysis of a specimen from a 14-year-old adolescent revealed high amounts of alcohol both in blood and in tissue, but ethyl glucuronide, a metabolite of alcohol, was not detected in the liver tissue. The authors concluded that postmortem alcohol in this patient was due to the action of the bacterial strain *Lactococcus garvieae* in blood of the deceased, which is capable of producing alcohol from glucose [45]. The dermal absorption of alcohol is not an issue for adults and cannot account for a significant blood alcohol level in a person attempting to use such defense. However, keratinized stratum, which gives adult skin a barrier, begins to develop in the seventh month of fetal life, and poorly keratinized skin of a preterm infant may be a poor barrier for ethanol. Harpin and Rutter described a case of a premature infant born at 27 weeks of gestation who developed severe hemorrhagic skin necrosis on the back and buttocks after umbilical arterial catheterization, showing a blood ethanol level of 259 mg/dL and a blood methanol level of 26 mg/dL. Before catheterization, the infant's skin was cleaned with methylated spirit, which caused elevated blood ethanol and methanol due to percutaneous absorption [46].

CASE REPORT 3.3

A 3-year-old female with short bowel syndrome showed signs of alcohol intoxication, such as walking erratically and bizarre behavior, on repeated occasions. The girl was receiving *Lactobacillus*-containing carbohydrate-rich fruit drink two to four times per day. A breath analyzer showed an alcohol level that was equivalent to 22 mmol/L (101.2 mg/dL) of alcohol in blood. The carbohydrate-rich fruit drink was discontinued for 1 week, but when it was reinstated, signs of alcohol intoxication reappeared with a blood ethanol concentration of 15 mmol/L (69 mg/dL). Cultures of gastric fluid showed growth of *Candida kefyr*. The fruit drink was discontinued, and the girl was treated with fluconazole. Following treatment, her symptoms quickly resolved within 1 week. One month later, her symptoms reappeared, and breath analysis showed a high amount of ethanol. Culture of her gastric fluid was positive for *Saccharomyces cerevisiae*, and she was again treated with fluconazole and her symptoms disappeared immediately. After this episode, a diet with lower carbohydrate content was chosen and no symptoms of alcohol intoxication were observed during 2 years of follow-up [47].

3.5 URINE ALCOHOL DETERMINATION

Urine ethanol determination is less common than the breath alcohol test or blood alcohol test. Usually, alcohol may not be included in pre-employment drug testing protocol. However, following high-profile transportation incident, including the 1989 Exxon Valdez oil spill in Alaska, the 1990 conviction of three Northwest Airlines pilots, and the 1991 New York subway crash in which alcohol was involved, the U.S. Congress enacted the Omnibus Transportation Employee Testing Act, which mandated alcohol testing. Breath or saliva alcohol tests are conducted more commonly in such drug and alcohol testing programs [48]. In some workplace drug testing, the urine alcohol test may be conducted to monitor abstinence of a person from alcohol. The urine alcohol test may also be used by social service agencies, the criminal justice system, or alcohol rehabilitation programs. Fraser reported that only 1.6% of specimens collected by a social service agency in Nova Scotia, Canada, from parents with a history of drug abuse, showed positive alcohol results. In contrast, cannabinoid was present in 11% of specimens, benzodiazepines in 14.5%, cocaine metabolite in 5%, and codeine/morphine in 71.1% [49]. In general, enzymatic assays used for blood alcohol determination can be used with or without modification for determination of urine alcohol. However, as with blood alcohol, urine alcohol, especially in legal situations, should be determined by GC.

After absorption from the stomach and intestine, alcohol enters the bloodstream and is distributed in all body organs depending on water content. Most of the alcohol is metabolized, and up to 2% of alcohol is excreted unchanged in urine. During early absorption of alcohol, the urine to blood alcohol ratio is less than unity, whereas in the late absorption/distribution phase of alcohol, the ratio is usually 1.0–1.2 and on reaching the postabsorptive phase the ratio averages 1.2–1.4, indicating that in the postabsorptive phase urine concentration is higher than blood concentration. After consuming a moderate amount of alcohol, approximately 2% of the dose could be recovered in the urine after 7 hr. Urine should be collected with sodium fluoride as a preservative to prevent *in vitro* formation of alcohol due to the presence of glucose and yeast such as *Candida albicans*. In general, the ratio of urine to blood alcohol in drunk drivers does not depend on creatinine content of the urine and therefore the relative dilution of the urine specimens. The change in urine alcohol levels in two successive voids can help to resolve whether a large amount of alcohol has been consumed recently [50]. However, it is possible that urine is stored in the bladder for a variable period of time, during which the blood alcohol level changes, thus complicating the interpretation of urine alcohol level. In general, during the absorptive phase, the urine to blood alcohol ratio tends to be lower, but it increases during the postabsorptive phase; if alcohol-free urine is present in the bladder before drinking is initiated, the urine to blood ratio of alcohol may be abnormally low. If two consecutive urine voids can be collected 60 min apart, blood and urine alcohol may correlate better, and this can be applied to alcohol-impaired drivers who claim drinking alcohol after driving, which is known as the "hip-flask" defense in the United Kingdom.

As mentioned previously, for determination of urine alcohol, preservative should be used to prevent *in vitro* formation of alcohol post collection. Uncontrolled diabetes mellitus may cause glycosuria. If yeast infection is present, *in vitro* production of alcohol may result after collection due to the presence of *C. albicans* in the specimen, which can produce alcohol from glucose after collection of the urine specimen. Storing urine at 4°C and using

CASE REPORT 3.4

In a 61-year-old male suspected of driving under the influence of alcohol, two blood samples were collected at 4 and 5:06 PM, and one urine specimen was collected at 3:45 PM. The two blood specimens showed positive blood alcohol, with a value of 75 mg/dL at 4 PM and 58 mg/dL at 5:06 PM, indicating a decrease of blood alcohol at a rate of 16 mg/dL per hour, which was reasonable for a person on the postabsorptive part of the alcohol curve. However, the urine specimen collected at 3:45 PM was negative (urine alcohol was 4 mg/dL, but concentrations <10 mg/dL are considered negative). The creatinine level was 370 mg/dL, indicating that urine was concentrated. The time of driving according to the police report was 3 PM; the driver had purchased a bottle of vodka and claimed that he consumed 80–100 mL immediately after driving. Positive blood alcohol but negative urine alcohol indicated that his bladder contained a pool of alcohol-free urine prior to consuming alcohol, and alcohol in freshly produced ureter urine was diluted because the urine specimen was collected only 45 min after the presumed consumption of alcohol [51].

1% sodium fluoride or potassium fluoride as a preservative can minimize this problem. Alternatively, alcohol biomarkers such as ethyl glucuronide may be tested in urine. Positive urine alcohol and negative ethyl glucuronide in urine indicates *in vitro* production of alcohol after collection. Helander *et al.* reported cases within a maritime alcohol drug testing program in which one subject showed high urine alcohol (108 mg/dL) but the specimen contained low levels of ethyl glucuronide and ethyl sulfate, indicating the possibility of post collection production of alcohol. *Candida albicans* was present in the specimen. In four other subjects, urine ethanol was positive, but both ethyl glucuronide and ethyl sulfate were negative again, indicating post collection formation of ethanol [52]. See Chapter 8 for more detail.

3.6 SALIVA ALCOHOL DETERMINATION

Urine, sweat, and saliva are body fluids derived from plasma and contain ethanol, which may be correlated with serum ethanol level. Saliva (oral fluid) can be collected noninvasively and can be used for the determination of ethanol concentration, although the breath alcohol test is more commonly used compared to saliva ethanol testing. Nevertheless, saliva ethanol may be used to monitor abstinence, although monitoring ethyl glucuronide is superior to direct alcohol measurement to show abstinence due to the longer window of detection of ethyl glucuronide compared to saliva alcohol. In general, ethanol level in saliva correlates with blood ethanol level, although the concentration of ethanol in saliva is higher than that in venous blood during absorption of alcohol from the gut and slightly lower than that in blood during the elimination phase of alcohol [53]. Onsite alcohol testing devices such as the quantitative ethanol detector (QRD) saliva alcohol test can be used to measure salivary ethanol level. The test is based on enzymatic oxidation of alcohol by alcohol dehydrogenase. Saliva is collected using a cotton swab, which can be inserted into the analyzer. Each test may be read by a color bar, and the test has a built-in quality control check that indicates if the test is performed properly. The test can measure saliva ethanol level up to 150 mg/dL. Jones commented that whereas acetone, 2-butanone, and ethylene glycol had no effect on alcohol determination using the QED device, propanol at 100 mg/dL produced an apparent ethanol level of 60 mg/dL and isopropyl alcohol at the same concentration produced an apparent value of 20 mg/dL. However, ethanol levels determined by the QED device correlated well with blood and breath alcohol levels, indicating that this device could be used for quick analysis of ethanol in saliva [54]. In contrast, Bendtsen *et al.* did not observe a good correlation between QED salivary ethanol and breath or blood ethanol levels [55]. The Alco-Screen saliva dipstick is an inexpensive, easy-to-use colorimetric test that provides a semiquantitative estimation of alcohol. Schwartz *et al.* compared salivary alcohol levels with blood alcohol levels in 53 patients suspected of ingesting alcohol. The authors concluded that although correlation was good with a blood alcohol level of approximately 100 mg/dL, for lower alcohol levels between 20 and 50 mg/dL, the performance of the Alco-Screen test was unsatisfactory. The authors concluded that Alco-Screen can be used as a screening test, but definitive diagnosis of alcohol intoxication requires confirmatory blood or breath alcohol analysis [56].

3.7 TRANSDERMAL ALCOHOL SENSORS

Some individuals with alcohol problems are viewed as posing a risk to the general public; as such, a court may order them to completely abstain from alcohol. Sometimes this sentence is enforced by alcohol ignition interlock devices that effectively prevent driving a car after consuming alcohol. Ignition interlock devices, also known as breath alcohol ignition devices, function like a breath analyzer and use ethanol-specific fuel cell technology. Periodic calibration is performed, and usually the offender pays for installation, calibration, and maintenance of the device. Most U.S. states allow judges to order the installation of the device as a condition of probation for repeat offenders, but it may also be required for first-time offenders in some states. Currently, in all states, installation of ignition interlock devices may be considered as an alternative to jail time by judges. The time an offender must use ignition interlock devices varies from state to state. However, ignition interlock devices can monitor alcohol use by an individual prior to driving only. Depending on the offense, the court may order the confinement of a problem drinker to his or her home using electronic proximity monitoring coupled with regular breath tests. In addition, alcohol biomarkers may be tested, but all such approaches require periodic specimen collection. An alternative to these approaches is direct transdermal monitoring of alcohol, which is used in the criminal justice system to ensure abstinence from alcohol by offenders.

Approximately 1% of ingested alcohol is excreted through the skin, mostly through "insensible perspiration," which is vapor that escapes through the skin when a person sweats but cannot be detected by the olfactory system. Early research focused on detecting transdermally excreted alcohol using a sweat patch. The patch was applied to the user's skin for a period of several days, and it absorbed liquid sweat excreted through the skin. The patch was then removed, followed by determination of alcohol content in the laboratory using a suitable method, such as GC. Later researchers demonstrated good correlation between ethanol concentrations in vapor above the skin and breath as well as blood alcohol. Therefore, transdermal alcohol could be measured by putting a portable electrochemical sensor directly above the skin, and the value was used to estimate blood alcohol level. Later, the SCRAM (Secure Continuous Remote Alcohol Monitor) bracelet was marketed as an ankle-worn device by Alcohol Monitoring Systems (Littleton, CO) for transdermal monitoring of alcohol. The Wrist Transdermal Alcohol Sensor (WrisTAS) was developed by Giner (Newton, MA) [57]. The Transdermal Alcohol Detection device (TAD), which is also an ankle-worn device, is commercially available for transdermal alcohol monitoring from BI Incorporated (Boulder, CO). However, there is a lag time between observation of peak blood alcohol and peak transdermal alcohol. In one study, the authors observed a lag time of 33 min between peak blood and transdermal alcohol after subjects consumed a dose of 15 mL of alcohol (alcohol was diluted in 150 mL of fluid). Moreover, after consuming 60 mL of alcohol (diluted with fluid), the lag time was increased to 53 min [58].

SCRAM was the first device available commercially for transdermal alcohol monitoring; the updated model SCRAM2 is now used. This device is locked to the ankle and should be worn continuously (it weighs ~8 oz). The SCRAM device is capable of measuring alcohol in insensible perspiration using a fuel cell technology which is a technique similar to that widely used in breath analyzers. The device measures alcohol concentration every 60 min unless a measurable alcohol level is detected, in which case it takes samples every 30 min until the readings decline below 0.02% (20 mg/dL). The device also records the temperature and skin reflectance using infrared light to provide data that detect attempts to remove or tamper with the device. The device automatically transmits information stored in the ankle bracelet via a modem to a secure web server through a telephone line. The instrument is programmed to transmit data once per day when the offender is likely to be at home, such as during the middle of the night, so that he or she is close to a modem. When data are transmitted to a secure website, a designated person such as a parole officer can review the data for appropriate action to take. Barnett *et al.* studied the performance of the SCRAM device using 66 heavy drinking adults who wore the device between 1 and 28 days and reported their alcohol consumption daily by web-based survey. On days when bracelets were functional, 502 of 690 drinking episodes (72.8%) were detected by remote sensing data, and no gender difference was observed. However, for consumption of fewer than five drinks, women's drinking episodes were more likely to be detected than men's drinking episodes due to higher transdermal alcohol level in women, but for five or more drinks, no gender difference was observed. In multivariant analysis, no variable other than the number of drinks significantly predicted alcohol detection [59]. Sakai *et al.* reported that the SCRAM device consistently detected consumption of approximately two standard drinks, but individual readings did not always correlate with breath alcohol levels, and the device showed discriminative validity for semiquantitative consumption of alcohol in subjects [60].

Marques and McKnight evaluated the performance of two types of transdermal devices—SCRAM and WrisTAS—using 22 paid research subjects. WrisTAS is worn using a Velcro strap and is approximately the size of a wristwatch. SCRAM should be worn during showers but should not be immersed in water. In contrast, WrisTAS should be removed during showers and does not have a lock like SCRAM. WrisTAS uses a different technology (hydrated proton exchange membrane) for detecting transdermal alcohol. In their study, the authors observed that although SCRAM correctly detected 57% of drinking episodes, the true positive detection rate by WrisTAS was 24%. When subjects consumed a sufficient amount of alcohol to reach a blood alcohol level of 80 mg/dL or more, SCRAM correctly detected 88% of those events [61]. The WrisTAS is not currently commercially available, but the technology is used in the commercially available TAD ankle bracelet. The transdermal alcohol sensor transmits data allowing continuous monitoring of *in vivo* alcohol use without intrusive daily contact from a human monitor. It is recommended that transdermal alcohol sensor devices be used to monitor sobriety of individuals and positive/negative semiquantitative values should be used for interpretation of results. Currently, transdermal devices do have some limitations, and more research is needed for further improvement of such devices [62].

3.8 CONCLUSIONS

If consumed in moderation, alcohol has many health benefits, but alcohol abuse is detrimental to health. Drivers under the influence of alcohol are responsible for a majority of fatalities from accidents. In addition, alcohol poisoning may require immediate medical care. Although breath analyzers are widely used by law enforcement officers to identify people driving under the influence of alcohol, the results of breath analyzers may be affected by interference. Blood alcohol determination in a toxicology laboratory using GC provides more accurate results compared to blood alcohol level calculated from breath alcohol level. In addition, GC is superior to the enzymatic method of serum alcohol determination. Transdermal alcohol sensors are used in the criminal justice system to monitor the sobriety of individuals. However, at best, these devices produce acceptable semiquantitative results for interpretation.

References

[1] Flowers NT, Naimi TS, Brewer RD, Elder RW, et al. Patterns of alcohol consumption and alcohol-impaired driving in the United States. Alcohol Clin Exp Res 2008;32(4):639–44.

[2] Jones AW, Norberg A, Hahn RG. Concentration-time profiles of ethanol in arterial and venous blood and end-expired breath during and after intravenous infusion. J Forensic Sci 1997;42:1088–94.

[3] Pavlic M, Grubwieser P, Libiseller K, Rabl W. Elimination rates of breath alcohol. Forensic Sci Int 2007;171:16–21.

[4] Swift R. Direct measurement of alcohol and its metabolites. Addiction 2003;98(Suppl. 2):73–80.

[5] Hlastala MP, Anderson JC. The impact of breathing pattern and lung size on the alcohol breath test. Ann Biomed Eng 2007;35:264–72.

[6] George SC, Babb AL, Hlastala MP. Modeling the concentration of ethanol in the exhaled breath following pretest breathing maneuvers. Ann Biomed Eng 1995;23:48–60.

[7] Kechagias S, Jonsson KA, Franzen T, Anderson L, et al. Reliability of breath alcohol analysis in individuals with gastroesophageal reflux disease. J Forensic Sci 1999;44:814–18.

[8] Fessler CC, Tulleners FA, Howitt DG, Richards JR. Determination of mouth alcohol using the Dräger Evidential Portable Alcohol System. Sci Justice 2008;48:16–23.

[9] Harding PM, McMurray MC, Laessig RH, Smiley II DO, et al. The effect of dentures and denture adhesives on mouth alcohol retention. J Forensic Sci 1992;37:999–1007.

[10] Logan BK, Distefano S, Case GA. Evaluation of the effect of asthma inhalers and nasal decongestant sprays on a breath alcohol test. J Forensic Sci 1998;43:197–9.

[11] Lutmer B, Zurfluh C, Long C. Potential effect of alcohol content in energy drinks on breath alcohol testing. J Anal Toxicol 2009;33:167–9.

[12] Laakso O, Pennanen T, Himbwerg K, Kuitunen T, et al. Effect of eight solvents on ethanol analysis by Drager 7110 evidentiary breath analyzer. J Forensic Sci 2004;49:1113–16.

[13] Jones AW, Rossner S. False positive breath alcohol test after a ketogenic diet. Int J Obesity (Lond) 2007;31:559–61.

[14] Aderjan R, Schmitt G, Wu M. Glue solvent as the cause of breath alcohol value of 1.96 permille. Blutalkohol 1992;29:360–4 [in German].

[15] Caldwell JP, Kim ND. The response of the Intoxilyzer 5000 to five potential interfering substances. J Forensic Sci 1997;42:1080–7.

[16] Caravati EM, Anderson KT. Breath alcohol analyzer mistakes methanol poisoning for alcohol intoxication. Ann Emerg Med 2010;55:198–200.

[17] Barnholl Jr MT, Herbert D, Wells Jr. DJ. Comparison of hospital laboratory serum alcohol levels obtained by an enzymatic method with whole blood levels forensically determined by gas chromatography. J Anal Toxicol 2007;31:23–30.

[18] Rainey P. Relation between serum and whole blood ethanol concentrations. Clin Chem 1999;39:2288–92.

[19] Thompson WT, Malhotra D, Schammel DP, Blackwell W, et al. False positive ethanol in clinical and postmortem sera by enzymatic assay: elimination of interference by measuring alcohol in protein free ultrafiltrate. Clin Chem 1994;40:1594–5.

[20] Badcock NR, O'Reilly DA. False-positive EMIT-st ethanol screen with post-mortem infant plasma. Clin Chem 1992;38:434.

[21] Nine JS, Moraca M, Virji MA, Rao KN. Serum ethanol determination: comparison of lactate and lactate dehydrogenase interference in three enzymatic assays. J Anal Toxicol 1995;19:192–6.

[22] Dunne JR, Tracy JK, Scalea TM, Napolitano L. Lactate and base deficit in trauma: does alcohol or drug use impair predictive accuracy? J Trauma 2005;58:959–66.

[23] Winek CL, Wahba WW, Windisch R, Winek CL. Serum alcohol concentrations in trauma patients determined by immunoassays versus gas chromatography. Forensic Sci Int 2004;139:1–3.

[24] Powers RH, Dean DE. Evaluation of potential lactate/lactate dehydrogenase interference with an enzymatic alcohol analysis. J Anal Toxicol 2009;33:561–3.

[25] Gharapetian A, Holmes DT, Urquhart N, Rosenberg F. Dehydrogenases interference with enzymatic ethanol assays: forgotten but not gone. Clin Chem 2008;54:1251–2.

[26] Jones TE. False-positive ethanol blood concentrations leading to clinical confusion on Christmas day. Clin Biochem 2011;44:1355–7.

[27] Jain NC. Direct blood injection method for gas chromatographic determination of alcohols and other volatile compounds. Clin Chem 1971;17:82–5.

[28] Smith NB. Determination of volatile alcohols and acetone in serum by non-polar capillary gas chromatograph after direct sample injection. Clin Chem 1984;30:1672–4.

[29] Maleki R, Farhadi K, Martin AA. Analysis of ethanol and methanol in human fluids by headspaces solid phase microextraction coupled with capillary gas chromatography. Anal Sci 2006;22:1253–5.

[30] Tiscione NB, Alford I, Yeatman DT, Shan X. Ethanol analysis by headspace gas chromatography with simultaneous flame-ionization and mass spectrometry detection. J Anal Toxicol 2011;35:501–11.

[31] Wasfi A, Al-Awadhi AH, Al-Hatali ZN, Al-Rayami FJ, et al. Rapid and sensitive static headspace gas chromatography–mass spectrometry method for the analysis of ethanol and abused inhalants in blood. J Chromatogr B Analyt Technol Biomed Life Sci 2004; 799:331–6.

[32] Cordell RL, Pandya H, Hubbard M, Turner MA, et al. GC–MS analysis of ethanol and other volatile compounds in micro-volume blood samples—Quantifying neonatal exposures. Ann Bioanal Chem 2013;405:4139–47.

[33] Xiao HT, He L, Tong RS, Yu JY, et al. Rapid and sensitive headspace gas chromatography–mass spectrometry method for the analysis of ethanol in the whole blood. J Clin Lab Anal 2014;28:386–90.

[34] Penetar DM, McNeil JF, Ryan ET, Lukas SE. Comparison among plasma, serum and whole blood ethanol concentrations: impact of storage conditions and collection tubes. J Anal Toxicol 2008;32:505–10.

[35] Winek CL, Paul LJ. Effect of short-term storage conditions on alcohol concentrations in blood from living human subjects. Clin Chem 1983;29:1959–60.

[36] Sutlovic D, Versic-Bratincevic M, Definis-Gojanovic M. Blood alcohol stability in postmortem blood samples. Am J Forensic Med Pathol 2014;35:55–8.

[37] Shan X, Tiscione NB, Alford I, Yeatman DT. A study of blood alcohol stability in forensic antemortem blood samples. Forensic Sci Int 2011;211:47–50.

[38] Labianca DA, Simpson G. Statistical analysis of blood to breath alcohol ratio data in the logarithm-transformed and non-transformed modes. Eur J Clin Chem Clin Biochem 1996;34:111–17.

[39] Simpson G. Do breath tests really underestimate blood alcohol concentrations? J Anal Toxicol 1989;13:120–3.

[40] Jones AW, Anderson L. Comparison of ethanol concentrations in venous blood and end-expired breath during a controlled drinking study. Forensic Sci Int 2003;132:18–25.

[41] Kriikku P, Wilhelm L, Jenckel S, Rintatalo J, et al. Comparison of breath alcohol screening test results with venous blood alcohol concentration in suspected drunken drivers. Forensic Sci Int 2014;239:57–61.

[42] Jones AW, Mardgh G, Anggard E. Determination of endogenous ethanol in blood and breath by gas chromatography–mass spectrometry. Pharmacol Biochem Behav 1983;18 (Suppl. 1):267–72.

[43] Madrid AM, Hurtado C, Gatica S, Chacon I, et al. Endogenous ethanol production in patients with liver cirrhosis, motor alteration and bacterial overgrowth. Rev Med Chil 2002;130:1329–34.

[44] Logan BK, Jones AW. Endogenous ethanol "auto-brewery syndrome" as a drunk driving defense. Med Sci Law 2000;40:206–15.

[45] Appenzeller BM, Schuman M, Wennig R. Was a child poisoned by ethanol? Discrimination between antemortem and postmortem formation. Int J Legal Med 2008;122:429–34.

[46] Harpin V, Rutter N. Percutaneous alcohol absorption and skin necrosis in a preterm infant. Arch Dis Child 1982;57:477–9.

[47] Jansson-Nettelbladt E, Meurling S, Perini B, Soling J. Endogenous ethanol fermentation in a child with short bowel syndrome. Acta Paediatr 2006;95:502–4.

[48] Li G, Brady JE, DiMaggio C, Baker SP, et al. Validity of suspected alcohol and drug violations in aviation employees. Addiction 2010;105:1771–5.

[49] Fraser AD. Urine drug testing for social service agencies in Nova Scotia, Canada. J Forensic Sci 1998;43:194–6.

[50] Jones AW. Urine as a biological specimen for forensic analysis of alcohol and variability in the urine to blood relationship. Toxicol Rev 2006;25:15–35.

[51] Jones AW, Kugelberg FC. Relationship between blood and urine alcohol concentrations in apprehended drivers who claimed consumption of alcohol after driving and without supporting evidence. Forensic Sci Int 2010;194:97–102.

[52] Helander A, Hagelberg CA, Beck O, Petrini B. Unreliable alcohol testing in a shipping safety program. Forensic Sci Int 2009;189:e45–7.

[53] Gubala W, Zuba D. Gender differences in the pharmacokinetics of ethanol in saliva and blood after oral ingestion. Pol J Pharmacol 2003;55:639–44.

[54] Jones AW. Measuring ethanol in saliva with QED enzymatic test device: comparison of results with blood and breath alcohol concentrations. J Anal Toxicol 1995;19:169–74.

[55] Bendtsen P, Hultberg J, Carlsson M, Jones AW. Monitoring ethanol exposure in a clinical setting by analysis of blood, breath, saliva, and urine. Alcohol Clin Exp Res 1999;23:1446–51.

[56] Schwartz RH, O'Donnell RM, Thorne MM, Getson PR, et al. Evaluation of colorimetric dipstick to detect alcohol in saliva: a pilot study. Ann Emerg Med 1989;18:1001–3.

[57] Hawthorne JS, Wojcik MH. Transdermal alcohol measurement: a review of literature. Can Soc Forensic Sci J 2006;39:65–71.

[58] Webster GD, Gabler HC. Feasibility of transdermal ethanol sensing for the detection of intoxicated drivers. Annu Proc Assoc Adv Automot Med 2007;51:449–64.

[59] Barnett NP, Meade EB, Glynn TR. Predictors of detection of alcohol use episodes using a transdermal alcohol sensor. Exp Clin Psychopharmacol 2014;22:86–96.

[60] Sakai JT, Mikulich-Gilbertson SK, Long RJ, Crowley TJ. Validity of transdermal alcohol monitoring: fixed and self-regulated dosing. Alcohol Clin Exp Res 2006;30:26–33.

[61] Marques PR, McKnight AS. Field and laboratory alcohol detection with 2 types of transdermal devices. Alcohol Clin Exp Res 2009;33:703–11.

[62] Leffingwell TR, Cooney NJ, Murphy JG, Luczak S, et al. Continuous objective monitoring of alcohol use: twenty-first century measurement using transdermal sensors. Alcohol Clin Exp Res 2013;37:16–22.

CHAPTER 4

Alcohol Biomarkers

An Overview

4.1 INTRODUCTION

The term *biomarkers* is frequently used to describe any statistically significant biochemical or molecular change between two populations. For alcohol biomarkers, one population comprises individuals who are either teetotalers or social drinkers and another population consists of individuals who consume alcohol in excess, including alcohol-dependent individuals. The biomarkers definition working group of the National Institutes of Health (NIH) defined biomarker as "a characteristic that is objectively measured and evaluated as an indicator of normal biological process, pathogenic processes, or pharmacological response to a therapeutic intervention." The difference between this definition and the broad definition of biomarker is that in the broad definition, a significant difference in mean between two populations is sufficient to qualify a molecule as a biomarker, whereas in the NIH definition, a biomarker should also provide information regarding individual subjects. With regard to alcohol biomarkers, a particular biomarker should be an accurate indicator of alcohol consumption by an individual. Similarly, a biomarker may be an indicator of the pathogenic process in alcohol-induced organ or tissue damage and dysfunction. There is no universal biomarker for all aspects of alcohol consumption. Instead, a biomarker may apply to particular aspects of consuming alcohol, such as heavy drinking versus social drinking, or a biomarker may differentiate between individuals who are teetotalers and those who are social drinkers [1].

Alcohol abuse is a serious public health issue not only in the United States but also worldwide. Two billion people worldwide consume alcohol, and there are 76.3 million with diagnosable alcohol use disorders. It is estimated that 3.8% of all global deaths and 4.6% of the global burden of disease and injury can be attributed to alcohol [2]. The World Health Organization's (WHO) "Global Status Report on Alcohol and Health 2014" estimated that

CONTENTS

4.1 Introduction .. 91
4.2 State Versus Trait Alcohol Biomarkers 92
4.3 Liver Enzymes as Alcohol Biomarkers 94
4.4 Mean Corpuscular Volume as Alcohol Biomarker 97
4.5 Carbohydrate-Deficient Transferrin as Alcohol Biomarker 98
4.5.1 Combined CDT–GGT as Alcohol Biomarker 100
4.6 β-Hexosaminidase as Alcohol Biomarker 101
4.7 Ethyl Glucuronide and Ethyl Sulfate as Alcohol Biomarkers 102

A. Dasgupta: Alcohol and Its Biomarkers. DOI: http://dx.doi.org/10.1016/B978-0-12-800339-8.00004-3

4.8 Fatty Acid Ethyl Ester as Alcohol Biomarker 104
4.9 Phosphatidylethanol as Alcohol Biomarker 106
4.10 Total Plasma Sialic Acid as Alcohol Biomarker 108
4.11 Sialic Acid Index of Apolipoprotein J 109
4.12 5-HTOL/5-HIAA as Alcohol Biomarker 109
4.13 Other Alcohol Biomarkers 110
4.14 Clinical Application of Alcohol Biomarkers 111
4.14.1 Diagnosis Using DSM-IV and DSM-5.. 111
4.14.2 Self-Assessment of Alcohol Use........... 111
4.14.3 Alcohol Biomarkers and AUDIT......................... 112
4.14.4 Application of Alcohol Biomarkers.. 113
4.14.5 Combining Alcohol Biomarkers.. 114
4.15 Conclusions....... 116
References 116

3.3 million deaths in 2012 worldwide were attributable to alcohol abuse. WHO estimates that there are 140 million alcoholics worldwide. It is also estimated that 18 million people in the United States are alcoholics or alcohol dependent. There are more male alcohol-dependent individuals than female. It is estimated that alcohol dependence and alcohol abuse result in an estimated cost of $220 billion annually [3]. Therefore, alcohol biomarkers play an important role in public health.

4.2 STATE VERSUS TRAIT ALCOHOL BIOMARKERS

A biomarker may be a state or a trait biomarker. When a biomarker generally refers to a condition such as alcohol consumption, it is termed a state biomarker because it may describe a pathological state such as alcohol-induced organ damage. When clinicians evaluate history of alcohol consumption by patients, they want to know not only about recent drinking patterns (acute alcohol consumption) but also about patients' drinking history, such as whether they are social drinkers versus heavy consumers of alcohol. Various state alcohol biomarkers can be used for evaluating patients in order to answer these questions along with self-described patterns of drinking by patients. Trait markers help physicians to identify whether patients have a genetic predisposition to alcohol abuse and alcoholism. Knowing which individuals are at risk of developing alcoholism can help to prevent alcohol problems altogether or enable these people to seek early treatment for alcohol-related problems [4].

A state alcohol biomarker may be an indirect biomarker of alcohol consumption or a direct biomarker. Blood and urine alcohol levels are a direct indication of alcohol consumption, and blood alcohol level is used to determine if a person is driving while intoxicated. In the United States, a whole blood alcohol level of 0.08% (80 mg/dL) is considered the legal limit for driving (see Chapter 1). Acetaldehyde, a metabolite of alcohol, reacts readily with various proteins, and hemoglobin–acetaldehyde adduct is a direct state alcohol biomarker. Ethyl glucuronide, a minor metabolite of alcohol, is also a direct alcohol biomarker. On the other hand, liver enzymes such as γ-glutamyl transferase (GGT), alanine aminotransferase (ALT), and aspartate aminotransferase (AST), which are elevated after heavy alcohol consumption because alcohol has toxic effects on the liver, are indirect alcohol biomarkers. Mean corpuscular volume (MCV), carbohydrate-deficient transferrin (CDT), and serum and urine hexosaminidase are also indirect state alcohol biomarkers. Neurotransmitters such as γ-aminobutyric acid and β-endorphin are examples of trait alcohol biomarkers. State and trait alcohol biomarkers are listed in Table 4.1, and short-term and long-term state alcohol biomarkers are listed in Box 4.1. Direct determination of alcohol in serum, breath, urine,

Table 4.1 Various State and Trait Alcohol Biomarkers

Type of Alcohol Biomarker	Examples
State alcohol biomarkers	Indirect Biomarkers ■ Liver enzymes (GGT mostly commonly used, but ALT and AST may also be used) ■ Mean corpuscular volume (MCV) ■ Carbohydrate-deficient transferrin ■ Serum or urine β-hexosaminidase ■ Total serum or plasma sialic acid ■ Sialic acid index of apolipoprotein J ■ 5-Hydroxytryptophol Direct Biomarkers ■ Ethyl glucuronide ■ Ethyl sulfate ■ Fatty acid ethyl esters ■ Phosphatidylethanol ■ Acetaldehyde–hemoglobin adduct
Trait alcohol biomarkers	■ Polymorphisms of genes encoding following enzymes/receptors ■ Alcohol dehydrogenase and aldehyde dehydrogenase enzymes (major effect) ■ Dopamine receptors and transporters ■ Monoamine oxidase ■ Catechol-*O*-methyltransferase (COMT) ■ γ-Aminobutyric acid (GABA) receptor ■ Serotonin receptor ■ Cannabinoid receptor ■ Acetyl choline receptor ■ Glutamate receptor ■ Adenylyl cyclase ■ Neuropeptide Y receptor

BOX 4.1 SHORT-TERM AND LONG-TERM STATE ALCOHOL BIOMARKERS

Short-Term Markers of Alcohol Use (Hours to Days)

- Serum/plasma/blood/saliva ethanol (see Chapter 2)
- Fatty acid esters (24 hr in serum but up to months in hair)
- 5-Hydroxytryptophol
- Ethyl glucuronide (2 or 3 days in urine but longer in hair)
- Ethyl sulfate (1 or 2 days in urine but longer in hair)

Long-Term Markers of Alcohol Use (One to Several Weeks)

- *N*-acetyl-β-hexosaminidase
- Mean corpuscular volume (MCV)
- γ-Glutamyl transferase (GGT)
- Carbohydrate-deficient transferrin (CDT)
- Combination of CDT and GGT
- Acetaldehyde–protein adducts
- Total sialic acid in serum/plasma
- Phosphatidylethanol
- Plasma sialic acid index of apolipoprotein J

or saliva can also be considered a direct short-term alcohol biomarker. Determination of alcohol in blood, urine, saliva, and breath is discussed in detail in Chapter 3, and alcohol metabolism is discussed in Chapter 2.

This chapter provides an overview of state alcohol biomarkers that are used routinely in clinical practice. Some trait markers hold promise for clinical use in the near future, and trait markers are discussed in Chapter 10. Various state markers are discussed in detail in Chapters 5–9. In Chapter 5, the application of various liver enzymes as alcohol biomarkers is discussed with an emphasis on γ-glutamyl transferase. Chapter 6 discusses the application of MCV and CDT as alcohol biomarkers. In Chapter 7, the application of β-hexosaminidase, acetaldehyde–protein adducts, and dolichol as alcohol biomarkers is addressed. Direct alcohol biomarkers such as ethyl glucuronide, ethyl sulfate, fatty acid ethyl esters, and phosphatidylethanol are discussed in Chapter 8. In Chapter 9, various other alcohol biomarkers are discussed, including total serum or plasma sialic acid, sialic acid index of plasma apolipoprotein J, 5-hydroxytryptophol, plasma cholesterol transfer protein, homocysteine, and circulating cytokines. Characteristics of various state alcohol biomarkers are summarized in Table 4.2.

4.3 LIVER ENZYMES AS ALCOHOL BIOMARKERS

Liver enzymes are one of the first described alcohol biomarkers. Breakdown of hepatocytes results in the release of aminotransferases (also referred to as transaminases) such as ALT and AST into the blood. ALT is a cytosolic enzyme and more specific for liver disease. AST is primarily a mitochondrial enzyme that is also found in heart, muscle, kidney, and brain. ALT has a longer half-life than that of AST. In acute liver injury, AST levels are higher than ALT levels; however, after 24–48 hr, ALT levels may be higher than those of AST. ALT is considered a more specific marker for liver injury. Because alcohol has toxic effects on the liver, liver enzymes are elevated in individuals who consume excessive amounts of alcohol. GGT is a membrane-bound glycoprotein located on the cell surface membrane of most cell types, including hepatocytes. Although GGT is present in many tissues, only liver GGT is detectable in serum. Of all liver enzymes, GGT is the most often used enzyme in clinical settings as an alcohol biomarker [5].

Increased activity of GGT in serum may be due to increased synthesis as a result of enzyme induction by alcohol or release from hepatocytes due to damage caused by excessive alcohol use. However, elevation of GGT in response to alcohol consumption varies widely among individuals, and levels correlate only moderately with the amount of alcohol consumed.

Table 4.2 Characteristics of Various State Alcohol Biomarkers

Alcohol Biomarker	Specimen	Detection Window	Indicated Use	Sensitivity (%)	Specificity (%)	Cutoff Value
GGT	Blood	3–4 weeks	Chronic heavy drinking	64	72	30 U/L
MCV	Blood	3–4 months	Chronic heavy drinking	48	52	100 fL
%CDT	Blood	2–3 weeks	Moderate to heavy drinking	84	92	2.4%
Serum β-hexosaminidase	Blood	7–10 days	Chronic heavy drinking	94	91	35%[a]
Ethyl glucuronide	Urine	2–3 days	Moderate drinking	76	93	100 ng/mL
Ethyl glucuronide	Hair	Several months	Moderate to heavy drinking	96	99	30 pg/mg
Ethyl sulfate	Urine	1–2 days	Moderate drinking	82	86	25 ng/mL
Fatty acid ethyl esters	Hair	Only 24 hr in serum but several months in hair, which is more frequently used	Showing abstinence or differentiating heavy drinking from social drinking depending on cutoff	90	100	0.29 ng/mg
Phosphatidylethanol	Blood	2–3 weeks	Heavy drinking	94.5	100	0.36 μmol/L
5-HTOL/5-HIAA	Urine	5–15 hr	Acute consumption of 50 g or more of alcohol	77	100	15 pmol/nmol
Total plasma sialic acid	Blood	3 weeks	Differentiating heavy drinking from social drinking	57.7 in females 47.8 in males	95.5 in females 81.3 in males	77.8 mg/dL in females 80 mg/dL in males
Sialic acid index of apolipoprotein J	Blood	6–8 weeks	Heavy drinking	90–92	100	Not firmly established

[a]*35% cutoff value represents β-Hex-B% (Hex-B represents heat-stable β-hexosaminidase activity; therefore, Hex-B% is the ratio of heat-stable β-hexosaminidase to total β-hexosaminidase activity multiplied by 100).*

In general, sustained alcohol consumption is needed for elevation in GGT levels, but if a person has a history of alcohol abuse, GGT may be elevated if drinking resumes. Consumption of 60 g or more of alcohol for 3–6 weeks is generally necessary to observe significant increases in serum GGT levels. Regular consumption is more likely to increase GGT levels rather than episodic drinking. The sensitivity and specificity of GGT vary widely among different populations and also depending on cutoff level. Arndt *et al.* reported a sensitivity of 64% and specificity of 72% at a 30 U/L cutoff [6]. However, other authors have reported relatively poor sensitivity of GGT because only 30–50% of patients who are excessive drinkers can be identified by elevated GGT in the general community or family practice setting. In this setting, specificity may vary from 40 to 90%. However, in residential alcohol rehabilitation programs, sensitivity may be higher (50–90%) with reasonable specificity (65–90%). A wide range of values (35–80 U/L) has been proposed as a cutoff for GGT. Therefore, it is difficult to compare sensitivity and specificity described in different reports because the reference range of GGT may be different [7]. Gender-specific cutoff (33 U/L for females and 56 U/L for males) had also been proposed. Although average GGT levels are significantly higher in current and former drinkers compared to lifetime abstainers, one study observed that in men, daily drinking showed the highest levels of GGT, whereas in women the highest levels of GGT were observed in weekend drinkers. Women who consumed alcohol without food exhibited higher GGT levels compared to women who consumed alcohol with food. However, no such relationship was observed in males [8].

There are many limitations to using GGT as an alcohol biomarker. Serum GGT levels are not usually elevated in adolescents and young adults after heavy drinking. GGT levels are likely to be elevated in individuals who are aged at least 30 years or older and who consume alcohol on a regular basis. Obesity and hepatitis C infection may increase serum GGT levels [7]. Cigarette smoking and diabetes may also cause elevated serum GGT levels (see Chapter 5). False-positive results may be encountered in patients receiving therapy with barbiturates, phenytoin, phenazone, dextropropoxyphene, monoamine oxidase inhibitors, tricyclic antidepressants, warfarin, thiazide diuretics, or anabolic steroids. In addition, damage to the liver due to viral infection such as hepatitis or ischemic damage to the liver may also significantly increase serum GGT levels [9]. Danielsson *et al.* reported that in men, elevation of serum GGT level induced by heavy drinking (>280 g/week) was significantly reduced by coffee consumption exceeding four cups per day. A similar trend was also observed in women [10]. Limitations of GGT as an alcohol biomarker are summarized in Table 4.3.

Table 4.3 Limitations of GGT, MCV, %CDT, and β-Hexosaminidase as Alcohol Biomarkers

Alcohol Biomarker	Comments
GGT	■ False-positive results may be encountered in patients receiving therapy with phenytoin, phenazone, dextropropoxyphene, monoamine oxidase inhibitors, tricyclic antidepressants, warfarin, and other drugs. See Chapter 5 for details. ■ Obesity, cigarette smoking, and diabetes may increase serum GGT levels. ■ Viral hepatitis, nonalcoholic liver diseases, and ischemic damage to liver may increase serum GGT levels. ■ Coffee drinking (exceeding four cups per day) may lower serum GGT levels in individuals who consume a high amount of alcohol.
MCV	■ MCV may be increased in various anemias, including megaloblastic anemia. ■ Various drugs, such as antibiotics, anticancer agents, antibiotics, anticonvulsants, and antidiabetic and anti-inflammatory agents, may also increase MCV, causing false-positive results.
%CDT	■ Genetic variant in transferrin may cause false-positive %CDT test results. ■ Liver diseases such as biliary cirrhosis and end-stage liver disease may elevate both %CDT and the absolute value of CDT. ■ People with chronic illnesses, such as chronic pulmonary disease and rheumatoid arthritis, and pre-liver transplant patients, may show elevated %CDT as well as absolute value of CDT. ■ Total CDT levels may be affected by factors that increase transferrin levels, such as iron deficiency, chronic illness, and menopausal status. ■ False-negative results may be associated with female gender, episodic lower level alcohol use, and acute trauma with blood loss. ■ Certain drug therapies may affect %CDT as well as absolute value of CDT. Anticonvulsants and ACE inhibitors may elevate both %CDT and CDT, whereas loop diuretics may lower %CDT and absolute value of CDT; further studies are needed to establish such effects.
β-Hexosaminidase	■ Hepatitis, liver metastasis, diabetes, rheumatoid arthritis, inflammatory bowel disease, rheumatoid arthritis, and myocardial and cerebral infarction may increase serum β-hexosaminidase levels. ■ Use of oral contraceptives and pregnancy may also increase serum β-hexosaminidase levels, causing false-positive results.

4.4 MEAN CORPUSCULAR VOLUME AS ALCOHOL BIOMARKER

MCV is a value routinely calculated by automated hematology analyzers during complete blood count analysis. Like GGT, MCV is a biomarker of excessive alcohol use. Increased MCV due to macrocytosis in the presence of excess alcohol may occur with normal folate levels, although in up to 30% of alcohol-dependent patients, some reduction of folate level may be observed due to dietary deficiency or impaired absorption due to excess use of alcohol. Because the life span of erythrocytes is 120 days, it may take several months before MCV returns to a normal level after abstinence. MCV has limited value as a single alcohol biomarker in screening because it has poor sensitivity, usually less than 50%. Specificity also varies [7]. In general, sensitivity and specificity of MCV as an alcohol biomarker are lower than those of GGT [11]. Macrocytosis is usually defined as an MCV value greater than 100 fL, and several authors have used 100 fL as the cutoff value of MCV to study the relationship between alcohol consumption and MCV. However, a lower cutoff, such as 96 fL, has also been suggested. A major cause of macrocytosis is megaloblastic anemia due to vitamin B_{12} and/or folate deficiency. In general, MCV is less than 110 fL in chronic alcohol abusers [12]. However, in megaloblastic anemia, MCV can be higher than 110 fL, sometimes reaching 130 fL or greater. Limitations of MCV as an alcohol biomarker are listed in Table 4.3 (see Chapter 6 for more details).

4.5 CARBOHYDRATE-DEFICIENT TRANSFERRIN AS ALCOHOL BIOMARKER

Transferrin, a glycoprotein synthesized in the liver, functions as an iron transporter. Transferrin exists in various isoforms, which are named depending on the number of terminal sialic acid residues in the transferrin molecule. The major isoform contains four sialic acid residues and is called tetrasialotransferrin. This isoform represents 64–80% of total transferrin molecules in the serum [13]. In general, transferrin molecules that are deficient in sialic acid (containing zero to two sialic acid molecules) are minor isoforms present in sera of normal individuals. These minor isoforms include asialo (no sialic acid), monosialo (one sialic acid molecule), and disialotransferrin (two sialic acid molecules). CDT is the collective term for a group of minor isoforms of human transferrin with a low degree of glycosylation. After a period of chronic heavy alcohol consumption, CDT isoforms usually increase in serum, making CDT a useful alcohol biomarker in screening for alcohol abuse and monitoring the progress of patients enrolled in alcohol and drug rehabilitation programs. The precise mechanism behind the increase in CDT in serum after chronic heavy alcohol use is not fully elucidated. Sialyltransferase present in liver plays an important role in the incorporation of sialic acid in the transferrin molecule. After chronic alcohol intake, liver membrane sialyltransferase activity decreases, which may be related to the incorporation of fewer sialic acid molecules in transferrin. In addition, the sialidase enzyme is responsible for removing sialic acid from transferrin. Plasma sialidase activity increases after heavy alcohol consumption [14].

In general, heavy alcohol consumption (50–80 g/day) for a period of at least 1 week leads to increased concentrations of CDT in serum; after abstinence, CDT returns to a normal level, with a half-life of 14 days.

In common practice, the ratio of CDT to total transferrin is calculated, and the value is presented as a percentage (%CDT). However, the absolute value of CDT may also be used as an alcohol biomarker. CDT was the first test to receive approval by the U.S. Food and Drug Administration as an alcohol biomarker; it has been in use in the U.S. health care system since 2001 [15]. Various cutoff values have been proposed for both CDT and %CDT. Whereas CDT is gender dependent (women show higher values than men) and has gender-specific cutoffs, %CDT is not gender dependent and one cutoff can be used for both males and females. Malcom *et al.* used a serum CDT cutoff of 17 U/L for males and 25 U/L for females to evaluate the usefulness of CDT as an index of heavy alcohol use during postmortem examination. They concluded that elevated CDT levels appeared to indicate antemortem alcohol abuse prior to death [16]. Sorvajarvi *et al.* defined the normal range of CDT as up to 20 U/L for men and up to 26 U/L for women [17]. Arndt *et al.* observed that at a cutoff of 2.4% for CDT sensitivity and specificity were 84 and 92%, respectively. Sensitivity and specificity of %CDT were superior to those of other alcohol biomarkers, such as GGT and MCV [6]. However, both sensitivity and specificity of %CDT may change with cutoff level. In a study that included 396 women and 403 men and that used the receiver operator curve characteristic to define the cutoff value to determine the best fit for subjects who consumed more than 90 drinks per month, Fleming and Mundt concluded that 2.5% represented the best cutoff value for %CDT. Median %CDT was 2.0 for abstainers, 2.1 for moderate drinkers, and 2.7 for heavy drinkers. The overall sensitivity and specificity of the %CDT test were 61 and 85%, respectively [18]. Hock *et al.* proposed a significantly higher cutoff value of 3.0% for %CDT based on the 95th percentile value of social drinkers [19].

There are issues of both false positive and false-negative test results with %CDT as well as the absolute value of CDT when used as an alcohol biomarker because levels may be influenced by other conditions not related to alcohol consumption. Genetic variant in transferrin may cause false-positive %CDT test results. Liver diseases, including end-stage liver disease, may elevate both %CDT and the absolute value of CDT. In addition, total CDT levels may be affected by factors that increase transferrin levels, such as iron deficiency, chronic illness, and menopausal status. False-negative results may be associated with female gender, episodic lower-level alcohol use, and acute trauma with blood loss. Certain drug therapies may affect %CDT as well as the absolute value of CDT. Anticonvulsants and angiotensin-converting enzyme (ACE) inhibitors may elevate both %CDT and CDT, whereas loop diuretics may lower %CDT and the absolute value of CDT; however, additional studies are needed to establish such correlation [20]. See Chapter 6 for an in-depth discussion of CDT.

CASE REPORT 4.1

In 2007, a patient with a known history of alcohol abuse (who consumed 90–150 g of ethanol—approximately 7–11 drinks—per day) showed a %CDT value of 3.3% along with elevated levels of AST (137 U/L), ALT (120 U/L), GGT (434 U/L), and MCV (101 fL). In December 2007, after double coronary artery bypass surgery, he was treated with amlodipine, atorvastatin, isosorbide mononitrate, carvedilol, ticlopidine, and pantoprazole. After 6 months of reduced alcohol consumption following surgery, he resumed alcohol abuse. However, despite witnessed alcohol abuse, in July 2009 the patient showed a normal %CDT value of 0.8%. Two months later, he was admitted to the trauma service with hip fracture, and his blood alcohol was 80 mg/dL 16 hr after admission. In January 2011, he again showed a normal %CDT value of 1.1% but had significantly elevated GGT (555 U/L) and MCV (101 fL) with a concurrently high urinary ethyl glucuronide concentration (>50 mg/L). He was admitted to an alcohol rehabilitation center, and in May 2011, he was transferred to a private medical center. This study indicates that drug therapy may hamper the identification of chronic alcohol abuse using CDT as an alcohol biomarker [21].

4.5.1 Combined CDT–GGT as Alcohol Biomarker

Combining GGT and CDT values to mathematically derive a new parameter may improve the sensitivity and specificity of this calculated parameter as an alcohol biomarker. The mathematical formula to derive this parameter is as follows:

$$\text{GGT}-\text{CDT}=[0.8\times\ln(\text{GGT})]+[1.3\times\ln(\text{CDT})]$$

In this case, both the GGT level and the CDT level are expressed as units per liter. Sillanaukee and Olsson proposed a cutoff value of 6.5 using the 95th percentile of the data from the control group for this combined marker. The authors further reported that sensitivity and specificity of the GGT–CDT marker were 79 and 93%, respectively, which were superior to sensitivity of 65% and specificity of 94% observed with CDT. The sensitivity and specificity of GGT were 59 and 91%, respectively. The authors concluded that combined GGT–CDT is more sensitive and specific than CDT alone as an alcohol biomarker [22]. A similar approach can be adopted to calculate the GGT–%CDT parameter:

$$\text{GGT}-\%\text{CDT}=[0.8\times\ln(\text{GGT})]+[1.3\times\ln(\%\text{CDT})]$$

However, the cutoff level should be different from that of the GGT–CDT parameter. Based on receiver operator curve analysis, Anttila *et al.* established a cutoff value of 4.0 for GGT–%CDT. The sensitivity and specificity of GGT–%CDT in alcoholics with liver disease were 94 and 100%, respectively. In alcoholics without liver disease, the respective sensitivity and specificity were 90 and 100% [23].

4.6 β-HEXOSAMINIDASE AS ALCOHOL BIOMARKER

N-acetyl-β-hexosaminidase (β-hexosaminidase) is a complex group of glycoprotein lysosomal isoenzymes that releases *N*-acetylglucosamine and *N*-acetylgalactosamine from the nonreducing end of oligosaccharide chains of glycoproteins, glycolipids, and glycosaminoglycans. Two major isoenzymes of β-hexosaminidase have been characterized—isoenzyme A (one α and one β chain) and isoenzyme B (two β chains)—whereas isoenzyme S (two α chains) represents only approximately 0.02% of isoenzyme activity [24]. However, there are other isoforms of β-hexosaminidase present in serum that can be separated by isoelectrofocusing. Isoforms P (P isoform is elevated in pregnancy, hence the name) and I_1 and I_2 (intermediate heat-stable forms normally present in serum) all consist of two β subunits. Listed in order of decreasing isoelectric points, the isoenzymes of β-hexosaminidase can be arranged as B, I_1, I_2, P, A, and S. Hexosaminidase C has been purified from human placenta, and this isoform is more active in patients deficient in β-hexosaminidase A and B activity. Hexosaminidase C has a distinct isoelectric point and is found predominantly in the brain [25].

Chronic alcohol consumption results in increased levels of β-hexosaminidase in both serum and urine. It has been assumed that consuming more than 60 g of alcohol ($>$4.5 standard drinks) per day for 10 days or more results in an increased level of β-hexosaminidase in serum. Karkkainen *et al.* reported that the mean β-hexosaminidase level was 35.0 U/L among drunken men ($n=25$), but the mean value was 16.8 U/L among healthy males ($n=16$) who were social drinkers. Interestingly, the mean β-hexosaminidase level was 19.8 U/L among teetotalers. The authors further reported that sensitivity of β-hexosaminidase among heavy drinkers was 85.7% and specificity of β-hexosaminidase as an alcohol biomarker was 97.6% [26]. Therefore, elevated serum β-hexosaminidase concentration is a marker for heavy drinking. β-Hexosaminidase in serum exists as heat labile (A and S) and heat stable (B, I_1, I_2, and P). The heat-stable form is often referred to as β-Hex-B. The ratio of β-Hex-B and total serum β-hexosaminidase activity expressed as a percentage (β-Hex-B%) is frequently used as an alcohol biomarker. In one study, the mean half-life of β-Hex-B was 6.5 days and that of CDT was 8.6 days, and combining both markers further increased sensitivity as an alcohol biomarker [27]. Stowell *et al.* reported that mean β-Hex-B% was 52.4% for alcoholics, 40.2% for heavy drinkers, 29.0% for moderate drinkers, and 27.5% for nondrinkers, indicating that β-Hex-B% is a good biomarker for alcohol abuse. Also, this parameter correlated well with serum CDT values. The authors concluded that the cutoff value of serum β-Hex-B% was 35%, and any value greater than this indicates heavy alcohol consumption ($>$60 g per week). At that cutoff value, this alcohol biomarker had a specificity of

91% and sensitivity of 94%. This marker is more sensitive than GGT, AST, ALT, or MCV as an alcohol biomarker. Moreover, serum β-Hex-B% is slightly more sensitive than CDT [28].

β-Hexosaminidase is a large molecule that is not filtered during glomerular filtration. However, it is abundantly present in the cells of proximal tubules, and a small amount is excreted in urine of normal subjects due to the exocytosis process. Elevated total β-hexosaminidase levels (two- or threefold greater than normal value) in urine have been reported in chronic alcohol abusers. Wehr *et al.* suggested that urinary β-hexosaminidase can be used to monitor sobriety in alcohol-dependent individuals [29]. Various conditions may falsely elevate serum and urine β-hexosaminidase levels (see Table 4.3). See Chapter 7 for further discussion.

4.7 ETHYL GLUCURONIDE AND ETHYL SULFATE AS ALCOHOL BIOMARKERS

Alcohol biomarkers such as GGT, MCV, CDT, and β-hexosaminidase are indirect alcohol biomarkers. In contrast, ethyl glucuronide and ethyl sulfate are direct alcohol biomarkers because they are minor alcohol metabolites. Ethyl glucuronide and ethyl sulfate are minor products of phase II ethanol metabolism, representing less than 0.1% of total ethanol disposition. Ethyl glucuronide is formed by conjugation with glucuronic acid catalyzed by the enzyme UDP-glucuronosyltransferase, whereas ethyl sulfate formation is catalyzed by sulfotransferase. The presence of ethyl glucuronide and ethyl sulfate indicates recent alcohol consumption when alcohol is no longer present in the urine. Consumption of a relatively small quantity of alcohol, such as 7 g, may result in a detectable ethyl glucuronide level in urine for up to 6 hr. Detection time is longer after consumption of higher amounts of alcohol. Both ethyl glucuronide and ethyl sulfate could be detected in urine after 2–5 days after abstinence in alcohol-dependent patients who were admitted to an alcohol rehabilitation center. The cutoff concentration of urinary glucuronide may vary from laboratory to laboratory; cutoff values of 100, 500, and 1000 ng/mL have been reported [30]. However, a cutoff value of 100 ng/mL (0.1 mg/L) for ethyl glucuronide may be more appropriate to establish abstinence. A cutoff value of 500 ng/mL for ethyl glucuronide is widely used by many laboratories, although higher cutoff levels have also been proposed, especially in forensic situations. The proposed cutoff value for ethyl sulfate ranges from 50 to 200 ng/mL [31].

In recent years, much attention has focused on determining the concentration of ethyl glucuronide in hair. Usually, a scalp hair specimen is collected by cutting hair using scissors as closely as possible to the skin, and an approximately 3-cm proximal segment is used for analysis. In general, the ethyl glucuronide level in hair in 95% of abstainers studied was less than 1.0 pg/mg of hair, whereas 30% of abstainers had an ethyl glucuronide level below the detection limit of the highly sensitive liquid chromatography combined with tandem mass spectrometry assay (LC-MS/MS; detection limit: 0.5 pg/mg of hair) [32]. Various cutoff concentrations have been proposed for analysis of ethyl glucuronide in hair, where the value is expressed as picograms per milligram of hair. Morini *et al.* stated that 27 pg/mg provided the best compromise between sensitivity (92%) and specificity (96%). The authors further commented that hair color, gender, age, body mass index, smoking, and cosmetic treatment of hair did not influence hair analysis for ethyl glucuronide [33]. Based on a meta-analysis of 15 records, Boscolo-Berto *et al.* noted that a cutoff of 30 pg/mg limits the false-negative effect in differentiating heavy from social drinking, whereas the 7 pg/mg cutoff may only be used for suspecting an active alcohol use and not proving complete abstinence [34].

Ethyl glucuronide in meconium is also measured to investigate possible exposure of fetuses to maternal alcohol use. Based on a study of 557 women with singleton births and available data including meconium specimens, Goecke *et al.* reported that only ethyl glucuronide in meconium showed an association with alcohol consumption history [35]. Bana *et al.* used a cutoff of 50 ng/g of meconium for ethyl glucuronide and 1000 ng/g of meconium for fatty acid ethyl esters in their study and reported that 34.6% of women consumed alcohol during pregnancy, whereas 17% of women showed positive results with both markers [36]. For hair ethyl glucuronide, sensitivity of 96% and specificity of 99% have been reported at a cutoff concentration of 30 pg/mg of hair to identify individuals who were consuming alcohol chronically at an amount exceeding 60 g per day [37]. For urinary glucuronide at a cutoff of 100 ng/mL, sensitivity and specificity were 76 and 93%, respectively. The authors also determined that sensitivity and specificity of urinary ethyl sulfate at a 25 ng/mL cutoff were 82 and 86%, respectively, for identifying alcohol consumption 3–7 days prior to clinic visits [38].

Both false-positive and false-negative results have been reported with ethyl glucuronide and ethyl sulfate. False-positive test results with ethyl glucuronide and sulfate may be due to incidental exposure to alcohol-containing products such as mouthwash and hand sanitizers, especially if lower cutoff concentrations are used. Consuming nonalcoholic beer and wine in larger amounts may also produce false-positive results because such products may contain a small amount of alcohol. Eating baker's yeast with sugar, drinking large amounts of apple juice, or even eating ripe bananas may cause detectable amounts of ethyl glucuronide and ethyl sulfate in urine. Urinary tract infection may produce false-negative test results due to degradation of

Table 4.4 Limitations of Direct Alcohol Biomarkers

Direct Alcohol Biomarkers	Comments
Ethyl glucuronide and ethyl sulfate in urine	■ Incidental exposure to ethanol other than intentional consumption may cause false-positive test results with ethyl glucuronide and/or ethyl sulfate if a lower cutoff is used. Extensive use of ethanol-containing mouthwash or hand sanitizer and drinking large amounts (2.5 L) of nonalcoholic beer or wine (750 mL) may cause false-positive results. ■ Therapy with chloral hydrate or use of propanol/isopropanol-containing hand sanitizer may cause false-positive results only if DRI immunoassay for ethyl glucuronide is used. LC/MS/MS-based assays are not affected. ■ False-negative ethyl glucuronide result may be encountered in a patient with urinary tract infection because *E. coli* present in urine may degrade ethyl glucuronide. Ethyl sulfate is not affected. Sodium fluoride should be used as preservative to avoid such degradation.
Fatty acid ethyl esters in hair	■ Regular use of hair products containing as little as 10% alcohol can impact fatty acid ethyl ester values in hair. In this case, ethyl glucuronide should be tested because ethyl glucuronide values in hair seem to be unaffected by use of hair products containing alcohol.
Phosphatidylethanol in blood	■ Phosphatidylethanol may be formed *in vitro* in the presence of alcohol. ■ Phospholipase D is capable of producing phosphatidylmethanol in a person overdosed with methanol. Such a peak usually occurs near the peak of phosphatidylethanol in the chromatogram during LC/MS/MS analysis.

ethyl glucuronide in urine by β-glucuronidase enzyme present in *Escherichia coli*. Ethyl sulfate is not affected (Table 4.4). In 2006, the U.S. Substance Abuse and Mental Health Services Administration issued an advisory because of concern of false-positive test results with ethyl glucuronide testing and warned against use of ethyl glucuronide as the sole evidence in determining abstinence in criminal justice, regulatory, or legal settings [30]. See Chapter 8 for more in-depth discussion on this topic.

4.8 FATTY ACID ETHYL ESTER AS ALCOHOL BIOMARKER

Fatty acid ethyl esters are minor metabolites of ethanol that are formed after alcohol consumption in virtually all tissues due to interaction of ethanol with free fatty acids as well as triglycerides, lipoproteins, and phospholipids. This pathway is an enzyme-mediated esterification of fatty acid or fatty acetyl-CoA by ethanol. The fatty acid ethyl ester synthase plays a role in the formation of fatty acid ethyl esters. In serum, fatty acid ethyl esters appear after alcohol consumption and are bound to albumin and also found in the core of lipoproteins along with other neutral lipids. Fatty acid ethyl esters can be used as a marker to determine both acute and chronic ingestion of alcohol. The concentrations of fatty acid ethyl esters in serum parallel the ethanol level in serum after acute ingestion of alcohol. If blood alcohol is negative but fatty acid ethyl ester test is positive, then it can be assumed that alcohol consumption has occurred within the past 24 hours [39].

Although serum can be used to determine concentrations of fatty acid ethyl esters, the recent focus has been the determination of hair concentration of fatty acid ethyl esters in both clinical and forensic investigations. However, hair from a strictly abstinence person may show a very small amount of fatty acid ethyl ester, which may be related to trace amounts of endogenous ethanol production or nutrition or the use of hair cosmetics. Hair specimens collected from the pubic region, armpit, chest, arm, or thigh show comparable fatty acid ethyl esters to specimens collected from scalp hair. Although approximately 15–20 fatty acid ethyl esters can be detected in a specimen, the most common ones used for calculation in varying amounts include ethyl laurate, ethyl myristate, ethyl palmitate, ethyl palmitoleate, ethyl stearate, ethyl oleate, ethyl linoleate, ethyl linolenate, ethyl arachidonate, and ethyl docosahexaenoate. Sometimes only the sum of the concentrations of four fatty acid ethyl esters (ethyl myristate, ethyl oleate, ethyl palmitate, and ethyl stearate) is used to determine the fatty acid ethyl ester concentration in a specimen. The sum of these fatty acid ethyl esters in hair specimens varies from less than 0.2 ng/mg of hair in strict teetotalers to more than 30 ng/mg in samples from alcoholic death [40]. Bertol *et al.* also proposed a cutoff of 0.5 ng/mg for fatty acid ethyl ester concentration in hair (3-cm segment analyzed) to differentiate between social drinking and excessive alcohol consumption (>60 g per day) [41]. Wurst *et al.* reported that at a 0.29 ng/mg cutoff for fatty acid ethyl esters in hair, the sensitivity and specificity were 100 and 90%, respectively, in individuals who chronically abused alcohol [42]. Combining fatty acid ethyl ester and ethyl glucuronide in hair to differentiate social drinking from heavy drinking has also been proposed. Pragst *et al.* commented that the cutoff values of 0.5 ng/mg for fatty acid ethyl ester in hair and 30 pg/mg for ethyl glucuronide in a proximal hair segment 0–3 cm long seem to be an optimal compromise for discriminating heavy consumption of alcohol from social drinking [43].

Meconium analysis of fatty acid ethyl esters is a valid method for identifying heavy prenatal ethanol exposure. However, small amounts of fatty acid ethyl esters are found in the meconium of neonates without any maternal use of alcohol. This may originate from endogenous ethanol or a trace amount of alcohol coming from food. Depending on the combination of fatty acid ethyl esters used for calculation, various cutoff values have been proposed; a cutoff of 2 nmol/g (approximately 600 ng/g) has been widely used. See Chapter 8 for further discussion.

False-positive results have been reported with fatty acid ethyl ester analysis in hair. Hartwig *et al.* evaluated the effect of hair care products on fatty acid ethyl ester concentrations in hair and observed false-positive results due to frequent treatment with hair lotions containing high amounts of alcohol. Some hair wax products also contain fatty acid ethyl esters, but use of such hair wax should not increase the concentration of fatty acid ethyl ester in hair. In doubtful cases, pubic hair should be analyzed for comparison [44]. Gareri *et al.* studied the effect of hair care products on fatty acid ethyl ester concentrations in hair and concluded that regular use of hair products containing as little as 10% alcohol can impact fatty acid ethyl ester values in hair. In this case, ethyl glucuronide should be tested because ethyl glucuronide values in hair seem to be unaffected by use of hair products containing alcohol [45].

CASE REPORT 4.2

A male infant was born at 39 weeks of gestation by uncomplicated vaginal delivery. Birth weight was 2770 g, length was 46 cm, and head circumference was 32 cm. The infant was jittery at birth. The mother was 36 years old, and she tested positive for hepatitis C. Her two other children were in foster care, and she admitted that she had been consuming alcohol, primarily beer, on weekends for many years. She also reported that during pregnancy, she smoked two packs of cigarettes every 2 or 3 days and drank beer occasionally until 6 months of pregnancy, when she consumed high amounts of alcohol. She reduced her alcohol consumption significantly 2 weeks before delivery but did not stop drinking. Analysis of the infant's meconium showed a very high level of total fatty acid ethyl ester (13,126 ng/g). In contrast, analysis of meconium of three infants not exposed to alcohol *in utero* according to maternal report showed a mean value of 410 ng/g. It is not completely understood whether fatty acid ethyl esters are formed mostly in the mother and cross the placenta or whether they are formed in the fetus. However, the placenta has all the necessary enzymes to produce fatty acid ethyl esters. In addition, fatty acid ethyl esters may be produced in meconium in the presence of ethanol. For example, when the authors supplemented meconium *in vitro* with ethanol, an increased level of linoleic acid ethyl ester was observed. However, no fatty acid ethyl ester was produced when amniotic fluid was supplemented *in vitro* with ethanol. The authors commented that fatty acid ethyl ester is an alcohol biomarker for *in utero* alcohol exposure [46].

4.9 PHOSPHATIDYLETHANOL AS ALCOHOL BIOMARKER

Phosphatidylethanol is a group of phospholipids formed through phospholipase D-mediated enzymatic reaction between ethanol and phosphatidylcholine in cell membranes. Phosphatidylethanol is measured in whole blood, in which it is mostly associated with erythrocyte cell membranes.

Phosphatidylethanol is not a single molecule but, rather, a group of glycerophospholipid homologs with a common phosphoethanol head group and two long fatty acid chains attached to a glycerol backbone (*sn*-1 and *sn*-2 positions). There exist many combinations of fatty acid chain lengths and numbers of double bonds in the fatty acids that are attached to the *sn*-1 and *sn*-2 positions of the glycerol backbone, but phosphatidylethanol 16:0/18:1 (palmitic acid/oleic acid) and phosphatidylethanol 16:0/18:2 (palmitic acid/linoleic acid) are two major molecular species extracted from human erythrocytes [47]. Although a combination of phosphatidylethanol molecular species 16:0/18:1 and 16:0/18:2 may be sufficient to represent total phosphatidylethanol level in blood, some investigators have used a combination of more molecular species for their studies, such as 16:0/18:1, 16:0/18;2, 16:0/20:4, 18:1/18:1, and 18:1/18:2. However, other investigators prefer to use 16:0/18:1 molecular species, which is the most abundant molecular species for quantification. Moreover, reference standards are commercially available.

Consumption of one alcoholic drink is not sufficient to produce a detectable level of phosphatidylethanol in blood. In general, consumption of 50 g or more alcohol per day for several weeks is necessary to produce detectable amounts of phosphatidylethanol in blood. However, once positive, it remains positive for 2 or 3 weeks, whereas blood alcohol may be undetectable several hours after consumption of the last drink. The half-life of phosphatidylethanol in blood is 4 days, and the amount of alcohol consumed correlates with blood phosphatidylethanol level. This marker is almost 100% specific because its formation is totally dependent on the presence of ethanol. In addition, this marker is more sensitive than other traditional alcohol biomarkers, including CDT, GGT, and MCV [48].

No cutoff concentration has been firmly established for clinical application of phosphatidylethanol, but values between 0.2 and 0.7 μmol/L have been proposed based on the limit of quantitation of the high-performance liquid chromatography method. In Sweden, 0.7 μmol/L (~492 ng/mL) has been proposed, at which value people who consume 50 g of alcohol per day or more can be identified; at a cutoff of 0.2 μmol/L (~140 ng/mL), alcohol consumption of 40 g or less may be assumed [47]. Phosphatidylethanol in blood has the highest specificity of 100% at a cutoff value of 0.36 μmol/L, whereas the sensitivity at this cutoff is 94.5% to differentiate between drinkers and sober patients [49]. However, phosphatidylethanol may be formed *in vitro* if ethanol is present. Moreover, a patient overdosed with methanol may show the presence of a phosphatidylmethanol peak, which is eluted very close to the phosphatidylethanol peak in the chromatogram (see Table 4.4). See Chapter 8 for further details.

4.10 TOTAL PLASMA SIALIC ACID AS ALCOHOL BIOMARKER

Sialic acids comprise a family of more than 50 naturally occurring carbohydrates that are derivatives of the nine-carbon sugar neuraminic acid (5-amino-3,5-dideoxy-D-glycero-D-galactononulsonic acid). One branch of the sialic acid family is *N*-acetylated to form *N*-acetylneuraminic acids, which are the most common form of sialic acid in humans [50]. In human serum, the majority of sialic acid is *N*-acetylneuraminic acid; other forms are found only in trace amounts. Total sialic acid in serum is the sum of protein-bound sialic acid, lipid-bound sialic acid, and free sialic acid, although free sialic acid represents only a very small fraction of total sialic acid. The molecular weights of sialic acids vary with their substitutions, but the average molecular weight is assumed to be 328.2 for sialic acid in human serum or plasma. The normal concentration of total sialic acid in serum or plasma is 1.58–2.22 mmol/L (52–73 mg/dL), and that of free sialic acid is 0.5–3 μmol/L (164–985 ng/mL) [51]. Chronic alcoholic consumption inhibits glycosylation of many proteins, such as transferrin, fibrinogen, and complement proteins. Because alcohol interferes with glycosylation, it is expected that total sialic acid concentration in serum should increase in individuals who consume alcohol on a regular basis. However, total sialic acid in serum or plasma is an indirect alcohol biomarker. Sillanaukee *et al.* studied 38 social drinkers and 77 alcoholics and suggested a cutoff concentration of 77.8 mg/dL in females and 80 mg/dL in males. Using these cutoffs, the sensitivity and specificity of sialic acid as an alcohol biomarker were 57.7 and 95.5%, respectively, in women and 47.8 and 81.3%, respectively, in men [51]. Total sialic acid in plasma may be increased in a variety of diseases, which are listed in Table 4.5. See Chapter 9 for more details.

Table 4.5 Limitations of Total Plasma Sialic Acid, Sialic Acid Index of Apolipoprotein J, and 5-HTOL/5-HIAA Ratio as Alcohol Biomarkers

Alcohol Biomarker	Comments
Total plasma sialic acid level	■ Various cancers, renal disease, diabetes, bacterial infection, inflammatory diseases, and sympathetic ophthalmitis may increase plasma total sialic acid level. ■ Inherited disorders of sialic acid may cause elevated levels. ■ Elderly people may have higher plasma total sialic acid level.
Sialic acid index of apolipoprotein J	■ No specific cases of false-positive results have been reported.
5-HTOL/5-HIAA	■ Therapy with aldehyde dehydrogenase inhibitors such as disulfiram (Antabuse) may cause a high 5-HTOL/5-HIAA ratio in urine.

4.11 SIALIC ACID INDEX OF APOLIPOPROTEIN J

Apolipoprotein J (Apo J) or clusterin is a highly sialylated glycoprotein that is a normal component of plasma high-density lipoprotein particles. Approximately 30% of the Apo J molecule is carbohydrate, and there may be up to 28 mol of sialic acid per mole of Apo J [52]. The synthesis of the mature Apo J molecule requires the addition of sugars to the molecule in a sequential manner and termination of the attachment with sialic acid molecules. Therefore, in alcoholics, the activity of enzymes such as sialyltransferase may be reduced, thus inhibiting the incorporation of sialic acid molecules in the mature Apo J molecule. Gong *et al.* reported downregulation of the liver sialyltransferase gene in human alcoholics, which caused defective glycosylation of a number of proteins, including Apo E and Apo J [53]. As a result, the sialic acid index of plasma Apo J could be used as an alcohol biomarker, where moles of sialic acid per mole of Apo J should be reduced in alcoholics compared to healthy individuals. One study reported that in human subjects, intake of alcohol for 30 days resulted in an almost 50% decrease in the sialic acid index of Apo J. Patients who consumed 50–60 g of alcohol per day showed a mean sialic index of Apo J (moles of sialic acid per mole of Apo J) of 14 ($n = 15$), whereas in controls the value was 28 ($n = 15$). The values returned to normal after 6–8 weeks of abstinence [54]. Specificity of the sialic acid index of Apo J is approximately 100%, sensitivity is approximately 90–92%, and half-life is 4 or 5 weeks [55].

4.12 5-HTOL/5-HIAA AS ALCOHOL BIOMARKER

Normally, serotonin (5-hydroxytryptamine), a neuromodulator, is metabolized to 5-hydroxyindole-3-acetaldehyde by the action of the monoamine oxidase enzyme (monoamine oxidase A has the highest affinity for serotonin). Then 5-hydroxyindole-3-acetaldehyde is either oxidized to 5-hydroxyindole-3-acetic acid (5-HIAA) or reduced to 5-hydroxytryptophol (5-HTOL). Oxidation to 5-HIAA is the major metabolic pathway of serotonin, whereas 5-HTOL is a minor metabolic pathway. However, alcohol consumption causes a shift in the serotonin metabolism from 5-HIAA to 5-HTOL due to competitive inhibition of aldehyde dehydrogenase by acetaldehyde, the major metabolite of alcohol. In addition, increased levels of NADH due to alcohol metabolism also favor 5-HTOL formation [56]. Whereas 5-HIAA is secreted in the urine unconjugated, 5-HTOL is excreted in the urine as glucuronide conjugate (GTOL). Sulfate conjugate of 5-HTOL is also found in the urine. The ratio of 5-HTOL to 5-HIAA in urine is usually determined as picomoles of 5-HTOL per nanomoles of 5-HIAA to evaluate alcohol consumption because the ratio increases significantly in people who consume alcohol. The advantages of using this ratio are that it compensates for urine dilution and accounts for dietary sources of serotonin. In general, it is assumed that the detection window of the 5-HTOL/5-HIAA ratio in urine is approximately 5−15 hr longer than that of ethanol in urine, and this biomarker is considered a 24-hr alcohol biomarker. Although consumption of low amounts of ethanol (<10 g) may not increase the 5-HTOL/5-HIAA ratio in urine, consumption of 50 g or more should increase the ratio significantly. Higher ratios are indicative of more ethanol consumption, almost in a dose-dependent manner. The specificity of this alcohol biomarker is almost 100% [57]. The sensitivity for consuming 50 g or more of alcohol is 77% at a cutoff value of 15 pmol/nmol [58]. However, Voltaire *et al.* proposed a cutoff of 20 pmol/nmol [59]. Instead of the 5-HTOL/5-HIAA ratio, some authors have used the GTOL/5-HIAA ratio as an alcohol biomarker, which provides similar results as those obtained using the 5-HTOL/5-HIAA ratio as an alcohol biomarker. Apart from alcohol ingestion, treatment with an aldehyde dehydrogenase inhibitor such as disulfiram is the only known cause of an abnormally high 5-HTOL/5-HIAA ratio in urine [60]. See Chapter 9 for more details. Limitations of the 5-HTOL/5-HIAA ratio as an alcohol biomarker are summarized in Table 4.5.

4.13 OTHER ALCOHOL BIOMARKERS

In addition to the biomarkers previously described, there are other alcohol biomarkers that are less often used in the clinical setting. Acetaldehyde, the major metabolite of ethanol, is a highly reactive molecule that reacts with various proteins forming adducts. Hemoglobin adduct of acetaldehyde can be used as an alcohol biomarker. Dolichol is also a potential alcohol biomarker (see Chapter 7). Plasma cholesteryl ester transfer protein (CETP) plays an important role in reverse cholesterol transport—the process in which cholesterol is transported from peripheral tissue back to the liver. Alcohol consumption lowers the activity of CETP; thus, CETP has a potential application as an alcohol biomarker. Homocysteine is an excitatory amino acid that markedly enhances vulnerability to neural cells to oxidative injury. Chronic alcoholism can increase the serum or plasma concentration of homocysteine. Cytokines play an important role in regulating various processes, including inflammation, cell death, cell proliferation, and cell mitigation. Circulating cytokines such as tumor necrosis factor-α, interleukin-1 (IL-1), and IL-6 are elevated in both chronic and acute alcohol-induced liver disease. In addition, IL-8, IL-12, and monocyte chemoattractant protein-1 (MCP-1) may have potential as alcohol biomarkers. See Chapter 9 for more details.

4.14 CLINICAL APPLICATION OF ALCOHOL BIOMARKERS

Although moderate alcohol consumption has health benefits, heavy alcohol consumption is detrimental to health (see Chapter 1). Alcohol use disorder (AUD) is a medical condition affecting an estimated 18 million Americans. Therefore, early identification of individuals who may be consuming excessive alcohol is important so that intervention may be instituted to prevent alcohol-related injuries and illness. Because there is no universal alcohol biomarker that can be used as a gold standard, it is generally assumed that the structured interview provides the best reference standard on which a screening test such as an alcohol biomarker can be validated. In general, a structured interview by a clinician or a health care professional, assessment of alcohol use using a questionnaire, and laboratory tests including alcohol biomarkers can be used to identify individuals who are at high risk of developing AUD.

4.14.1 Diagnosis Using *DSM-IV* and *DSM-5*

AUD can be diagnosed in a structured interview setting using the criteria in the fourth edition of the *Diagnostic and Statistical Manual of Mental Disorders* (*DSM-IV*; American Psychiatric Association). The *DSM-5*, an updated version of *DSM-IV*, is now available, but there is considerable overlap between *DSM-IV* and *DSM-5*. *DSM-IV* describes two distinct disorders—alcohol abuse and alcohol dependence—with specific criteria for each. However, in *DSM-5*, both disorders are integrated into one disorder termed alcohol use disorder (AUD) with subclassification of mild, moderate, and severe disorders. Eleven criteria of AUD are listed in *DSM-5*. The presence of at least 2 of the 11 criteria in the past 12 months indicates that a patient is suffering from AUD. If only 2 or 3 criteria are present, the diagnosis is mild AUD; if 4 or 5 criteria are present, it is moderate AUD; and if 6 or more criteria are present, the diagnosis is severe AUD. In one study using *DSM-5* criteria, the prevalence of AUD was 10.8% among 34,653 survey participants. According to *DSM-IV* criteria, 9.7% in the same population would have the AUD diagnosis [61].

4.14.2 Self-Assessment of Alcohol Use

In addition to the structured interview format, self-assessment of alcohol use by patients can be used for diagnosis of AUD. Clinically, self-assessment questionnaires such as AUDIT, MAST, and CAGE are widely used. CAGE is based on self-assessment of four factors (C: Have you ever felt you should cut down on your drinking? A: Have people annoyed you by criticizing your drinking? G: Have you ever felt bad or guilty about your drinking? E: Eye opener—Have you ever had a drink first thing in the morning to steady your nerves or to get rid of a hangover?). MAST (Michigan Alcoholism Screening Test) is a long test containing 24 questions. AUDIT (Alcohol Use Disorder Identification Test) was developed in collaboration with WHO as a simple screening test for early detection of AUD. AUDIT consists of 10 questions about alcohol consumption (questions 1–3), dependence (questions 4–6), and alcohol-related harm (questions 7–10). In general, AUDIT is better than CAGE in clinical settings. Subsequently, a three-question version of AUDIT called AUDIT-C (Alcohol Use Disorder Identification Test—Consumption) was developed that is simpler than AUDIT because it contains only the first three questions of the AUDIT questionnaire. This simple version is useful for screening patients by trauma physicians, and the American College of Surgeons has mandated that all trauma patients should undergo routine AUD screening. The optimum cutoff score of AUDIT is 8 (maximum possible score 40), whereas the corresponding cutoff score of AUDIT-C is 5 (maximum possible score 12), providing a sensitivity and specificity of 88 and 91%, respectively, for both tests. Both AUDIT and AUDIT-C have excellent overall accuracy for the detection of hazardous alcohol consumption. There is no significant difference in performance between AUDIT and AUDIT-C [62]. However, Bradley *et al.* suggested an AUDIT-C cutoff score of 4 in men (sensitivity, 86%; specificity, 89%) and 3 in women (sensitivity, 73%; specificity, 91%) to identify alcohol misuse [63]. Berner *et al.* found that AUDIT was largely heterogeneous because sensitivity as well as specificity varied significantly between different studies. Therefore, AUDIT should be restricted to the primary care population, inpatients, and elderly patients [64]. Dawson *et al.* compared the performance of AUDIT-C in screening for *DSM-IV* and *DSM-5* criteria in the diagnosis of AUD and found that AUDIT-C score at a cutoff point of 4 optimally screens for *DSM-IV* criteria and also optimizes screening for *DSM-5* criteria. Therefore, clinicians should not have to undertake any major overhaul of their current screening procedure as a result of *DSM-5* revision and should benefit from fewer false-positive results [65].

4.14.3 Alcohol Biomarkers and AUDIT

AUDIT score is capable of providing a history of alcohol use in the past year, whereas no alcohol biomarker has this capability. Most alcohol biomarkers can provide information regarding recent use of alcohol, and hair testing may provide information up to a few months. Nevertheless, self-reported alcohol use such as AUDIT score has limitations. Patients may be reluctant to provide an accurate representation of past drinking history due to the social stigma associated with heavy consumption of alcohol. Therefore, AUDIT score and alcohol biomarkers may be complimentary in identifying patients who may be consuming excess alcohol. The validity of self-reporting may be improved if it is used in combination with a laboratory test such as alcohol biomarker. Using an AUDIT cutoff score of 8 and CDT cutoff scores of 20 U/L in males and 27 U/L in females, Hermansson *et al.* reported that out of 570 subjects who participated in a workplace health examination, 105 subjects (18.4%) screened positive according to AUDIT score, CDT cutoff, or both. The AUDIT score varied between 8 and 12 in subjects who screened positive by AUDIT, whereas CDT levels were 22–65 U/L in males and 30–36 U/L in females who screened positive. If GGT was included, then 125 subjects (22%) screened positive. However, if only AUDIT was used in the screening process, the proportion of all positives would have decreased by nearly half. The authors concluded that AUDIT and CDT are complementary for alcohol screening in a routine workplace health examination [66]. Wurst *et al.* found that self-reported ethanol intake in the past 28 days correlated with AUDIT score, with the direct ethanol metabolite (ethyl glucuronide and ethyl sulfate), and with MCV. However, results from biomarker tests could indicate cases of under- as well as overreporting of alcohol consumption [67]. In another report, the authors commented that although hair ethyl glucuronide correlated with AUDIT score, hair ethyl glucuronide identified 10 more cases of positive alcohol [68].

CASE REPORT 4.3

A 39-year-old woman who initially presented with bleeding gums due to excessive warfarin therapy, which she took prophylactically due to deep vein thrombosis, was seen in the psychiatry clinic for depression. She had a history of substance abuse and binge drinking. However, she reported that she had stopped drinking and had remained abstinent for the past year because of fear of further damaging her kidneys. Laboratory tests showed AST of 13 U/L, ALT of 19 U/L, but an elevated GGT level of 104 U/L. Her MCV was also elevated to 101 fL. The combined elevated GGT and MCV has a sensitivity of 95% for alcohol abuse, and her laboratory tests indicated that she had recently consumed a substantial amount of alcohol [69].

4.14.4 Application of Alcohol Biomarkers

Because patients may not disclose recent alcohol consumption or may underreport alcohol consumption, it is important to conduct alcohol biomarker testing in patients suspected of alcohol abuse. Currently, alcohol biomarkers are used in various clinical settings, which are listed in Table 4.6. In general, only 10–13% of patients with AUD seek treatment, and the vast majority of patients do not seek any specific help. Therefore, it is likely that many more patients with drinking problems will see a primary care physician. Although a primary care physician may not routinely test for specific alcohol biomarkers in all patients, during routine laboratory investigation elevated liver enzymes such as GGT may alert a clinician regarding potential alcohol abuse by a patient. In this case, testing of CDT may be useful. Kapoor *et al.* reported that

Table 4.6 Clinical Application of Alcohol Biomarkers

Clinical Settings	Indication for Use	Possible Biomarkers Used
Primary care	Screening for alcohol abuse/potential relapse	GGT and CDT are most commonly used by ethyl glucuronide, and other markers may also be used.
Criminal justice	Abstinence	Blood and urine alcohol levels along with ethyl glucuronide, ethyl sulfate, and hair fatty acid ethyl ester are commonly used. Transdermal alcohol sensor[a] devices may also be used.
Drug/alcohol rehabilitation program	Abstinence	Ethyl glucuronide and ethyl sulfate, along with blood and urine alcohol, are used. Transdermal alcohol sensor devices may also be used.
Health and safety screening program	Screening for alcohol abuse	Ethyl glucuronide, ethyl sulfate, GGT, and CDT are used.
Workplace drug/alcohol testing	Abstinence	Blood and/or urine alcohol along with ethyl glucuronide and ethyl sulfate, if necessary, are used.
Emergency medicine/ trauma/transplant service	Abstinence	GGT, CDT, ethyl glucuronide, ethyl sulfate, and possibly hair fatty acid ethyl ester are used.
Pregnant women	Abstinence	Ethyl glucuronide, ethyl sulfate in urine, or meconium is useful. In addition, testing of fatty acid ethyl ester in maternal hair may provide additional information.

[a]*Transdermal alcohol sensor devices are discussed in Chapter 3.*

adding CDT screening to a patient's self-report of alcohol consumption in the primary care setting results in significant savings in health care costs [70]. Alcohol biomarker testings are widely used in criminal justice along with drug and possible urine alcohol determination. In addition, transdermal alcohol sensor devices (see Chapter 3) are widely used in criminal justice. Ethyl glucuronide and ethyl sulfate, as well as hair fatty acid ethyl ester, are used as alcohol biomarkers in criminal justice. A less commonly used biomarker is the 5-HTOL/5-HIAA ratio. In the workplace, it is important to monitor abstinence of problematic drinkers in high-risk job settings such as the transportation sector. Alcohol biomarker testing is also useful in pregnant women if alcohol consumption is suspected. In screening of pregnant women, testing for ethyl glucuronide, ethyl sulfate, and fatty acid ethyl ester may be useful. Testing of these biomarkers in meconium may also have additional benefits [71].

4.14.5 Combining Alcohol Biomarkers

To increase sensitivity and specificity of alcohol biomarkers, more than one biomarker may be combined. A common approach is to combine GGT and CDT using the mathematical equation that was discussed previously in this chapter and is also discussed in Chapter 5. In general, trisialotransferrin is not measured during the determination of CDT in serum. Tamigniau *et al.*

reported that CDT showed a poor sensitivity of 63% or lower in the identification of patients with heavy alcohol consumption. Combining trisialotransferrin with CDT significantly improved sensitivity and specificity. When MCV and GGT were added to this combination, performance was further improved [72]. Kip *et al.* studied 74 blood alcohol-negative male patients who presented to the emergency room with either thoracic or gastrointestinal complains using AUDIT score, phosphatidylethanol in whole blood, serum and urine ethyl glucuronide, %CDT, GGT, and MCV. The authors classified patients into two groups—one with an AUDIT score less than 8 ($n = 52$) and the other with an AUDIT score of 8 or higher ($n = 22$). The authors observed that out of 52 patients with an AUDIT score of 8 or less (cutoff score of 8 and higher indicates alcohol abuse), 13 patients tested positive with both urine ethyl glucuronide and whole blood phosphatidylethanol indicating recent alcohol consumption. In addition, %CDT was useful in identifying patients with a cutoff AUDIT score of less than 8 but who consumed alcohol despite claiming sobriety. The authors concluded that determination of phosphatidylethanol and urine ethyl glucuronide provided additional evidence in screening for AUD in the emergency room setting [73]. Karagulle *et al.* reported that combining homocysteine with CDT levels can be a useful method to identify patients at high risk of developing alcohol withdrawal seizures [74].

Another approach to increase sensitivity and specificity is to combine 8–15 biochemical and clinical indicators to develop an algorithm for identifying patients who consume excessive amounts of alcohol. It is assumed that in the general practice setting, maximally only 60% of patients with issues of alcohol abuse are identified, and such an algorithm may be helpful to identify more patients with problematic alcohol consumption. The main reasons for underdiagnosis are denial by certain patients and insufficient sensitivity of a selected biomarker to detect a less severe form of alcoholism. The Bayesian Alcoholism Test (BAT) can facilitate the confirmation of the diagnosis of hazardous and harmful alcohol use. In BAT, a computer program combines information about alcoholism and data on a particular patient (from selected blood test results and clinical signs) to yield a probability of hazardous and harmful alcohol use in the patient. Using clinical data and laboratory tests (MCV, CDT, GGT, ALT, AST, and alkaline phosphate), Korzec *et al.* developed the BAT, which had a higher sensitivity (94%) compared to CDT (63%) and GGT (73%). The authors concluded that BAT has better diagnostic properties than do CDT and GGT for confirming hazardous and harmful alcohol use [75]. Although the original BAT algorithm used a total of 15 clinical (hepatitis risk, diabetes mellitus, body mass index, palpable liver, spider naevi, level of response to alcohol, smoking, and responses to CAGE questions) and biochemical (GGT, ALT, AST, ALT/AST ratio, CDT, alkaline phosphatase, and MCV) parameters, 8 parameters can also be used in the BAT data set to calculate probability of harmful (>80 g/day), hazardous (40–80 g/day), and moderate alcohol consumption (<40 g/day). In general, the use of BAT with 8 parameters is also superior to CDT, GGT, or AST for predicting harmful or hazardous alcohol use [76].

4.15 CONCLUSIONS

Alcohol abuse is a significant problem worldwide, and early detection of individuals who may be prone to such abuse is essential. Alcohol abuse is associated with violent behavior, traffic accident fatalities, higher disease burden, and poor outcome in pregnancy. Therapeutic intervention is necessary to deal with alcohol abuse, and recently even yoga has been suggested as a feasible and adjunct therapy for alcohol dependence [77]. Structured interview using *DSM-IV* or *DSM-5* criteria, self-assessment of alcohol intake such as AUDIT score, and alcohol biomarkers are available to help clinicians identify patients who are consuming harmful or hazardous amounts of alcohol as well as patients who have AUD. Whereas structured interview and AUDIT can provide information regarding past year alcohol consumption pattern, alcohol biomarkers can provide information only on recent alcohol consumption or abstinence. BAT can facilitate the confirmation of the diagnosis of hazardous and harmful alcohol use.

References

[1] Freeman WM, Vrana KE. Future prospects for biomarkers of alcohol consumption and alcohol induced disorders. Alcohol Clin Exp Res 2010;34:946–54.

[2] Waszkiewicz N, Szajda SD, Kepka A, Szulc A, et al. Glycoconjugates in the detection of alcohol abuse. Biochem Soc Trans 2011;39:365–9.

[3] Adewale A, Ifudu O. Kidney injury, fluid, electrolyte and acid base abnormalities in alcoholics. Niger Med J 2014;55:93–8.

[4] Peterson K. Biomarkers of alcohol use. Alcohol Res Health 2004–2005;28:30–7.

[5] Whitfield JB. Gamma glutamyltransferase. Crit Rev Clin Lab Sci 2001;38:263–355.

[6] Arndt T, Behnken L, Martens B, Hackler R. Evaluation of the cut-off for serum carbohydrate deficient transferrin as a marker of chronic alcohol abuse determination by ChronAlco ID assay. J Lab Med 1999;23:507–10.

[7] Conigrave KM, Davies P, Haber P, Whitfield JB. Traditional markers of excessive alcohol use. Addiction 2003;98(Suppl. 2):31–43.

[8] Stranges S, Freudenheim JL, Muti P, Farinaro E, et al. Differential effects of alcohol drinking pattern on liver enzymes in men and woman. Alcohol Clin Exp Res 2004;28:949–56.

[9] Hannuksela ML, Liisanantti M, Nissinen A, Savolainen M. Biochemical markers of alcoholism. Clin Chem Lab Med 2007;45:953–61.

[10] Danielsson J, Kangastupa P, Laatikainen T, Aalto M, et al. Dose and gender specific interactions between coffee consumption and serum GGT activity in alcohol consumers. Alcohol Alcohol 2013;48:303–7.

[11] Salaspuro M. Carbohydrate deficient transferrin compared to other markers of alcoholism: a systematic review. Alcohol 1999;19:261–71.

[12] Kaferle J, Strzoda CE. Evaluation of macrocytosis. Am Fam Physician 2009;79:203–8.

[13] Martensson O, Harlin A, Brandt R, Seppa K, et al. Transferrin isoform distribution: gender and alcohol consumption. Alcohol Clin Exp Res 1997;21:1710–15.

[14] Sillanaukee P, Strid N, Allen JP, Litten RZ. Possible reasons why heavy drinking increases carbohydrate deficient transferrin. Alcohol Clin Exp Rep 2001;25:34–40.

[15] Arndt T. Carbohydrate deficient transferrin a marker of chronic alcohol abuse: a critical review of preanalysis, analysis and interpretation. Clin Chem 2001;47:13–27.

[16] Malcom R, Anton RF, Conradi SF, Sutherland S. Carbohydrate deficient transferrin and alcohol use in medical examiner's case. Alcohol 1999;17:7–11.

[17] Sorvajarvi K, Blake JE, Israel Y, Niemela O. Sensitivity and specificity of carbohydrate deficient transferrin as a marker of alcohol abuse are significantly influenced by alterations in serum transferrin: comparison of two methods. Alcohol Clin Exp Res 1996;20:449–54.

[18] Fleming M, Mundt M. Carbohydrate-deficient transferrin: validity of a new alcohol biomarker in sample patients with diabetes and hypertension. J Am Board Fam Pract 2004;17:245–55.

[19] Hock B, Schwarz M, Domke I, Grunert VP, et al. Validity of carbohydrate-deficient transferrin gamma glutamyl transferase (gamma-GT) and mean corpuscular erythrocyte volume (MCV) as biomarkers of chronic alcohol abuse: a study in patients with alcohol dependence and liver disorders of non-alcoholic and alcoholic origin. Addiction 2005;100:1477–86.

[20] Fleming MF, Anton RF, Spies CD. A review of genetic, biological, pharmacological, and clinical factors that affect carbohydrate deficient transferrin levels. Alcohol Clin Exp Res 2004;28:1347–55.

[21] Vidali M, Bianchi V, Bagnati M, Atzeni N, et al. False negativity to carbohydrate-deficient transferrin and drugs: a clinical case. Biochem Med (Zagreb) 2014;24:175–9.

[22] Sillanaukee P, Olsson U. Improved diagnostic classification of alcohol abusers by combining carbohydrate deficient transferrin and γ-glutamyl transferase. Clin Chem 2001; 47:681–5.

[23] Anttila P, Jarvi K, Latvala J, Blake JE, et al. A new modified γ-%CDT method improves the detection of problem drinking: studies in alcoholics with or without liver disease. Clin Chim Acta 2003;338:45–51.

[24] Borzym-Kluczyk M, Radziejewska I, Olszewska E, Szajda S, et al. Statistical evaluation of isoform pattern of N-acetyl-β-hexosaminidase from human renal cancer tissue separated by isoelectrofocusing. Clin Biochem 2007;40:403–6.

[25] Beutler E, Kuhl W. The tissue distribution of hexosaminidase S and hexosaminidase C. Ann Hum Genet 1977;41:163–7.

[26] Karkkainen P, Poikolainen K, Salaspuro M. Serum beta-hexosaminidase as a marker of heavy drinking. Alcohol Clin Exp Res 1990;14:187–90.

[27] Hultberg B, Isaksson A, Berglund M, Alling C. Increases and time course variations in beta-hexosaminidase isoenzyme B and carbohydrate-deficient transferrin in serum of alcoholics are similar. Alcohol Clin Exp Res 1995;19:452–6.

[28] Stowell L, Stowell A, Garrett N, Robinson G. Comparison of serum β-hexosaminidase isoenzyme B activity with serum carbohydrate deficient transferrin and other markers of alcohol abuse. Alcohol Alcohol 1997;32:713–14.

[29] Wehr H, Habrat B, Czartoryska B, Gorska D, et al. Urinary beta-hexosaminidase activity as a marker for the monitoring of sobriety. Psychiatr Pol 1995;29:689–96 [in Polish].

[30] Ingall GB. Alcohol biomarkers. Clin Lab Med 2012;32:391–406.

[31] Albermann ME, Musshoff F, Doberentz E, Heese P, et al. Preliminary investigations on ethyl glucuronide and ethyl sulfate cut-offs for detecting alcohol consumption on the basis of an ingestion experiment and on data from withdrawal treatment. Int J Legal Med 2012;126:757–64.

[32] Pirro V, Di Corcia D, Seganti F, Salomone A, et al. Determination of ethyl glucuronide levels in hair for assessment of alcohol abstinence. Forensic Sci Int 2013;232:229–36.

[33] Morini L, Politi L, Polettini A. Ethyl glucuronide in hair: a sensitive and specific marker of chronic heavy drinking. Addiction 2009;104:8915–20.

[34] Boscolo-Berto R, Viel G, Montisci M, Terranova C, et al. Ethyl glucuronide concentration in hair for detecting heavy drinking and/or abstinence: a meta-analysis. Int J Legal Med 2013;127:611–19.

[35] Goecke TW, Burger P, Fasching PA, Bakdash A, et al. Meconium indicators of maternal alcohol abuse during pregnancy and association with patients' characteristics. Biomed Res Int 2014;2014:702848.

[36] Bana A, Tabernero MJ, Perez-Munuzuri A, Lopez-Suarez O, et al. Prenatal exposure and its repercussion on newborns. J Neonatal Perinatal Med 2014;7:47–54.

[37] Boscolo-Berto R, Favretto D, Cecchetto G, Vincenti M, et al. Sensitivity and specificity of ETG in hair as a marker of chronic excessive drinking: pooled analysis of raw data and meta-analysis of diagnostic accuracy studies. Ther Drug Monit 2014;36:560–75.

[38] Stewart SH, Koch DG, Burgess DM, Willner IR, et al. Sensitivity and specificity of urinary glucuronide and ethyl sulfate in liver disease patients. Alcohol Clin Exp Res 2013;37: 150–5.

[39] Laposata M. Fatty acid ethyl esters: short-term and long-term serum markers of ethanol intake. Clin Chem 1997;43:1527–34.

[40] Sube S, Selavka CM, Mieczkowski T, Pragst F. Fatty acid ethyl ester concentrations in hair and self-reported alcohol consumption in 644 cases from different origins. Forensic Sci Int 2010;196:111–17.

[41] Bertol E, Bravo ED, Vaiano F, Mari F, et al. Fatty acid ethyl esters in hair: correlation with self-reported ethanol intake in 160 subjects and influence of estroprogestin therapy. Drug Test Anal 2014;6:930–5.

[42] Wurst FM, Alexson S, Wolfersdorf M, Bechtel G, et al. Concentration of fatty acid ethyl esters in hair of alcoholics: comparison to other biological markers and self-reported ethanol intake. Alcohol Alcohol 2004;39:33–8.

[43] Pragst F, Rothe M, Moench B, Hastedt M, et al. Combined use of fatty acid ethyl esters and ethyl glucuronide in hair for diagnosis of alcohol abuse: interpretation and advantages. Forensic Sci Int 2010;196:101–10.

[44] Hartwig S, Auwarter V, Pragst F. Effect of hair care cosmetics on the concentrations of fatty acid ethyl esters in hair as markers of chronically elevated alcohol consumption. Forensic Sci Int 2003;131:90–7.

[45] Gareri J, Appenzeller B, Walasek P, Koren G. Impact of hair care products on FAEE hair concentrations in substance abuse program. Anal Bioanal Chem 2011;400:183–8.

[46] Klein J, Karaskov T, Koren G. Fatty acid ethyl esters: a novel biological marker for heavy in utero ethanol exposure: a case report. Ther Drug Monit 1999;21:644–6.

[47] Helander A, Zheng Y. Molecular species of the alcohol biomarker phosphatidylethanol in human blood measured by LC-MS. Clin Chem 2009;55:1395–405.

[48] Isaksson A, Walther L, Hansoon T, Andersson A, et al. Phosphatidylethanol in blood (B-PEth): a marker for alcohol use and abuse. Drug Test Anal 2011;4:195–200.

[49] Hartmann S, Aradottir S, Graf M, Wiesbeck M, et al. Phosphatidylethanol as a sensitive and specific biomarker: comparison with gamma-glutamyl transpeptidase, mean corpuscular volume and carbohydrate deficient transferrin. Addict Biol 2007;12:81–4.

[50] Wang B, Brand-Miller J. The role and potential of sialic acid in human nutrition. Eur J Clin Nutr 2003;57:1351–69.

[51] Sillanaukee P, Ponnio M, Jaaskelainen IP. Occurrence of sialic acids in healthy humans and different disorders. Eur J Clin Invest 1999;29:413–25.

[52] Der Silva HV, Stuart WD, Park YB, Mao SJ, et al. Purification and characterization of apolipoprotein J. J Biol Chem 1990;265:14292–7.

[53] Gong M, Castillo L, Redman RS, Garige M, et al. Down-regulation of liver Galβ1,4GlcNac α2,6-sialyltransferase gene by ethanol significantly correlates with alcoholic steatosis in humans. Metabolism 2008;57:1663–8.

[54] Ghosh P, Hale EA, Lakshman R. Plasma sialic acid index of apolipoprotein J (SIJ): a new novel alcohol intake marker. Alcohol 2001;25:173–9.

[55] Waszkiewicz N, Szajda SD, Kepka A, Szulc A, Zwierz K. Glycoconjugates in the detection of alcohol abuse. Biochem Soc Trans 2011;39:365–9.

[56] Svensson S, Some M, Lundsjo A, Helander A, et al. Activities of human alcohol dehydrogenase in the metabolic pathways of ethanol and serotonin. Eur J Biochem 1999;262:324–9.

[57] Beck O, Helander A. 5-Hydroxytryptophol as a marker for recent alcohol intake. Addiction 2003;98(Suppl. 2):63–72.

[58] Helander A, von Wachenfeldt J, Hiltunen A, Beck O, et al. Comparison of urinary 5-hydroxytryptophol, breath ethanol and self-report for detection of recent alcohol use during outpatient treatment: a study on methadone patients. Drug Alcohol Depend 1999;56:33–8.

[59] Voltaire A, Beck O, Borg S. Urinary 5-hydroxytryptophol: a possible marker of recent alcohol use. Alcohol Clin Exp Res 1992;16:281–5.

[60] Helander A. Monitoring relapse drinking during disulfiram therapy by assay of urinary 5-hydroxytryptophol. Alcohol Clin Exp Res 1998;22:111–14.

[61] Agrawal A, Heath AC, Lynskey M. DSM-IV to DSM-5: the impact of proposed revisions on diagnosis of alcohol use disorder. Addiction 2011;106:1935–43.

[62] Vitesnikova J, Dinh M, Leonard E, Boufous S, et al. Use of AUDIT-C as a tool to identify hazardous alcohol consumption in admitted trauma patients. Injury 2014;45:1440–4.

[63] Bradley KA, DeBenedetti AF, Volk RJ, Williams EC, et al. AUDIT-C as a brief screen for alcohol misuse in primary care. Alcohol Clin Exp Res 2007;31:1208–17.

[64] Berner MM, Kriston L, Bentele M, Harter M. The Alcohol Use Disorders Identification Test for detecting at-risk drinking: a systematic review and meta-analysis. J Stud Alcohol Drugs 2007;68:461–73.

[65] Dawson DA, Smith SM, Saha TD, Rubinsky AD, et al. Comparative performance of the AUDIT-C in screening for DSM-IV and DSM-V alcohol use disorders. Drug Alcohol Depend 2012;126:384–8.

[66] Hermansson U, Helander A, Huss A, Brandt L, et al. The Alcohol Use Disorders Identification Test (AUDIT) and carbohydrate-deficient transferrin (CDT) in a routine workplace health examination. Alcohol Clin Exp Res 2000;24:180–7.

[67] Wurst FM, Haber PS, Wiesbeck G, Watson B, et al. Assessment of alcohol consumption among hepatitis-C positive people receiving opioid maintenance treatment using direct ethanol metabolites and self-report: a pilot study. Alcohol Biol 2008;13:416–22.

[68] Wurst FM, Dursteler-MacFarland KM, Auwaerter V, Ergovic S, et al. Assessment of alcohol use among methadone maintenance patients by direct ethanol metabolites and self-report. Alcohol Clin Exp Res 2008;32:1552–7.

[69] Spiegel DR, Dhadwal N, Gill F. I'm sober, doctor, really: best biomarkers for underreported alcohol use. Current Psychiatry 2008;7:15–28.

[70] Kapoor A, Kraemer KL, Smith KJ, Roberts MS, et al. Cost-effectiveness of screening for unhealthy alcohol use with %carbohydrate-deficient transferrin: results from a literature-based decision analytical computer model. Alcohol Clin Exp Res 2009;33:1–10.

[71] Litten RZ, Bradley AM, Moss HB. Alcohol biomarkers in applied settings: recent advances and future research opportunities. Alcohol Clin Exp Res 2010;34:955–67.

[72] Tamigniau A, Wallemacq P, Maisin D. Could trisialotransferrin be used as an additional biomarker to CDT in order to improve detection of chronic excessive alcohol intake? Clin Biochem 2014;47:1203–8.

[73] Kip MJ, Spies CD, Neumann T, Nachbar Y, et al. The usefulness of direct ethanol metabolites in assessing alcohol intake in nonintoxicated male patients in an emergency room setting. Alcohol Clin Exp Res 2008;32:1284–91.

[74] Karagulle D, Heberlein A, Wilhelm J, Frieling H, et al. Biological markers for alcohol withdrawal seizures: a retrospective analysis. Eur Addict Res 2012;18:97–102.

[75] Korzec A, de Bruijn C, van Lambalgen M. The Bayesian alcoholism test had better diagnostic properties for confirming diagnosis of hazardous and harmful alcohol use. J Epidemiol 2005;58:1024–32.

[76] Korzec S, Korzec A, Conigrave K, Gisolf J, et al. Validation of the Bayesian alcoholism test compared to single biomarkers in detecting harmful drinking. Alcohol Alcohol 2009;44:398–402.

[77] Hallgren M, Rombereg K, Bakshi AS, Andreasson S. Yoga as an adjunct treatment for alcohol dependence: a pilot study. Complement Ther Med 2014;22:441–5.

CHAPTER 5

Liver Enzymes as Alcohol Biomarkers

5.1 INTRODUCTION

The liver is the largest internal organ of the body, weighing approximately 1.2–1.5 kg. The liver performs multiple functions that are essential for sustaining life, and it is the principal site for synthesis of all circulating proteins except γ globulins. A functioning normal liver produces 10–12 g of albumin daily, and the half-life of albumin is approximately 3 weeks. When liver function is compromised over a prolonged period of time, albumin synthesis is severely impaired and hypoalbuminemia is a common finding in chronic liver disease. However, a significant reduction in serum albumin levels might not be observed in patients with acute liver failure. In addition to albumin, all clotting factors with the exception of factor VIII are produced in the liver. Therefore, as expected when liver function is significantly impaired, low levels of clotting factors are produced by the liver. As a result, coagulation tests such as prothrombin time (PT) are prolonged. Conventional liver function tests consist of the determination of serum or plasma level of bilirubin as well as activities of various liver enzymes, including alanine aminotransferase (ALT), aspartate aminotransferase (AST), alkaline phosphatase (ALP), and γ-glutamyl transferase (GGT; also known as γ-glutamyl transpeptidase). In addition, serum or plasma concentrations of total protein and albumin are useful in assessing liver function. Normal values of various analytes used as liver function tests are summarized in Table 5.1.

Breakdown of hepatocytes results in the release of aminotransferases (also referred to as transaminases) such as ALT and AST into the blood. ALT is a cytosol enzyme and more specific for liver disease. AST is primarily a mitochondrial enzyme that is also found in heart, muscle, kidney, and brain. ALT has a longer half-life than AST. In acute liver injury, AST levels are higher than those of ALT; however, after 24–48 hr, ALT levels should be

CONTENTS

5.1 Introduction 121
5.2 Factors Affecting Liver Function Tests... 122
5.3 Effect of Moderate Alcohol Consumption on Liver Enzymes... 125
5.4 GGT as Alcohol Biomarker......... 126
5.4.1 Elevated GGT in Various Diseases and as a Risk Factor for Mortality and Certain Illnesses 130
5.4.2 GGT Fraction as Alcohol Biomarker.... 132
5.5 Laboratory Determinations of Liver Enzymes... 133
5.6 Conclusions 134
References 134

A. Dasgupta: Alcohol and Its Biomarkers. DOI: http://dx.doi.org/10.1016/B978-0-12-800339-8.00005-5

Table 5.1 Normal Values of Common Analytes Used as Liver Function Tests

Analyte	Normal (Reference Value)[a]
Alanine aminotransferase (ALT)	7–55 U/L
Aspartate aminotransferase (AST)	8–48 U/L
Alkaline phosphatase (ALP)	45–115 U/L
γ-Glutamyl transferase (GGT)	9–48 U/L
Albumin	3.5–5.0 g/dL
Total protein	6.3–7.9 mg/dL
Bilirubin	0.1–1.0 mg/dL
Prothrombin time (PT)	9.5–13.8 sec

[a]Reference values were adopted from those described by the Mayo Medical Clinic (Rochester, MN). Various enzyme activities (also referred to as enzyme levels) are expressed in units per liter, which are numerically equivalent to international units per liter (IU/L).

higher than AST levels. ALT is considered to be a more specific marker for liver injury. ALP is found in liver, bone, intestine, and placenta. ALP is located in the canalicular and sinusoidal membrane of the liver. Because alcohol has toxic effects on the liver, liver enzymes are elevated in individuals who consume excessive amounts of alcohol. GGT is a membrane-bound glycoprotein located on the cell surface membrane of most cell types, including hepatocytes. Although liver enzymes, especially GGT, may be elevated in people who consume alcohol, the activities of liver enzymes are also affected by many personal and environmental factors, including smoking. Various factors other than alcohol consumption that may affect liver enzymes are summarized in Table 5.2.

5.2 FACTORS AFFECTING LIVER FUNCTION TESTS

In addition to hepatic injury, hepatic diseases, and hepatitis infection, many other factors can cause abnormal serum levels of liver enzymes. Liver enzymes are measured in clinical laboratories as enzymatic activities (values are expressed as units per liter (U/L) or international units per liter (IU/L); these are numerically equivalent). However, enzyme activities may also refer to serum enzyme levels. For an in-depth discussion of elevated liver enzymes and hepatic disease, the reader is advised to consult a medical textbook. In this chapter, the relationship between liver enzymes and alcohol consumption is discussed, and a brief discussion of various factors other than hepatic injury that may cause elevated liver enzymes is also presented.

Table 5.2 Various Factors Other Than Alcohol Consumption and Liver Diseases That May Affect Serum or Plasma Concentrations of Liver Enzymes

Factor	Effects
Age	ALT values decrease with advancing age, whereas GGT values increase. Up to a 30% increase in mean GGT values is observed between the age groups of 18–25 and 56–65 years in both men and women.
Gender	Girls may have lower ALT levels than boys. Men have higher GGT values than women.
Childbirth	GGT values may increase 5–10 days after delivery with greater increases after cesarean section. Values then return to normal.
Race	GGT values are higher in African Americans than Caucasians.
Body mass index (BMI)	May increase both ALT and AST levels.
Cigarette smoking	May increase serum GGT levels.
Coffee consumption	Reduces levels of ALT, AST, and GGT in serum.
Strenuous exercise	Increases AST levels but may decrease ALT levels.
Diet	High-carbohydrate, high-calorie diet may increase serum transaminase levels.
Malnutrition	Elevated serum levels of liver enzymes.
Diurnal variation	ALT levels are lowest at night and highest during the afternoon.
Cardiac diseases	Heart failure and myocardial infarction may elevate liver enzyme activities in serum, especially GGT.
Musculoskeletal diseases	Significantly elevated AST levels in patients with muscular dystrophy, polymyositis, and rhabdomyolysis.
Endocrine disorders	Hyperthyroidism increases ALT and AST levels.
Wilson's disease	Elevated serum ALT and AST levels.

In general, 1–4% of asymptomatic patients exhibit abnormal liver function tests. Many factors may affect liver enzymes; for example, hyperthyroidism may elevate both AST and ALT levels. This may be due to excessive thyroid activity that leads to hepatic ischemia by increasing hepatic and splanchnic oxygen requirement [1]. Serum ALT levels decrease with age [2]. In general, girls have lower ALT levels than boys. Based on study of 1293 children, England *et al.* observed that ALT levels decreased with increasing age, whereas ALT levels were significantly lower in girls than in boys. Reference cutoffs before 18 months of age were 60 U/L for boys and 55 U/L for girls, decreasing to 40 U/L for boys and 35 U/L for girls after 18 months of age [3]. ALT and AST levels are 40–50% higher in individuals with high body mass index (BMI). Strenuous exercise can elevate AST values by up to threefold, but ALT values are approximately 20% lower in individuals who exercise [4]. Muscle injury is associated with significantly higher serum AST levels in conditions such as muscular dystrophy, polymyositis, and rhabdomyolysis [5]. Cardiac diseases may result in elevated liver enzyme levels in serum. Lofthus *et al.* demonstrated that acute ST-elevated myocardial infarction is associated with elevated levels of ALT and AST along with creatine kinase-MB isoenzyme (CK-MB) [6]. Wilson's disease is an autosomal recessive disease with genetic abnormality on chromosome 13 that results in defective copper metabolism. As a result, increased copper concentrations are observed in multiple organs, including the liver, central nervous system, and other organs. Serum ALT and AST are usually elevated in patients with Wilson's disease because copper disposition in the liver causes liver damage [7].

Blood transaminase activities and triglyceride concentrations are increased in individuals who consume a diet rich in carbohydrates and high in calories [8]. However, coffee consumption may reduce levels of ALT, AST, and GGT. An inverse relationship between coffee consumption and liver enzymes is more evident in individuals with a low BMI [9]. Klatsky *et al.* noted that there is an ingredient in coffee that protects the liver against cirrhosis, especially alcoholic cirrhosis [10]. Malnutrition can cause significant increases in serum transaminase levels [11].

CASE REPORT 5.1

A 26-year-old Caucasian woman with a history of anorexia nervosa had a BMI of 10.8 upon admission to the hospital. Her laboratory test results indicated highly elevated ALT (3930 IU/L), AST (9980 IU/L), amylase (1002 IU/L), lipase (1437 IU/L), lactate dehydrogenase (6830 IU/L), and creatine kinase (783 IU/L). Glomerular filtration rate was reduced to 35 mL/min, indicating dehydration and prerenal azotemia. No other cause of liver damage besides malnutrition was found. The patient recovered with hydration and nutritional support. She was discharged after staying in the hospital for 37 days. At discharge, all of her laboratory test results were within normal limits, including liver enzymes [12].

Various drugs and herbal supplements may cause liver damage resulting in elevated serum levels of various liver enzymes. Acetaminophen, an analgesic, is present in many over-the-counter medications as well as in prescription pain medications. Acetaminophen may cause liver damage and elevated levels of liver enzymes if taken in excess. Many drugs also have toxic effects on the liver. Herbal supplements such as kava, comfrey, germander, and chaparral may cause liver damage, and as a result, elevated levels of liver enzymes may be observed in serum [13]. Common drugs and herbal supplements that may cause liver damage are summarized in Box 5.1.

BOX 5.1 COMMON DRUGS AND HERBAL SUPPLEMENTS THAT MAY CAUSE ABNORMAL LIVER ENZYME LEVELS[a]

Drugs

- Acetaminophen
- Acetylsalicylate (aspirin)
- Diclofenac
- Indomethacin
- Tetracycline
- Erythromycin
- Isoniazid
- Allopurinol
- Phenytoin
- Valproic acid
- Amiodarone
- Quinidine
- Niacin
- Statins

Herbal Supplements

- Kava
- Comfrey
- Chaparral
- Germander
- Lipokinetix (weight-loss product)
- Mistletoe
- Pennyroyal oil

[a] *In addition to this list, there are many more drugs and herbal supplements that may cause liver damage.*

5.3 EFFECT OF MODERATE ALCOHOL CONSUMPTION ON LIVER ENZYMES

The prevalence of unexplained elevated levels of liver enzymes ranges from 2.8 to 5.4% in the general U.S. population, and 80–90% of unexplained elevated levels of liver enzymes are related to nonalcoholic fatty liver diseases, which include a broad spectrum of diseases from simple fatty liver disease to nonalcoholic liver steatohepatitis with variable degrees of fibrosis or hepatic inflammation. Factors such as alcohol consumption, cigarette smoking, metabolic syndrome, and obesity may cause elevated levels of various liver enzymes. Park *et al.* observed a supra-additive effect of moderate alcohol consumption and both cigarette smoking and metabolic syndrome on levels of serum liver enzymes. Moreover, moderate alcohol consumption alone may also increase levels of liver enzymes [14]. Whitehead *et al.* studied the effects of cigarette smoking and alcohol consumption on serum liver enzyme activities using 46,775 men and observed that cigarette smoking produced significant increases in GGT activities in all drinking categories, including teetotalers, especially in individuals who smoked more than 20 cigarettes per day. Although smoking alone had no significant effect on serum ALT and AST levels, both smoking and drinking resulted in increased levels of GGT due to induction of this liver enzyme by both nicotine and alcohol. The authors concluded that smoking as well as drinking habits of patients must be accounted for when evaluating liver enzyme values [15].

Although liver enzymes may be elevated in individuals who drink in moderation, obesity may intensify such effect. In a study of 2164 apparently healthy subjects (1028 men and 1136 women) who either did not consume any alcohol or consumed less than 40 g of ethanol each day (moderate drinkers), Alatalo *et al.* observed that the activities of ALT and GGT but not AST were slightly but statistically significantly (by Mann–Whitney U test) higher in moderate drinkers than in nondrinkers. For example, mean ALT, AST, and GGT activities in abstainers ($n = 669$) were 23, 24, and 24 U/L, respectively, whereas corresponding activities in moderate drinkers ($n = 1495$) were 25, 24, and 29 U/L, respectively. Obese people showed higher levels of liver enzymes, and such elevations were more pronounced in moderate drinkers. In the group of abstainers who were obese, the mean ALT, AST, and GGT activities were 52, 18, and 36% above the baseline, respectively. In moderate drinkers who were obese, the corresponding activities were 105, 41, and 123% higher than baseline values, respectively. The authors concluded that the effect of moderate alcohol consumption on serum liver enzyme activities increases with increasing BMI [16]. In another study, which included approximately 21,000 men, the authors observed a positive correlation between alcohol consumption and serum levels of serum ALT, AST, and GGT. Increased consumption of cigarettes was associated with increased mean level of GGT. However, all liver enzymes increased with increasing BMI, and the effect of obesity was particularly important in the case of ALT: The prevalence of increased ALT values in obese individuals (BMI $\geq$ 31 kg/m^2) was eight times that in people with normal body weight (BMI $\leq$ 25 kg/m^2) [17].

In a study of 1,166,847 Koreans (731,560 men and 435,287 women) aged 30–95 years, Sull *et al.* estimated that ALT increased by 18.8 U/L and AST by 7.1 U/L across the range of BMI values (<18.5 to >32 kg/m^2) in men, whereas such increases were estimated at 9.9 and 4.5 U/L, respectively, in women. In general, average ALT and AST levels were higher in men than in women. The prevalence of abnormal ALT and AST was 10.1 and 7.1%, respectively, in men, whereas such prevalence was 4.0 and 3.4%, respectively, in women. The prevalence of alcohol consumption was 76.8% in men and 14.3% in women. The levels of ALT and AST were higher in heavy drinkers than in nondrinkers, but for women the relationship between aminotransferase levels and BMI did not vary by alcohol consumption [18]. Therefore, elevated liver enzymes should be interpreted in light of drinking history and other factors, such as obesity, metabolic syndrome, and smoking habit.

5.4 GGT AS ALCOHOL BIOMARKER

Although ALT and AST may be significantly elevated in individuals who abuse alcohol, GGT is considered to be the most sensitive biomarker of alcohol consumption. GGT is found not only in the liver but also in the spleen,

Table 5.3 Sensitivity and Specificity of Liver Enzymes Used as Alcohol Biomarkers[a]

Liver Enzyme	Sensitivity (%)	Specificity (%)	Cutoff (U/L)
ALT	32	92	35
AST	68	80	35
GGT	64	72	30

[a]These values are based on Arndt et al. [20]. However, other authors have reported low sensitivity for both ALT and AST, with sensitivity of ALT between 18 and 58% and that of AST between 15 and 69%. Sensitivity of GGT between 34 and 80% and specificity between 11 and 95% have also been reported.

kidney, pancreas, biliary tree, heart, brain, and seminal vesicles, but only liver GGT is detectable in the serum. In general, GGT has higher sensitivity and specificity than ALT and AST as an alcohol biomarker [19]. The sensitivity and specificity of various liver enzymes as alcohol biomarkers are summarized in Table 5.3.

Serum GGT levels are widely used as an alcohol biomarker because the test is a part of routine chemistry analysis performed in most clinical laboratories. Moreover, the test is also inexpensive. In general, GGT activity is higher in men than in women. Interestingly, GGT activity is also higher in postmenopausal women than in premenopausal women, and GGT activity also tends to be higher in women who take oral contraceptives [21]. Drinking intensity (heavy binge drinking) affects GGT levels more than does drinking frequency. It is assumed that GGT is a marker for heavy alcohol consumption. The half-life of GGT is 14–26 days. However, in alcoholic patients with decompensated cirrhosis, the half-life of GGT may vary between 11 and 54 days. In general, with abstinence, GGT levels return to normal after 4 or 5 weeks. Various characteristics of GGT as an alcohol biomarker are summarized in Box 5.2. It is generally assumed that if GGT is used along with carbohydrate-deficient transferrin, another alcohol biomarker, better specificity can be obtained (see Chapter 6).

A single dose of alcohol has no effect on serum GGT. Repeated doses of approximately 40–50 g of alcohol on successive days may produce a small increase in GGT level in serum. In eight male volunteers, consumption of 63 g of alcohol per day for 5 weeks produced a significant increase in the GGT level from a mean baseline of 27 to 52 U/L. However, consumption of 40 g or less alcohol per day for 4–6 weeks may not produce any significant elevation in GGT. It is generally assumed that GGT levels increase significantly after alcohol intake of 80–200 g per day (five or more standard drinks per day) for several weeks (5 or 6 weeks). Therefore, GGT should be considered a state marker of heavy alcohol consumption.

BOX 5.2 CHARACTERISTICS OF GGT AS ALCOHOL BIOMARKER

- GGT is a state biomarker of heavy alcohol consumption.
- The proposed cutoff concentration varies, but it is usually 30–80 U/L; sensitivity and specificity vary with cutoff levels.
- More drinks per week may be necessary for males for elevation in GGT levels than for females.
- Younger people (<40 years of age) may need to consume more drinks for elevated GGT.
- Coffee consumption may lower GGT levels in individuals who consume alcohol.
- Half-life is 14–26 days.
- GGT levels return to normal after abstinence for 4 or 5 weeks.
- GGT along with carbohydrate-deficient transferrin has better specificity as an alcohol biomarker.

Various cutoffs for GGT concentrations have been proposed for its use as an alcohol biomarker. Hastedt *et al.* observed that the mean GGT concentration in nondrinkers was 22 U/L, whereas the mean GGT value in alcoholics was 209 U/L. Using the receiver operating curve approach, the authors stated that at a GGT cutoff value of 41 U/L, 83.3% sensitivity and 92.3% specificity can be achieved for GGT as an alcohol biomarker. Interestingly, specificity for ALT and AST as alcohol biomarkers is considered low, but the authors observed 46.3% sensitivity and 91% specificity for ALT at a cutoff concentration of 53 U/L. For AST at a cutoff concentration of 42.5 U/L, sensitivity and specificity were 61.1 and 91.0%, respectively. As expected, GGT had higher sensitivity and specificity compared to ALT and AST [22]. Iffland found that a GGT value of 70 U/L or greater was indicative of long-standing abuse of alcohol [23]. In a study of 6962 subjects (3974 males and 2988 females), Sillanaukee *et al.* reported that for males, mean GGT concentration was 37.6 U/L in nondrinkers, 39.9 U/L (1–105 g of alcohol/week) in light drinkers, 51.4 U/L (106–280 g of alcohol/week) in moderate drinkers, and 71.5 U/L (280–420 g of alcohol/week) in heavy drinkers. However, for females, mean GGT level was 24.1 U/L, 24.2 U/L, 32.5 U/L, and 143.2 U/L respectively. The authors further reported that males who consumed more than 420 g of alcohol per week showed a mean GGT level of 172.8 U/L. No women in the study drank that heavily [24].

Alatalo *et al.* reported that the average GGT level in men who were heavy drinkers was 193 U/L, and 119 U/L in women who were heavy drinkers. In moderate drinkers, average GGT was 34 U/L in men and 23 U/L in women. In abstainers, average GGT was 26 U/L in men and 22 U/L in women. In addition, in men who were heavy drinkers, mean AST and ALT levels were

65 and 71 U/L, respectively. In moderate drinkers, AST and ALT levels were 26 and 29 U/L, respectively. In nondrinking men, AST and ALT levels were 25 and 26 U/L, respectively, indicating that drinking may also be associated with elevated AST and ALT levels. In women who were heavy drinkers, corresponding AST and ALT levels were 55 and 51 U/L, respectively. In women who were moderate drinkers, corresponding AST and ALT levels were 21 and 20 U/L, respectively. In women who were nondrinkers, corresponding AST and ALT levels were 23 and 21 U/L, respectively. These data also indicated that alcohol consumption affects GGT levels more significantly than AST and ALT levels [25]. Tynjala *et al.* studied the relationship between alcohol consumption, age, and gender with serum GGT levels and observed that in men older than 40 years, regular consumption of 8 standard drinks or more resulted in significant increases in serum GGT, whereas for men younger than 40 years, 14 or more standard drinks per week might be needed for such increases in serum GGT. For women older than 40 years, 4 or more standard drinks per week might be needed for significant elevations in serum GGT level, and such threshold was 7 standard drinks for women younger than 40 years [26].

Nemesanszky *et al.* reported that after 4 weeks of abstinence, serum GGT levels in drinkers (daily alcohol intake of 60–80 g) declined from a mean initial value of 86 U/L to 33 U/L [27]. Coffee consumption has been linked to decreased serum level of GGT. In a study of 18,899 individuals (8807 men and 10,092 women) with a mean age of 48 years (range, 25–74 years), Danielsson *et al.* observed that in men, elevation of GGT by heavy drinking (> 280 g of alcohol per week or 20 standard drinks or more) was significantly reduced by coffee consumption exceeding four cups per day. As expected, highest GGT levels in both men and women were observed in heavy drinkers who did not consume any coffee [28].

In the healthy population, the epidemiological association of alcohol intake with GGT level is at least partly explained by genetic pleiotropy. In a study of

CASE REPORT 5.2

A 39-year-old white female with a history of chronic alcohol ingestion (she had consumed 6–12 beers daily for many years) was seen in the clinic. Her serum GGT level was markedly elevated to 3952 U/L. Her other biochemical tests included the following: AST, 380 U/L; ALT, 144 U/L; total bilirubin, 5.7 mg/dL; direct bilirubin, 2.9 mg/dL; and prothrombin time, 12 sec. Her serological tests were negative for hepatitis A, B, and C. Her liver biopsy showed histological features of alcoholic foamy degeneration. The patient significantly reduced her intake of alcohol. After 116 days, her GGT level was significantly reduced to 486 U/L. Her total bilirubin was reduced to 1.6 mg/dL, and her direct bilirubin was reduced to 0.3 mg/dL, indicating reversibility of biochemical abnormality in alcoholic foamy degeneration [29].

6465 subjects, Van Beek *et al.* observed that in men, 7.2% variance in GGT levels could be explained by genetic effect, and in women 4.4% variance could be explained by genetic effect. In comparison, environmental factors underlying alcohol intake explained only 2% of variance in GGT levels in males and only 1% in females [30].

5.4.1 Elevated GGT in Various Diseases and as a Risk Factor for Mortality and Certain Illnesses

As expected, alcoholic liver diseases are also associated with increased GGT levels. Selinger *et al.* reported that the mean serum GGT level in normal individuals was 32 U/L and that in patients with fatty liver not due to alcohol it was 43 U/L, but in patients with alcoholic liver disease, the mean GGT level was significantly elevated to 216 U/L. As expected, mean serum GGT level was elevated to 274 U/L in patients with biliary cirrhosis, 107 U/L in patients with chronic active hepatitis including postnecrotic cirrhosis, and 56 U/L in patients with persistent hepatitis [31]. The AST/ALT ratio can provide important diagnostic clues. The normal AST/ALT ratio is approximately 0.8. An AST/ALT ratio of 2.0 or higher or ALT level exceeding 300 U/L may be indicative of alcoholic liver disease. However, the AST/ALT ratio is usually 1.0 or less in nonalcoholic fatty liver disease. Patients with a history of alcohol abuse but no significant alcoholic hepatitis or cirrhosis of the liver usually have an AST/ALT ratio less than 1.0. However, in patients with viral hepatitis, an AST/ALT ratio greater than 1.0 may be indicative of underlying cirrhosis with a high specificity (94–100%) but low sensitivity (44–75%) [32]. Patients with Wilson's disease can show an AST/ALT ratio greater than 4.0 [33].

Elevated GGT concentration may be associated with a higher risk of coronary heart disease, type 2 diabetes, and stroke. In addition, individuals with elevated GGT concentration may experience higher mortality partly due to the association between elevated GGT concentration and higher risk of developing certain illnesses and partly because GGT is an independent predictor of mortality [34]. Research has indicated that GGT may have direct involvement in atherosclerotic plaque formation, suggesting a role of GGT in the pathogenesis of cardiovascular disease. Prospective studies have also shown GGT to be associated with the development of hypertension, metabolic syndrome, and also type 2 diabetes, which are risk factors for cardiovascular disease. Risk of stroke also increases with increasing GGT [35]. Ruhl and Everhart defined an elevated GGT level as greater than 51 U/L in men and greater than 33 U/L in women. For ALT, an elevated level in men was greater than 30 U/L and in women, 19 U/L. Using multivariant-adjusted analyses, the authors showed that in the U.S. population, elevated GGT was associated with mortality from all causes, liver disease, cancer, and diabetes, whereas elevated ALT was associated with only liver disease mortality [36]. Koehler *et al.* observed that GGT was associated with increased mortality from cardiovascular diseases; ALP and AST were associated with cancer-related mortality. Participants with GGT and ALP levels in the top 5% had the highest risk for all-cause mortality [37]. Matsha *et al.* reported that GGT levels were independently associated with insulin sensitivity and metabolic syndrome in a middle-aged African population [38].

Although the mechanism underlying the association between GGT level and metabolic syndrome is not fully understood, GGT has a central role in glutathione homeostatics by initiating the breakdown of extracellular glutathione and turnover of vascular glutathione. Because glutathione is an antioxidant protecting cells from oxidative stress, increases in GGT may reflect increased oxidative stress. The role of oxidative stress in metabolic syndrome has been speculated on. Bradley *et al.* concluded that GGT is strongly associated with both cardiovascular and metabolic risk factors, including prevalent metabolic disease [39]. Targher reviewed the association between high normal serum GGT levels (mostly within the reference range) and risk of mortality and major vascular (i.e., cardiovascular morbidity and mortality) and nonvascular outcomes (i.e., the incidence of type 2 diabetes, chronic kidney disease, and cancer). These risks are independent of alcohol consumption and other prognostic factors [40].

Abnormal GGT expression is found in several human tumors, including breast cancer, ovarian cancer, and cervical cancer. In addition, a possible role of GGT in tumor progression, invasion, drug resistance, and prognosis has been suggested. Thus, studies have shown an association between elevated GGT levels and increased risk of developing cancer. In patients with cervical cancer, elevated GGT levels may be associated with poor prognosis. In one study, the authors reported that mean GGT level was 29.9 U/L in stage I endometrial cancer, 31.0 U/L in stage II, 33.7 U/L in stage III, and 38.2 U/L in stage IV [41]. Edlinger *et al.* reported that elevated levels of GGT were associated with poor survival among endometrial cancer patients [42]. In a cohort of 545,460 people, Van Hemelrijck *et al.* observed a positive association between categories of GGT (<18, 18–36, 36–72, and >72 U/L) and overall cancer risk, with a hazard regression of 1.07 for the second category (GGT 18–36 U/L), 1.18 for the third category (GGT 36–72 U/L), and 1.32 for the fourth category (GGT >72 U/L) compared to the first category (GGT <18 U/L). The strength of the association may vary by glucose levels because hyperglycemia can induce oxidative stress, thus initiating damaging pathways of carcinogens [43].

Interestingly, construction workers may have elevated GGT and AST levels. In a study of 8043 male construction workers aged 25–64 years, the authors found elevated GGT (>28 U/L) in 32% of individuals, elevated ALT (>22 U/L) in 22%, and elevated AST levels (>18 U/L) in 12%. Factors most

BOX 5.3 ELEVATED GGT AND RISK OF CERTAIN DISEASES ASSOCIATED WITH ELEVATED GGT

- Cardiovascular diseases
- Hypertension
- Metabolic syndrome
- Type 2 diabetes
- Chronic kidney disease
- Stroke
- Cancer
- Increased mortality

strongly related to elevated levels of liver enzymes were self-reported alcohol consumption, diabetes, and hypertension. BMI was also strongly correlated with elevated GGT and ALT levels but not AST levels. Elevated levels of GGT and AST were strongly related to early retirement and all-cause mortality. Men with AST levels exceeding 18 U/L had a twofold risk of early retirement and three times higher risk of all-cause mortality compared to men with AST levels lower than 18 U/L [44]. Higher risk factors for certain diseases associated with elevated GGT levels are summarized in Box 5.3.

An increase in serum GGT activity may be due to induction of GGT not only by alcohol but also by several drugs. Increased levels of GGT in patients receiving classical anticonvulsants such as phenytoin and phenobarbital have been well reported in the literature. Carbamazepine and possibly valproate may have similar effects. Other drugs, such as barbiturates, certain benzodiazepines, tricyclic antidepressants, warfarin, and nonsteroidal anti-inflammatory drugs, may have similar effects. Statins in general may cause elevated liver enzymes in certain patients [34]. Clofibrate and oral contraceptives can decrease GGT levels. Drugs that may affect GGT levels are summarized in Box 5.4.

5.4.2 GGT Fraction as Alcohol Biomarker

Using molecular size exclusion chromatography followed by post-column GGT-specific reactions, four plasma GGT fractions can be isolated and quantified. These fractions are termed big GGT (b-GGT; molecular weight, 2000 kDa), medium GGT (m-GGT; molecular weight, 1000 kDa), small GGT (s-GGT; molecular weight, 250 kDa), and free GGT (f-GGT; molecular weight, 70 KDa). In general, f-GGT is the major fraction in the serum. For example, median total GGT in 100 males was 25.3 U/L, where f-GGT comprised 13.2 and b-GGT, m-GGT, and s-GGT comprised 2.4, 1.0, and 9.2, respectively. Median total GGT in 100 females was 14.4 U/L, where f-GGT comprised 8.9 and b-GGT, m-GGT, and s-GGT comprised 1.1, 0.5, and 3.9, respectively [45]. Franzini *et al.* reported that total GGT elevation in alcoholics was associated with a significant increase in s-GGT fraction and smaller increases in the other three fractions. In 27 control subjects (mean

BOX 5.4 DRUGS THAT MAY AFFECT GGT LEVELS

Increased GGT Level

- Phenytoin
- Phenobarbital
- Carbamazepine (possibly)
- Barbiturates
- Certain benzodiazepines
- Phenazone
- Propoxyphene
- Nonsteroidal anti-inflammatory drugs
- Monoamine oxidase inhibitors
- Tricyclic antidepressants
- Thiazide diuretics
- Anabolic steroids
- Warfarin

Decreased GGT Levels

- Clofibrate
- Oral contraceptives

alcohol consumption, <20 g/day), median total GGT was 23.9 U/L; b-GGT, m-GGT, s-GGT, and f-GGT fractions were 2.4, 1.0, 7.3, and 13.7, respectively. In 24 subjects who were nondrinkers, total GGT was 37.5 U/L; b-GGT, m-GGT, s-GGT, and f-GGT fractions were 2.8, 2.1, 15.3, and 14.4, respectively. In 27 alcoholics (median alcohol consumption, 209 g/day), total GGT was elevated to 133.3 U/L; b-GGT, m-GGT, s-GGT, and f-GGT fractions were 7.5, 5.2, 95.7, and 22.7, respectively. These data indicated that in alcoholics, there was a very significant increase in s-GGT. Moreover, the b-GGT/s-GGT ratio was higher in abstainers than in alcoholics. The authors concluded that analysis of GGT fraction pattern may improve specificity of GGT as an alcohol biomarker [46]. However, in nonalcoholic fatty liver disease, b-GGT increased significantly, showing the highest diagnostic accuracy. In chronic hepatitis, total GGT along with m-, s-, and f-GGT fractions showed significant increases, but b-GGT did not show any characteristic increase [47].

5.5 LABORATORY DETERMINATIONS OF LIVER ENZYMES

GGT assays are usually based on the use of L-γ-glutamyl-*p*-nitroanilide as the donor substrate and glycylglycine as the glutamyl acceptor. The *para*-nitroaniline produced in the reaction can be determined colorimetrically at 405 nm. However, L-γ-glutamyl-*p*-nitroanilide has limited solubility, and the International Federation of Clinical Chemistry recommends use of L-γ-glutamyl-3-carboxy-4-nitroanilide as the substrate. GGT catalyzes transfer of the glutamyl moiety from the substrate to glycylglycine, thereby releasing 5-amino-2-nitrobenzoate, which also absorbs at 405 nm. In GGT assay for application on the Vista 1500 analyzer (Siemens Diagnostics, Deerfield, IL), the rate of this reaction is measured at two different wavelengths (405 and 600 nm). The activity of GGT is expressed as units per liter.

AST assay is usually based on the principle that AST catalyzes the transamination from L-aspartate to α-ketoglutarate, forming L-glutamate and oxaloacetate. The oxaloacetate formed is then reduced by malate dehydrogenase and, in this reaction, simultaneous oxidation of reduced NADH. The change of absorption at 340 nm due to conversion of NADH (absorbs at 340 nm) to NAD (no absorption at 340 nm) is measured and is proportional to the activity of AST. Again, AST level in serum or plasma is expressed as units per liter. ALT catalyzes the transamination from L-alanine to α-ketoglutarate, also forming pyruvate and L-glutamate. The pyruvate formed is reduced to lactate by lactate dehydrogenase, and in this process reduced NADH is converted to NAD. The change of absorption at 340 nm is proportional to ALT activity, which is expressed as units per liter.

5.6 CONCLUSIONS

GGT is a liver enzyme that has long been used as a state marker of heavy alcohol use and is a superior biomarker compared to liver enzymes AST and ALT. Mitochondrial isoenzyme of AST (mAST) has been explored as a potential alcohol biomarker; in one study, the authors observed increased mAST levels in 92% of alcoholics, but mAST was increased in 48% of patients with nonalcoholic liver diseases. Although the mAST/total AST ratio differentiated alcoholic subjects from patients with nonalcoholic liver diseases with a sensitivity of 92%, the specificity was only 70%. The authors concluded that desialylated transferrin/total transferrin ratio was a superior alcohol biomarker (see Chapter 6) [48].

Although GGT is widely used as an alcohol biomarker, it is not a specific biomarker because the GGT value can be increased in various liver diseases, including nonalcoholic fatty liver disease. Moreover, the higher end of the reference range of GGT in an individual may also indicate a higher risk of that person for various illnesses, including hypertension, type 2 diabetes, and cardiovascular disease, and also an increased risk for all-cause mortality. Because the GGT test can be easily performed in a clinical laboratory and is inexpensive, GGT is still used as an alcohol biomarker. However, its specificity as an alcohol biomarker can be improved if the GGT value is considered along with other more specific alcohol biomarkers, such as carbohydrate-deficient transferrin.

References

[1] Bayraktar M, Van Thiel DH. Abnormalities in measures of liver function and injury in thyroid disorders. Hepatogastroenterology 1997;44:1614–18.

[2] Dong MH, Bettencourt R, Brenner DA, Barrett-Connor E, et al. Serum levels of alanine aminotransferase with age in longitudinal analysis. Clin Gastroenterol Hepatol 2012;10: 285–90.

[3] England K, Thorne C, Pembrey L, Tovo PA, et al. Age- and sex-related reference ranges of alanine aminotransferase in children: European Pediatric HCV Network. J Pediatr Gastroenterol Nutr 2009;49:71–7.

[4] Dufour DR, Lott JA, Nolte FS, Gretch DR, et al. Diagnosis and monitoring of hepatic injury II: recommendations for use of laboratory tests in screening, diagnosis, and monitoring of hepatic injury. Clin Chem 2000;46:2050–68.

[5] Krier M, Ahmad A. The asymptomatic outpatient with abnormal liver function tests. Clin Liver Dis 2009;13:167–77.

[6] Lofthus DM, Stevens SR, Armstrong PW, Granger CB, et al. Pattern of liver enzyme elevations in acute ST-elevation myocardial infarction. Coron Artery Dis 2012;23:22–30.

[7] Mahjoub F, Fereiduni R, Jahanzad I, Farahmand F, et al. Atomic absorption spectrometry in Wilson's disease and its comparison with other laboratory tests and paraclinical findings. Iran J Pediatr 2012;22:52–6.

[8] Purkins L, Love ER, Eve MD, Wooldridge CL, et al. The influence of diet upon liver function tests and serum lipids in healthy male volunteers resident in a phase I unit. Br J Clin Pharmacol 2004;57:199–208.

[9] Ikeda M, Maki T, Yin G, Kawate H, et al. Relation of coffee consumption and serum liver enzymes in Japanese men and women with reference to effect modification of alcohol use and body mass index. Scand J Clin Lab Invest 2010;70:171–9.

[10] Klatsky AL, Morton C, Udaltsova N, Friedman GD. Coffee, cirrhosis and transaminase enzymes. Arch Intern Med 2006;166:1190–5.

[11] Urso C, Brucculeri S, Caimi G. Marked elevation of transaminases and pancreatic enzymes in severe malnourished male with eating disorder. Clin Ter 2013;164:e387–91.

[12] Di Pascoli L, Lion A, Milazzo D, Caregaro L. Acute liver damage in anorexia nervosa. Int J Eat Disord 2004;36:114–17.

[13] Ernst E. Risks of herbal medicinal products. Pharmacoepidemiol Drug Saf 2004;13: 767–71.

[14] Park EY, Lim MK, Oh JK, Cho H, et al. Independent and supra additive effects of alcohol consumption, cigarette smoking and metabolic syndrome on the elevation of serum liver enzyme levels. PLOS ONE 2013;8:e63439.

[15] Whitehead TP, Robinson D, Allaway SL. The effects of cigarette smoking and alcohol consumption on serum enzyme activities: a dose-related study in men. Ann Clin Biochem 1996;33:530–5.

[16] Alatalo PI, Koivisto HM, Hietala JP, Puukka KS, et al. Effect of moderate alcohol consumption on liver enzymes increases with increasing body mass index. Am J Clin Nutr 2008;88:1097–103.

[17] Robinson D, Whitehead TP. Effect of body mass and other factors on serum liver enzyme levels in men attending for well population screening. Ann Clin Biochem 1989;26:393–400.

[18] Sull JW, Yun JE, Lee SY, Ohrr H, et al. Body mass index and serum aminotransferase levels in Korean men and women. J Clin Gastroenterol 2009;43:869–75.

[19] Hannuksela ML, Liisanantti MK, Nissinen AE, Savolainen MJ. Biochemical markers of alcoholism. Clin Chem Lab Med 2007;45:953–61.

[20] Arndt T, Behnken L, Martens B, Hackler R. Evaluation of the cutoff for serum carbohydrate-deficient transferrin as a marker of chronic alcohol abuse determination by ChronAlco ID assay. J Lab Med 1999;23:507–10.

[21] Sillanaukee P, Massot N, Jousilahti P, Vartiainen E, et al. Enhanced clinical utility of gamma-CDT in a general population. Alcohol Clin Exp Res 2000;24:1201–6.

[22] Hastedt M, Buchner M, Rothe M, Gapert R, et al. Detecting alcohol abuse: traditional blood alcohol markers compared to ethyl glucuronide (EtG) and fatty acid ethyl esters (FAEE) measurement in hair. Forensic Sci Med Pathol 2013;9:471–7.

[23] Iffland I. New way to use biochemical indicators of alcohol abuse to regrant licenses in a fair manner after drunken driving in Germany. Alcohol Alcohol 1996;31:619–20.

[24] Sillanaukee P, Massot N, Jousilahti P, Vartiainen E, et al. Enhanced clinical utility of γ-CDT in a general population. Alcohol Clin Exp Res 2000;24:1202–6.

[25] Alatalo P, Koivisto H, Puukka K, Hietala J, et al. Biomarkers of liver status in heavy drinkers, moderate drinkers and abstainers. Alcohol Alcohol 2008;44:199–203.

[26] Tynjala J, Kangastupa P, Laatikainen T, Aalto M, et al. Effect of age and gender on the relationship between alcohol consumption and serum GGT: time to recalibrate goals for normal range. Alcohol Alcohol 2012;47:558–62.

[27] Nemesanszky E, Lott JA, Arato M. Changes in serum enzymes in moderate drinkers after an alcohol challenge. Clin Chem 1988;34:525–7.

[28] Danielsson J, Kangastupa P, Laatikainen T, Aalto M, et al. Dose- and gender-dependent interactions between coffee consumption and serum GGT activity in alcohol consumers. Alcohol Alcohol 2013;48:303–7.

[29] Suri S, Mitros FA, Ahluwalia JP. Alcoholic foamy degeneration and a markedly elevated GGT: a case report and literature review. Dig Dis Sci 2003;48:1142–6.

[30] Van Beek JH, de Moor MH, Geels LM, Sinke MR, et al. The association of alcohol intake with gamma-glutamyl transferase (GGT) levels: evidence for correlated genetic effects. Drug Alcohol Depend 2014;134:99–105.

[31] Selinger MJ, Matloff DS, Kaplan MM. Gamma-glutamyl transferase activity in liver disease: serum elevation is independent of hepatic GGTP activity. Clin Chim Acta 1982;125: 283–90.

[32] Sorbi D, Boynton J, Lindor KD. The ratio of aspartate aminotransferase to alanine aminotransferase: potential value in differentiating nonalcoholic steatohepatitis from alcoholic liver disease. Am J Gastroenterol 1999;94:1018–22.

[33] Berman DH, Leventhal RI, Gavaler JS, Cadoff EM, et al. Clinical differentiation of fulminant Wilsonian hepatitis from other causes of hepatic failure. Gastroenterology 1991;100: 1129–34.

[34] Whitfield JB. Gamma-glutamyltransferase. Crit Rev Clin Lab Sci 2001;38:263–355.

[35] Wannamethee SG, Lennon L, Shaper A. The value of gamma-glutamyltransferase in cardiovascular risk prediction in men without diagnosed cardiovascular disease or diabetes. Atherosclerosis 2008;201:168–75.

[36] Ruhl CE, Everhart JE. Elevated serum alanine aminotransferase and gamma-glutamyltransferase and mortality in the United States population. Gastroenterology 2009;136:477–85.

[37] Koehler EM, Sanna D, Hansen BE, van Rooij FJ, et al. Serum liver enzymes are associated with all-cause mortality in an elderly population. Liver Int 2014;34:296–304.

[38] Matsha TE, Macharia M, Yako YY, Erasmus RT, et al. Gamma-glutamyltransferase, insulin resistance and cardiometabolic risk profile in a middle-aged African population. Eur J Prev Cardiol 2013 [E-pub ahead of print]. PMID 23945039.

[39] Bradley R, Fitzpatrick AL, Jenny NS, Lee DH, et al. Association between total serum GGT activity and metabolic risk: MESA. Biomarker Med 2013;7:709–21.

[40] Targher G. Elevated serum gamma-glutamyltransferase activity is associated with increased risk of mortality, incidence of type 2 diabetes, cardiovascular events, chronic kidney disease and cancer: a narrative review. Clin Chem Lab Med 2010;48:147–57.

[41] Seebacher V, Polterauer P, Grimm C, Rahhal J, et al. Prognostic significance of gamma-glutamyltransferase in patients with endometrial cancer: a multi-center trial. Br J Cancer 2012;106:1551–5.

[42] Edlinger M, Concin N, Concin H, Nagel G, et al. Lifestyle-related biomarkers and endometrial cancer survival: elevated gamma-glutamyltransferase as an important risk factor. Cancer Epidemiol 2013;37:156–61.

[43] Van Hemelrijck M, Jassem W, Walldius G, Fentiman IS, et al. Gamma-glutamyltransferase and risk of cancer in a cohort of 545,460 persons: the Swedish AMORIS study. Eur J Cancer 2011;47:2033–41.

[44] Arndt V, Brenner H, Rothenbacher D, Zschenderlein B, et al. Elevated liver enzyme activity in construction workers: prevalence and impact on early retirement and all-cause mortality. Int Arch Occup Environ Health 1998;71:405–12.

[45] Franzini M, Passino C, Ottaviano V, Fierabracci V, et al. Fractions of plasma gamma-glutamyltransferase in healthy individuals: reference value. Clin Chim Acta 2008;395: 188–9.

[46] Franzini M, Fornaciari I, Vico T, Monicini M, et al. High sensitivity gamma-glutamyltransferase fraction pattern in alcohol addicts and abstainers. Drug Alcohol Depend 2013;127:239–42.

[47] Franzini M, Fornaciari I, Fierabracci V, Elawadi HA, et al. Accuracy of b-GGT fraction for the diagnosis of non-alcoholic fatty liver disease. Liver Int 2012;32:629–34.

[48] Kwoh-Gain I, Fletcher LM, Price LW, Halliday JW. Desialylated transferrin and mitochondrial aspartate aminotransferase compared as laboratory markers of excessive alcohol consumption. Clin Chem 1990;36:841–5.

CHAPTER 6

Mean Corpuscular Volume and Carbohydrate-Deficient Transferrin as Alcohol Biomarkers

CONTENTS

6.1 Introduction. 139

6.2 Mean Corpuscular Volume as Alcohol Biomarker 139

6.2.1 Mechanism of Increased MCV in Alcoholics 141

6.2.2 Other Causes of Macrocytosis............... 141

6.3 Carbohydrate-Deficient Transferrin 142

6.3.1 Mechanism of Formation of CDT....... 144

6.3.2 Cutoff Values, Sensitivity, and Specificity of CDT....... 144

6.3.3 CDT and GGT as Combined Alcohol Biomarker 147

6.3.4 Application of CDT.............................. 148

6.3.5 Limitations of CDT as Alcohol Biomarker 151

6.4 Laboratory Determination of CDT 154

6.5 Conclusions. 158

References 158

6.1 INTRODUCTION

Alcohol and its metabolites have toxic effects on erythrocytes. Excessive alcohol consumption is known to increase mean corpuscular volume (MCV) by a pathogenic process known as macrocytosis. Transferrin is found in blood in several different molecular forms that differ in their number of carbohydrate groups (sialic acid and other carbohydrate residues) and that carry different electrical charges. The major form of transferrin is tetrasialotransferrin, which has four sialic acid (carbohydrate) moieties attached to the molecule and represents approximately 80% of all transferrin molecules in a healthy individual. Other transferrin molecules found in blood may have more (five to eight sialic acid moieties) or less (up to three sialic acid moieties) carbohydrates. Acetaldehyde formed during alcohol metabolism is known to interfere with the incorporation of the sialic acid moiety to transferrin, resulting in the formation of transferrin molecules with zero, one, or two sialic acid moieties attached to the final molecules (carbohydrate-deficient transferrin (CDT)). In this chapter, the usefulness and limitations of MCV and CDT as alcohol biomarkers are addressed.

6.2 MEAN CORPUSCULAR VOLUME AS ALCOHOL BIOMARKER

Mean corpuscular volume or mean cell volume of erythrocytes (MCV) is calculated by dividing the hematocrit by the total number of erythrocytes (red blood cell (RBC) count).

$$\text{MCV} = \frac{\text{Hematocrit (\%)}}{\text{RBC count (millions/}\mu\text{L)}} \times 10$$

A. Dasgupta: Alcohol and Its Biomarkers. DOI: http://dx.doi.org/10.1016/B978-0-12-800339-8.00006-7

MCV is printed in the report generated during complete blood count. The normal range is 82–95 fL. One of the most common causes of abnormal MCV is anemia. If MCV is below the normal range, then the anemia is considered as microcytic anemia, whereas if MCV is above the normal range, it is called macrocytic anemia. Anemia may also be present even if MCV is within the normal range (normocytic anemia). It may occur in an acute condition such as blood loss. MCV is also used as an alcohol biomarker because it increases in subjects with alcohol abuse (macrocytosis). As early as 1978, Whitehead *et al.* showed that consumption of 60 g (approximately four drinks) of alcohol daily increased MCV above the reference range in most individuals, with the highest MCV being 104 fL (range, 84–104 fL). The range of MCV was 84–100 fL for individuals who consumed up to three drinks daily and 84–98 fL for individuals who consumed only one drink daily [1]. With abstinence, MCV returns to normal values within 2–4 months.

Macrocytosis is usually defined as an MCV value greater than 100 fL, and alcohol dependence is one of the causes of macrocytosis other than anemia. In one study, the authors observed that approximately 3% of the general population had MCV values greater than 100 fL [2]. However, theoretically macrocytosis may be considered at an MCV value greater than 95 fL. In contrast, Rumsey *et al.* reported that 7% of the population they studied had an MCV greater than 96 fL, and approximately 1.7% had an MCV greater than 100 fL [3]. In another study, the authors observed that 138 of 3805 adult outpatients (3.7%) had an elevated MCV of greater than 98.5 fL (normal, 82–95 fL). The authors further evaluated 73 of these patients with a mean an MCV of 102.5 fL (another 55 patients were not evaluated further). Of these 73 patients, alcoholism was the cause of the elevated MCV in 47 patients [4]. Morgan *et al.* defined macrocytosis as an MCV value greater than 95 fL and studied 303 alcoholics with liver disease, 60 nonalcoholics with liver disease, and 35 control subjects. The authors observed that 70.3% of subjects with alcoholic liver disease showed macrocytosis, but macrocytosis was also found in 23.3% of subjects with nonalcoholic liver disease. An MCV value greater than 100 fL was observed in 49.5% of subjects with alcoholic liver disease but only in 3.3 subjects with nonalcoholic liver disease. Macrocytosis was more common in female alcoholics (86.3%) than in male alcoholics (63.0%). The authors concluded that an MCV greater than 100 fL in patients with liver disease is indicative of alcohol-related liver disease [5]. Koivisto *et al.* reported that an MCV up to 98 fL was observed in moderate drinkers (1–40 g alcohol per day); the reference interval of MCV calculated from mean and two standard deviations (mean ± 2SD) in these individuals was 82–98 fL, whereas the reference range calculated for abstainers was 82–96 fL. Use of 96 fL as the cutoff between alcoholics and nonalcoholics yielded a sensitivity of 44%, specificity of 98%, a positive predictive value of

96%, and a negative predictive value of 58% [6]. In general, MCV is less than 110 fL in chronic alcohol abusers [7]. However, in megaloblastic anemia, MCV can be greater than 110 fL, sometimes reaching 130 fL or higher.

6.2.1 Mechanism of Increased MCV in Alcoholics

The mechanism of increased MCV is probably related to hematotoxicity of both alcohol and its metabolite, acetaldehyde. Alcohol can permeate the cell membrane and alter lipid structures of the membrane. In addition, alcohol can alter erythrocyte metabolism, thus altering its stability [8]. Acetaldehyde, which is formed during alcohol metabolism by enzymatic reaction involving the alcohol dehydrogenase enzyme and nicotinamide adenine dinucleotide as a cofactor, is highly reactive and can form stable adducts with proteins and other constituents of the cell membrane. As a result, erythrocyte membrane structure may become more susceptible to damage such as hemolysis, thus shortening its half-life [9]. Patients who abuse alcohol and demonstrate macrocytosis often show the presence of circulating antibodies that recognize acetaldehyde-modified epitopes of protein, indicating that the acetaldehyde–protein adduct may play an important role in erythrocyte abnormalities seen in alcohol-dependent patients [6]. Studies of male Japanese alcoholics showed that patients with inactive aldehyde dehydrogenase 2 (ALDH-2) enzyme due to the presence of the *ALDH2*1/2*2* genotype had higher MCV than patients with normal ALDH-2 enzyme activity. ALDH-2 is a key enzyme involved in the removal of the toxic acetaldehyde metabolite of alcohol. Therefore, patients with inactive ALDH-2 enzyme may have a higher acetaldehyde concentration, thus further indicating the role of acetaldehyde in increasing MCV. In addition, MCV of 106 fL or greater indicates a high risk of esophageal squamous cell carcinoma in Japanese alcoholic men [10]. Another study showed that in Japanese men with MCV greater than 99 fL but no flushing experience after alcohol use, the risk of esophageal cancer was 2.5 times greater compared to that for Japanese men with an MCV of less than 93 fL [11].

6.2.2 Other Causes of Macrocytosis

In addition to alcohol abuse, macrocytosis is observed in megaloblastic anemia, commonly caused by vitamin B_{12} and folate deficiency. Vitamin B_{12} deficiency alone may not cause megaloblastic anemia. In megaloblastic anemia, erythrogenic precursors are larger than mature red blood cells because folate and vitamin B_{12} deficiencies result in defective RNA and DNA synthesis. A nonmegaloblastic process leads to round macrocytes or macroreticulocytes. Although increased MCV due to alcohol abuse is a nonmegaloblastic process, individuals who chronically abuse alcohol may be deficient in vitamin B_{12} and folate [12]. Hypothyroidism and other diseases may also cause

BOX 6.1 CAUSES OF MACROCYTOSIS (INCREASED MCV) OTHER THAN ALCOHOL ABUSE

- Vitamin B_{12} and folate deficiency
- Hypothyroidism
- Nonalcoholic liver disease
- Reticulocytosis
- Monoclonal gammopathy
- Bone marrow dysplasia
- Acute leukemia
- Aplastic anemia
- Anorexia nervosa
- Medications

Table 6.1 Common Medications That May Cause Macrocytosis

Drug Class	Individual Drugs
Antiretrovirals	Zidovudine, stavudine
Anticancer agents	Hydroxyurea, methotrexate, 5-fluorouracil, cyclophosphamide, azathioprine, mercaptopurine, cytosine arabinoside
Antibiotics	Bactrim (sulfamethoxazole and trimethoprim)
Anticonvulsants	Phenytoin, primidone, valproic acid
Anti-inflammatory	Sulfasalazine
Hypoglycemic agent	Metformin

macrocytosis. Keenan commented that macrocytosis defined by an MCV greater than 100 fL should be considered as an indicator of human disease. In Keenan's study, alcoholism was the main cause of macrocytosis. Anorexia nervosa may also cause macrocytosis [13]. Macrocytosis can be a manifestation of monoclonal gammopathy [14]. Unexplained macrocytosis may not be a benign finding and requires close follow-up because patients may develop worsening cytopenia or ultimately may be diagnosed with primary hematological malignancy [15]. Conditions that may cause macrocytosis other than alcohol abuse are summarized in Box 6.1. Various drugs may also cause increased MCV [16]. Common medications that may cause macrocytosis are listed in Table 6.1.

6.3 CARBOHYDRATE-DEFICIENT TRANSFERRIN

Transferrin, a glycoprotein responsible for iron transport, is synthesized in hepatocytes and secreted into the blood. Transferrin has a molecular weight of 79,570 Da [17]. The transferrin molecule consists of a polypeptide chain of 679 amino acids and is divided into two domains, the N-terminal domain (amino acids 1–336) and the C-terminal domain (amino acids 337–679). Carbohydrate moieties are present in the C-terminal domain at positions 413 and 611, representing two glycosylation sites for transferrin. Glycosylation is a post-translational modification of protein by enzymatic addition of oligosaccharide chains (N-glycans) to the structure. Sialic acid is the terminal carbohydrate moiety in the transferrin molecule, giving the molecule a negative charge. Other carbohydrates present as part of the N-glycan structure in the transferrin molecule include galactose, mannose, and *N*-acetylglucosamine. The transferrin molecule has two binding sites of iron and is considered the most important transport protein for ferric ion (Fe^{3+}) [18]. The human transferrin molecule also shows genetic polymorphism with at least 38 known different transferrin variants. These variants are due to substitution of amino acid in the polypeptide chain. However, transferrin C is the most common phenotype. Allelic B and D variants with a different primary structure occur in low frequencies in various populations [19]. Transferrin exists in various isoforms that are named depending on the number of terminal sialic acid residues on N-glycans. The major isoform contains four sialic acid residues and is called tetrasialotransferrin. This isoform represents 64–80% of total transferrin molecules in the serum [20]. Other isoforms detected in serum of healthy people are pentasialotransferrin (12–18%), hexasialotransferrin (1–3%), heptasialotransferrin (<1.5%), and octasialotransferrin (nondetectable or trace), which contain more sialic acids than does tetrasialotransferrin. In addition, isoforms of transferrin containing fewer sialic acids than does the major isoform are also detected in serum of healthy subjects. These isoforms include disialotransferrin (<2.5%) and trisialotransferrin (4.5–9%). Asialo and monosialotransferrin are usually not detectable or represent less than 0.5% of total transferrin in healthy subjects [21].

CDT is the collective name for a group of minor isoforms of human transferrin with a low degree of glycosylation. These minor isoforms include asialo, monosialo, and disialotransferrin. In 1976, Stibler and Kjellin used the electrophoresis technique to show increased concentrations of these isoforms in cerebrospinal fluid and serum of alcoholics [22]. Since then, other studies have documented elevated levels of these isoforms in serum of individuals who abuse alcohol. In general, heavy alcohol consumption (50–80 g/day) for a period of at least 1 week leads to increased concentrations of CDT in serum; after abstinence, CDT returns to normal levels with a half-life of 14 days. In common practice, the ratio of CDT to total transferrin is determined and is expressed as a percentage (%CDT). However, the absolute level of CDT may also be used as an alcohol biomarker. In 2001, CDT was the first test to receive approval by the U.S. Food and Drug Administration (FDA) as an alcohol biomarker for use in the U.S. health care system. The CDT test had been used by insurance companies since 1995 to screen individuals who might abuse alcohol. CDT is widely used in European countries to monitor abstinence and identify patients with high-risk drinking behavior in surgical as well as medical settings [21].

6.3.1 Mechanism of Formation of CDT

The mechanism of increased CDT in serum of alcoholics is still under investigation. However, one possible mechanism is that ethanol intake decreases activities of enzymes responsible for incorporation of sialic acid in transferrin molecules. Enzymes such as sialyltransferase, galactosyltransferase, and *N*-acetylglucosamine transferase, found mostly in the hepatic Golgi complexes, are responsible for addition of sialic acid and other carbohydrate molecules to the transferrin molecule as part of post-translational glycosylation. Activities of these enzymes are inhibited by ethanol and/or its metabolite, acetaldehyde, thus interfering with the incorporation of sialic acid moieties to transferrin molecules. In addition, alcohol consumption may increase the activity of enzymes such as sialidase, which removes the carbohydrate moiety from the transferrin molecule. Animal experiments have found that chronic ethanol-treated rats show marked inhibition of liver sialyltransferase, which is responsible for incorporation of sialic acid into the transferrin molecule. In contrast, hepatic plasma membrane sialidase activity is increased. Acetaldehyde derived from ethanol may also play a role in inhibiting sialyltransferase activity [23].

6.3.2 Cutoff Values, Sensitivity, and Specificity of CDT

In general, concentrations of CDT are higher in women than in men. This may be due to the fact that women may have subclinical iron deficiency that may cause an elevated transferrin level, although the exact mechanism of such difference is not fully understood. However, the serum CDT to total transferrin ratio is not gender dependent [24]. Both the level of CDT and the amount of CDT expressed as a percentage of total transferrin (%CDT) have been used as alcohol biomarkers; however, if the absolute value of CDT is used, there is a gender difference in the reference range. Malcom *et al.* used a serum CDT cutoff of 17 U/L for males and 25 U/L for females to evaluate the usefulness of CDT as an index of heavy alcohol use during postmortem examination. Fifteen of 17 cases with evidence of alcohol abuse (as evidenced by microscopic alcoholic fatty changes in the liver, cirrhosis, or a clinical report that the subject abused alcohol) showed CDT levels above the cutoff values, giving CDT assay a sensitivity of 88%. Eight cases had no evidence of alcohol abuse, but 3 cases showed CDT levels above the cutoff, giving the CDT test a specificity of 63%. The authors concluded that elevated CDT levels appeared to indicate antemortem alcohol abuse [25]. Sorvajarvi *et al.* defined the normal range of CDT as up to 20 U/L for men and up to 26 U/L for women and commented that although earlier studies reported sensitivity of CDT 90% and higher and specificity of 90–100% as alcohol biomarkers, other investigators found sensitivity of only 20–45% in heavy drinkers. The authors compared CDT as well as %CDT values in alcohol-dependent patients, heavy drinkers, and controls and found that sensitivity of CDT was 49% in heavy drinkers but higher (90%) in alcohol-dependent patients. The mean CDT in control men was 14.5 U/L, whereas in control females it was 21.3 U/L. In contrast, the mean CDT was 34.6 and 39.2 U/L in alcohol-dependent male and female patients, respectively. In heavy drinkers, mean CDT was 26.2 U/L in males and 28.6 U/L in females. Similar trends were also observed in the % CDT index. In alcohol-dependent males and females, mean %CDT was 3.9 and 2.4%, respectively. In contrast, mean %CDT was only 0.1% in both control males and females. The authors concluded that CDT assay has higher sensitivity than %CDT. In addition, the absolute CDT result should be interpreted with caution in patients with increased transferrin levels, such as anemic patients, those who use oral contraceptives, and pregnant women. On the other hand, a low transferrin level associated with acute phase reaction could lead to false-negative results [26]. Therefore, %CDT, which is not gender specific, may have a slight advantage over absolute CDT levels as an alcohol biomarker, especially in situations in which abnormal amounts of transferrin may be present [27].

Various cutoff values have been proposed for %CDT to differentiate a heavy alcohol user from a moderate drinker or a nondrinker, and such cutoff levels are often method dependent and also depend on which transferrin isoforms are used to define CDT. In some methods, especially those based on high-performance liquid chromatography (HPLC) or electrophoresis, disialotransferrin or a combination of asialo- and disialotransferrin may be used in the calculation of CDT. In immunoassay-based methods, a combination of asialo, monosialo, and disialotransferrin is used to calculate CDT, and then %CDT values are derived by dividing CDT by total transferrin value. Therefore, capillary electrophoresis and HPLC-based methods usually use a lower cutoff value than immunoassay-based methods. In general, cutoff values may range from 1.3 to 2.6% based on the method. In addition, significantly different sensitivity and specificity of %CDT as an alcohol biomarker has been reported by different investigators.

In a study of 396 women and 403 men, Fleming and Mundt reported that 5.6% of these subjects (45 of 799) consumed more than 14 drinks per week. In addition, 2% of the cohort abused alcohol, and 3% were alcohol dependent. The authors used receiver operating curve characteristic to define the cutoff value to determine the best fit for subjects who consumed more than

90 drinks per month and concluded that 2.5% represented the best cutoff value. In general, median %CDT was 2.0 for abstainers, 2.1 for moderate drinkers, and 2.7 for heavy drinkers (90 or more drinks per week). The overall sensitivity and specificity of the %CDT test were 61 and 85% respectively. However, sensitivity was only 40% in diabetic patients, and specificity was 89%. In hypertensive patients, sensitivity and specificity were 58 and 87%, respectively. In general, %CDT as an alcohol biomarker has higher sensitivity in patients who are neither diabetic nor hypertensive [28]. Hock *et al.* proposed a significantly higher cutoff value of 3.0% for %CDT based on the 95th percentile value of social drinkers [29]. Madhubala *et al.* compared sensitivity and specificity of %CDT with other alcohol biomarkers, including aspartate aminotransferase (AST), alanine aminotransferase (ALT), γ-glutamyl transferase (GGT), and mean corpuscular volume (MCV), using 25 alcoholic subjects and 25 controls. The alcoholic subjects consumed 44.2 ± 12.2 drinks per week, whereas control subjects did not consume any alcohol. Using a cutoff value of 2.4%, the authors observed that %CDT had a sensitivity of 84% and a specificity of 92%. The %CDT values were elevated in 92% of subjects who abused alcohol compared to only 72% of alcoholics who showed elevated levels of GGT [30]. Sensitivity and specificity of various alcohol biomarkers as stated by these authors are summarized in Table 6.2.

Appenzeller *et al.* studied various cutoffs of %CDT, including 1.5 (recommend by the kit manufacturer), 2, 2.5, and 3.0%. They observed that at 3% cutoff, specificity greater than 95% can be achieved. In addition, almost all specimens with %CDT of approximately 3% showed the presence of asialotransferrin. For drivers with blood alcohol less than 50 mg/dL, almost no specimens showed %CDT greater than 3%. In contrast, 47% of specimens with a blood alcohol content of between 300 and 350 mg/dL showed %CDT greater than 3%. Moreover, 67% of subjects with blood alcohol higher than 350 mg/dL showed %CDT greater than 3%. Only 2.9% of drivers with blood alcohol below the legal limit (80 mg/dL) showed %CDT greater than

Table 6.2 Sensitivity and Specificity of Various Alcohol Biomarkers

Biomarker	Cutoff	Sensitivity (%)	Specificity (%)
% Carbohydrate-deficient transferrin (CDT)	2.4%	84	92
γ-Glutamyl transferase (GGT)	30 U/L	64	72
Mean corpuscular volume (MCV)	100 fL	48	52
Alanine aminotransferase (ALT)	35 U/L	32	92
Aspartate aminotransferase (AST)	35 U/L	68	80

Source: *Data from Arndt* et al. [24].

3% [31]. In another study that involved 40 drunk drivers and 51 controls, Bortolotti *et al.* observed statistically significant increases in %CDT in drunk drivers compared to controls using the Student *t*-test. The blood alcohol of drunk drivers varied from 1.00 to 4.21 g/L (average, 2.34 g/L (234 mg/dL)), whereas urine alcohol levels of control subjects were negative (blood alcohol level was not determined in control subjects). The %CDT in drunk drivers varied from 0.79 to 15.86%, with an average value of 3.88%. In contrast, %CDT in control subjects varied from 0.53 to 2.24%, with an average value of 1.38%. The authors used a %CDT cutoff of 2% in their laboratory. Using this criterion, 27 of 40 drunk drivers (67.5%) showed elevated %CDT. In contrast, only 2 of 51 control subjects (3.9%) showed %CDT greater than the 2% cutoff value. The authors concluded that %CDT can identify high prevalence of chronic alcohol abusers among drunken drivers [32]. In another study, the authors commented that the usual cut-off of immunonephelometric assay for determination of %CDT should be reduced from the screening cutoff of 2.5%, as stated by the manufacturer, to 2.32% in order to achieve better sensitivity [33]. In a study involving 502 patients admitted consecutively in a medical department during a 4-week period, the authors determined that clinical sensitivity and specificity of CDT for detection of alcohol consumption of more than 50 g per day were 69 and 92%, respectively [34].

6.3.3 CDT and GGT as Combined Alcohol Biomarker

Combining GGT and CDT values to derive a new parameter mathematically may improve the sensitivity and specificity of this calculated parameter as an alcohol biomarker. The following mathematical formula is used to derive this parameter:

$$\text{GGT} - \text{CDT} = [0.8 \times \ln(\text{GGT})] + [1.3 \times \ln(\text{CDT})]$$

In this case, both GGT level and CDT level are expressed as units per liter. Sillanaukee and Olsson proposed a cutoff value of 6.5 using the 95th percentile of the data from the control group for this combined marker. In social drinkers, the mean value of GGT–CDT observed by the authors was 5.5 for both males and females. However, for male and female alcoholics, the GGT–CDT mean values were 7.5 and 7.2, respectively. For CDT values, male alcoholics showed a mean value of 27.9 U/L, and female alcoholics showed a mean value of 33.0 U/L. In contrast, in controls, the mean value observed in males was 16.2 U/L, and that observed in females was 12.7 U/L. The authors further reported that sensitivity and specificity of the GGT–CDT marker were 79 and 93%, respectively, which were superior to sensitivity of 65% and specificity of 94% observed with CDT. The sensitivity and specificity of GGT were 59 and 91%, respectively. The authors concluded that combined GGT–CDT is more sensitive and specific than CDT alone as an alcohol biomarker [35].

A similar approach can be adopted to calculate the GGT–%CDT parameter:

$$\text{GGT–\%CDT} = [0.8 \times \ln(\text{GGT})] + [1.3 \times \ln(\text{\%CDT})]$$

However, the cutoff level should be different from that of the GGT–CDT parameter. Based on receiver operator curve analysis, Anttila *et al.* established a cutoff value of 4.0 for GGT–%CDT. In a study involving 34 alcoholics with biopsy-proven liver disease, 31 alcoholics without liver disease, and 45 controls, GGT–%CDT mean values were 5.7, 5.1, and 3.4 respectively. As expected, all alcoholics (with or without liver disease) showed a value well over the 4.0 cutoff level. The mean %CDT levels in alcoholics with liver disease, alcoholics without liver disease, and controls were 3.8, 4.0, and 2.0 respectively. The sensitivity and specificity of GGT–%CDT in alcoholics with liver disease were 94 and 100%, respectively. In alcoholics without liver disease, the respective sensitivity and specificity were 90 and 100%, respectively [36].

6.3.4 Application of CDT

CDT has application in both clinical medicine and forensic investigation as an alcohol biomarker. In general, it is assumed that CDT is a biomarker of chronic alcohol intake of more than 60 g per day. CDT is a superior biomarker compared to GGT and MCV. The major advantage of CDT as an alcohol biomarker is that CDT values are not significantly affected by medications except in immunocompromised patients, who may show low CDT values. In general, CDT values appear to be less elevated in women after alcohol consumption than in men. However, due to the relatively low sensitivity of CDT, it is not recommended for screening of the general population for identification of individuals who may abuse alcohol. Various features of CDT are summarized in Box 6.2.

A major application of CDT or %CDT as an alcohol biomarker is in the clinical management of alcohol-dependent patients. This marker can be used for monitoring abstinence or relapse in heavy drinkers attending clinics for detoxification. It can be used to identify heavy drinkers in medical outpatient or psychiatric outpatient clinics. This marker can also be used for screening presurgical patients.

CASE REPORT 6.1

A 53-year-old white male was referred to the substance abuse clinic after he was sentenced by a court to drug and alcohol counseling for a driving while intoxicated (DWI) conviction. The patient admitted that he was a recovering alcoholic, but on the night of the incident he consumed approximately six drinks after being sober for 3 years. At the clinic, he was diagnosed with major depressive disorder and was prescribed 60 mg fluoxetine, but he did not want intensive alcohol treatment because he was not drinking. His %CDT was 1.7%, indicating that he was not drinking. At his next appointment, his GGT was within normal limits, but %CDT was increased to 2.5%. Two weeks later, his GGT was still within normal limits, but his %CDT was further increased to 3.1%. He denied any alcohol consumption, but at a later appointment his %CDT showed a high value of 3.6%. At that time, he admitted drinking for the past 2 months on a regular basis and did not think laboratory tests were so accurate that he could be caught. At that time, he was admitted for an intensive outpatient alcohol detoxification program, which he completed successfully [37].

BOX 6.2 CHARACTERISTICS OF CDT AS ALCOHOL BIOMARKER

- CDT is a combination of minor isoforms: asialo-, monosialo-, and disialotransferrin.
- Either CDT alone (expressed as units per liter) or %CDT can be used as an alcohol biomarker.
- Usually, consumption of 50–60 g of alcohol per day chronically (for at least 2 or 3 weeks) increases CDT.
- The half-life of CDT is usually 14–17 days; values return to normal 3 or 4 weeks after abstinence.
- Women may have a higher CDT than men.
- Whereas absolute CDT value is gender dependent; %CDT is not.
- %CDT may be slightly superior to CDT as an alcohol biomarker because transferrin values may be elevated by various factors, such as anemia, use of oral contraceptives, and pregnancy.
- CDT values may be less elevated in women than in men after chronic alcohol consumption.
- The upper limit of normal for CDT in males is usually 20 U/L and in females 26 U/L, but different authors may use slightly different cutoffs.
- The cutoff value of %CDT varies among reports but is usually between 2 and 3%; however, many authors describe a cutoff value of 2.5 or 2.6%.
- Sensitivity and specificity of CDT or %CDT vary widely as reported by various authors, but in general %CDT is a better marker than GGT or MCV.
- Combined GGT–%CDT marker, based on the formula $[0.8 \times \ln(GGT)] + [1.3 \times \ln(\%CDT)]$, is superior to %CDT as an alcohol biomarker.
- CDT values are not influenced significantly by medication, but immunocompromised patients may have lower CDT values.

In at least eight European countries, alcohol biomarkers are used as part of the clinical evaluation of convicted drunk drivers for abstinence and regranting driver's licenses. In Switzerland, Italy, and Austria, repeat offenders are sent to therapy, and alcohol biomarkers including CDT are measured quarterly for 1 year to monitor abstinence. After 1 year with documented abstinence, the driver's license is reinstated. A CDT test was introduced in Belgium in 2008 within the framework of driver's license reinstatement. The blood samples are drawn bimonthly, and CDT along with other biomarkers are monitored to evaluate a driver's adherence to the abstinence program for 1 year. Based on a 2½-year study, Maenhout *et al.* reported that CDT is a powerful tool for driver's license renewal or regranting [38]. Of all routinely used alcohol biomarkers, %CDT is the main predictor of recidivism of drunk driving in previously convicted drunk driving offenders [39].

CDT has been studied as a marker of chronic alcohol use in death investigations. CDT is stable in postmortem blood for up to 36 hr and can be used as a marker of antemortem alcohol use prior to time of death in medical examiner cases [25]. Popovic *et al.* noted that CDT analysis can be performed in postmortem specimens up to 76 hr after collection. In addition, CDT analysis in death investigations for antemortem alcohol use had a sensitivity of 59% and specificity of 71%. However, high incidences of false-positive CDT results were observed in individuals with liver cirrhosis and liver failure that were nonalcoholic in origin [40].

Glycosylation is the most common post-translational modification of protein, and disturbance in glycosylation may be congenital or acquired. Approximately 50 congenital disorders of glycosylation of protein have been described. There are two main types of protein glycosylation, namely N-glycosylation and O-glycosylation. Most diseases are due to defects in the N-glycosylation pathways [41]. Congenital defect of glycosylation Ia is caused by phosphomannose mutase deficiency and, except for hereditary multiple exostoses, is the most commonly encountered type of defect. It affects mainly the nervous system but may also affect other organs, such as the heart, liver, gastrointestinal tract, and gonads. The incomplete addition of sialic acid to transferrin molecules as evidenced by an elevated level of CDT is the most commonly used test for identification of this type of disorder. Using a %CDT cutoff of 2.5%, Perez-Cerda *et al.* screened 7910 children with suspected metabolic disorders and identified 50 patients tentatively suffering from congenital disorders of glycosylation. In half of these cases, deficiencies were genetically confirmed, and in another 15 cases abnormalities were related to secondary alterations [42].

The CDT test can be used to monitor the progress of therapy in patients with galactosemia or fructosemia. Reversible hypoglycosylation of N-glycoprotein mimicking congenital disorder of glycosylation of protein type Ia as evidenced by transferrin isoform pattern is also present in hereditary fructose intolerance as well as galactosemia. The possible mechanism is inhibition of phosphomannose by fructose 1-phosphate in the liver, kidney, and intestine. Usually, with therapy, hypoglycosylation of protein is reversed, as evidenced by normalization of hypoglycosylated transferrin molecules commonly referred to as CDT. In a study of 10 patients with hereditary fructose intolerance and 17 patients with galactosemia, Pronicka *et al.* demonstrated that following treatment, elevated levels of %CDT were significantly reduced. For example, in patients with hereditary fructose intolerance, the mean pretherapy %CDT level of 27.3% was significantly reduced to a mean of 9.3% following a fructose-free diet. Similarly, in patients with galactosemia, mean %CDT level was reduced from 43.8 to 11.2% following dietary treatment [43].

6.3.5 Limitations of CDT as Alcohol Biomarker

There are several limitations to using CDT as an alcohol biomarker. For example, there is controversy regarding the level at which chronic consumption of alcohol should lead to elevated CDT levels or %CDT. In general, it is assumed that consuming 50–60 g of alcohol (~4 or 5 standard drinks) per day for 2 or 3 weeks should lead to elevated levels of CDT. However, some studies have suggested that consuming even 80 g of alcohol per day is not sufficient to increase CDT values over the reference range in certain individuals [44]. In contrast, Kim *et al.* reported that Korean male drinkers who experienced facial flushing after drinking due to a genetic defect leading to reduced enzymatic activity of the aldehyde dehydrogenase enzyme showed higher %CDT levels (cutoff of 2.47%) compared to male drinkers who did not experience the facial flushing reaction after drinking. In the flushing group, drinking as little as 47 g of alcohol per week (3.38 drinks) induced elevated %CDT, and the test had sensitivity and specificity of 77.8 and 70.4%, respectively. In contrast, in normal Korean male drinkers, the least amount of alcohol needed per week for elevated %CDT was 158 g (11.25 drinks). The respective sensitivity and specificity of %CDT were 62.2 and 69.6% [45]. Limitations of CDT as an alcohol biomarker are listed in Box 6.3.

BOX 6.3 FACTORS THAT MAY AFFECT %CDT TEST

False-Positive %CDT

- Congenital disorder of glycosylation
- Hereditary fructose intolerance
- Galactosemia
- End-stage liver disease including nonalcoholic liver diseases such as hepatocellular carcinoma
- Low ferritin level (both in anemic and in nonanemic patients)
- Cystic fibrosis
- Chronic obstructive pulmonary disease
- Combined pancreas and kidney transplantation
- Patient with achondroplasia
- Pregnancy, but elevated CDT level returns to normal in the postpartum stage
- Anorexia nervosa and other psychiatric illness in which a patient is in a negative catabolic state (female patients)

False-Negative %CDT

- Significant iron overload such as genetic hemochromatosis may reduce CDT level; intense iron removal therapy in these patients may increase CDT level significantly.
- Analytical interferences

Variable Effect Due to Interference with the Test

- Genetic variance of transferrin molecule

Although controversial, it is generally assumed that age, gender, ethnicity, body mass index, and smoking habit do not significantly affect the interpretation of %CDT results [46]. Nevertheless, some investigators have reported that low body mass index may result in higher CDT levels, whereas high body mass index may be associated with lower CDT levels. The authors further commented in the review article that liver diseases including end-stage liver disease may elevate both %CDT and the absolute value of CDT. In addition, total CDT levels may be affected by factors that increase transferrin levels, such as iron deficiency, chronic illness, and menopausal status. False-negative results may be associated with female gender, episodic lower level alcohol use, and acute trauma with blood loss. Certain drug therapies may affect %CDT as well as the absolute value of CDT. Anticonvulsants and angiotensin-converting enzyme (ACE) inhibitors may elevate both %CDT and CDT, whereas loop diuretics may lower %CDT and the absolute value of CDT. However, evidence that certain drugs may affect %CDT is not strong, and further studies are needed to establish such effects of certain drugs on %CDT or the absolute value of CDT [47]. Fagan *et al.* reported that heavy drinkers with a body mass index indicative of being overweight or obesity had significantly lower %CDT compared to that of lean heavy drinkers [48]. Szabo *et al.* reported three cases of bodybuilders who were ethanol abstinent but showed elevated %CDT. These bodybuilders were taking protein-containing food supplements [49].

In general, CDT levels are elevated in patients with end-stage liver disease even in the absence of alcohol abuse. In one study, the authors used HPLC to identify six patients with end-stage liver disease who showed elevated disialotransferrin levels. In addition, the authors identified one patient taking enzyme-inducing drugs (phenytoin and carbamazepine) who showed elevated disialotransferrin [50]. DiMartini *et al.* studied 79 patients with end-stage liver disease who did not consume any alcohol. They observed that nearly 50% of these patients had elevated %CDT (at a cutoff concentration of 2.6% or greater) and concluded that elevated %CDT may not indicate alcohol consumption in patients with advanced liver disease [51]. Although %CDT does not increase significantly in patients with chronic viral hepatitis, it is increased in viral liver cirrhosis. In addition, %CDT values are increased in patients with hepatocellular carcinoma [52].

False-positive CDT levels in patients with low ferritin have been reported. Both anemic and nonanemic patients with low ferritin levels may show elevated CDT levels. Low iron status or high iron demand result in increased transferrin synthesis in these patients; as a result, a proportional increase in CDT isoforms may occur [53]. Patients with cystic fibrosis may have an increased level of serum CDT due to abnormalities in protein glycosylation and sialylation [54]. Increased serum levels of CDT has also been reported in patients with chronic obstructive pulmonary disease [55]. Arndt *et al.* reported elevated serum CDT levels and %CDT in patients with combined kidney and pancreas transplantation. In contrast, diabetic patients and patients who received only kidney transplantation showed normal values of CDT and % CDT [56]. Increased CDT level has also been reported in a patient with achondroplasia [57]. During pregnancy, the relative disialo-, pentasialo-, and hexasialotransferrin levels increase along with reduced levels of tetrasialotransferrin and trisialotransferrin. As a result, false-positive %CDT based on disialotransferrin may be observed in pregnant women, with the most pronounced effect in the third trimester. In postpartum specimens, all transferrin isoforms return to normal levels [58].

Patients with eating disorders, especially anorexia nervosa, may show elevated CDT levels. In a study of 24 nonalcoholic subjects, Reif *et al.* observed that 57% of patients with anorexia nervosa showed elevated levels of CDT. In contrast, bulimia nervosa patients showed normal CDT levels. Patients with initially elevated CDT tended to be more seriously ill, but during therapy body mass index of anorexia nervosa patients normalized with parallel normalization of CDT levels [59]. Women in catabolic (negative metabolic) state may also have elevated CDT levels. For example, a woman with recent weight loss may have elevated CDT, but this phenomenon is more common in female psychiatry patients who are in a negative catabolic state but not consuming alcohol on a regular basis. In addition to anorexia nervosa, these patients may also suffer from dementia, schizophrenia, and related disorders leading to significant weight loss [60].

CDT levels are highly affected by body iron status. Patients with iron deficiency anemia usually show elevated levels of CDT, but such levels decrease with iron supplementation. Patients with genetic hemochromatosis often experience iron overload, and these patients may show false-negative results due to decreased CDT levels. In one study, the authors reported that the average CDT level in patients with genetic hemochromatosis was 9.6 U/L, whereas in controls it was 15.7 U/L. However, in patients with iron deficiency anemia, the average CDT level was higher (28.1 U/L) than that of controls. Therefore, CDT levels are dependent on iron status of the body [61]. Jensen *et al.* reported that in 11 patients with genetic hemochromatosis, CDT levels were significantly increased from a mean value of 8.5 U/L to a mean value of 16.6 U/L during intensive iron removal but not during maintenance iron removal therapy. Iron mobilization from the liver could be responsible for such increased CDT levels [62].

Various transferrin genetic variants have been reported, but they usually occur at low frequencies. Transferrin C, especially transferrin C1 and C2, is the predominant form, but other forms, such as B, can also be encountered. De Wolf *et al.* reported that a novel C2 transferrin variant, T139M, interferes with HPLC and capillary zone electrophoresis methods of analysis of CDT but does not affect the immunochemistry method [63]. The genetic transferrin B variant may cause a false-negative CDT test result, whereas the genetic D variant may cause a false-positive CDT result. Compared to HPLC, the immunoassay method for determining %CDT often produces a low result with the transferrin BC variant but a high result with transferrin CD and C2C3 variants [64].

CASE REPORT 6.2

A 23-year-old healthy professional soccer player was involved in a car accident related to alcohol consumption. His %CDT results were elevated (e.g., 3.2 and 3.65% measured by the ChronAlcol.D assay, which involves the ion exchange microcolumn separation method) above the reference range (cutoff, 2.5%), indicating alcohol abuse. The young man suffered from increasing social isolation due to the perception of chronic alcohol abuse. He was also referred for further evaluation for potential liver disease. However, he denied chronic alcohol abuse, and no evidence of liver disease was found. At this point, his serum transferrin pattern was further evaluated using isoelectric focusing, which revealed a pattern inconsistent with chronic alcohol abuse but led to the detection of a genetically determined transferrin D variant. It was established that initial CDT results were falsely elevated due to the presence of the transferrin D variant. Determination of this transferrin variant heterozygosity resulted in his social rehabilitation and reinstatement of his driver's license [65]

6.4 LABORATORY DETERMINATION OF CDT

As mentioned previously, asialo-, monosialo-, and disialotransferrin are collectively called CDT. It is controversial whether trisialotransferrin should also be considered as CDT, but most methods for CDT do not measure trisialotransferrin. Although the International Federation of Clinical Chemistry (IFCC) Working Group on Standardization of CDT recommended that disialotransferrin should be the primary target for measurement of CDT and standardization of CDT methods should be based on this single analyte, other isoforms, such as asialo- and monosialotransferrin, are measured along with disialotransferrin in various CDT methods. In general, %CDT is more widely used as an alcohol biomarker than is absolute CDT value.

CDT can be determined by various methods, including gel electrophoresis with isoelectric focusing, ion exchange chromatography, HPLC, capillary electrophoresis, and immunoassay. Currently, HPLC, capillary electrophoresis, and immunoassay are the methods most commonly used in the clinical laboratory for the determination of CDT, and the only immunoassay (immunonephelometric assay) on the market is the N-Latex CDT assay (Siemens Diagnostics). In the past, several immunoassays were available that required indirect separation of CDT from other isoforms by anion exchange minicolumns followed by immunoassay determination of various isoforms. However, many of these indirect immunoassays have been withdrawn from the market. Current CDT methods based on HPLC or capillary electrophoresis offer the advantage of graphic visualization of the band. HPLC methods rely on selective absorbances of the iron–transferrin complex at 460–470 nm, whereas the capillary electrophoresis method relies on ultraviolet detection at 200 nm due to peptide bonds present in transferrin isoforms. However, standardization of CDT methods is of utmost importance, as the IFCC working group on CDT standardization has emphasized [66].

In the past, many investigators used gel electrophoresis with isoelectric focusing (also known as zone electrophoresis or immunofixation electrophoresis (IFE)) to study CDT. In this method, molecules are separated by differences in their isoelectric point. IFE involves adding an ampholyte solution into immobilized pH gradient gel. During electrophoresis, a protein that is in a pH region lower than its isoelectric point should have positive charge and will migrate toward the cathode and during this process will come across a pH gradient that corresponds to the isoelectric point. Then the protein should have no net charge and should have no more migration. In this way, various proteins with different isoelectric points can be separated into distinct zones. CDT has an isoelectric point of 5.7, whereas that of a normal transferrin molecule is 5.4. After iron saturation, transferrin isoforms can be analyzed by isoelectric focusing. Appropriate staining technique can be used for visualization of bands. CDT can also be determined by isocratic microanion exchange chromatography at pH 5.65. In this case, normal transferrin isoforms are retained by the column. However, the rare transferrin D variant may elute with the CDT fraction [67]

The N-Latex CDT direct immunonephelometric assay is based on monoclonal antibody that specifically recognizes asialo-, monosialo-, and disialotransferrin in combination with a simultaneous assay for determination of total transferrin level (Siemens Diagnostics). Polystyrene particles coated with monoclonal antibody against CDT are agglutinated by CDT-coated polystyrene particles. If present in a patient's serum, CDT inhibits this reaction in a dose-dependent manner, which allows quantification of CDT using nephelometry over an 18-min time period. Because the degree of iron saturation of transferrin influences the binding affinity of antibody, in the first incubation step, transferrin-bound iron is removed by a chelating agent. The simultaneous determination of transferrin allows an automatic calculation of %CDT. The assay is automated and does not require any sample pretreatment. The measurement range is 0.77–25% CDT. The 97.5th percentile for %CDT cutoff was 2.35% as observed by Delanghe *et al.* in the cohort they studied. In addition, the authors found no interference in the assay from transferrin variants [68]. However, Maenhout *et al.* reported that the automated immunonephelometric method underestimates the %CDT in the presence of mutant transferrin compared to %CDT measured by the capillary zone electrophoresis method [69]. Although it is generally assumed that %CDT is a marker for heavy alcohol use (50–80 g/day), Whitfield *et al.*, using N-Latex CDT assay, observed that the probability of a %CDT greater than 2% increased with modest alcohol intake of 8–14 drinks per week in men and %CDT increased steadily with higher amounts of alcohol consumption. In women, mean %CDT was highest with 29–35 drinks per week and did not increase further with higher levels of alcohol consumption. Therefore, with higher alcohol consumption, males generally showed higher %CDT values than females [70]. In the past, many immunoassay methods required separation of CDT prior to quantitation by anion exchange minicolumn. The older ion exchange method reported CDT as units per liter, which corresponds to approximately 1 mg of transferrin per liter.

Capillary zone electrophoresis (CZE) is also used for determination of CDT. CZE requires a small amount of specimen, and assay can be performed using a fused silica capillary column at alkaline pH and solute detection can be carried out at 200-nm wavelength. This assay can also be automated. Joneli *et al.* reported that during a 10-year period, the CZE method used in their laboratory had acceptable performance and good precision. Moreover, results obtained by CZE compared well with those of HPLC and CZE assays performed in other laboratories [71]. An automated capillary electrophoresis system for analysis of CDT is available from Sebia (Capillarys2 analyzer for serum proteins; Sebia, Norcross, GA). All necessary reagents are supplied by the manufacturer with the test kit along with instructions to perform the test. Specimens are automatically diluted with an iron solution and injected at the anodic end of the capillary column, where transferrin isoforms are separated in an alkaline buffer in the presence of high voltage. Transferrin isoforms are separated based on their electrophoretic mobility, mainly depending on pH and electro-osmotic flow. Detection is achieved at the cathodic end at 200-nm wavelength. The relative amount of CDT is expressed as the sum of disialotransferrin and, when present, asialotransferrin, which is calculated automatically. The result can be expressed as percentage of transferrin (%CDT). In case of analytical interference in the capillary electrophoresis peak profile that may cause difficulty with CDT quantitation, the specimen is reanalyzed after cleanup with an immunosubtraction solution (provided by Sebia) that precipitates immunoglobulins. If the analytic problem persists after sample cleanup, an alternative method such as HPLC should be used. HPLC is the confirmatory method, with recommended cutoff of 1.3% [72]. Gonzalo *et al.* considered a value only greater than 2% to be positive using Bio-Rad HPLC and greater than 3% as positive using N-Latex CDT assay, which are higher cutoffs compared to those proposed by other authors. The authors commented that for patients with liver cirrhosis, neither method can correctly predict alcohol consumption because in 54% of patients, neither the Bio-Rad method nor the N-Latex % carbohydrate assay provided possible interpretation [73].

HPLC is considered as the reference method for %CDT determination. In the HPLC protocol described by Bergstrom and Helander, serum transferrin was saturated with iron using ferric nitrotriloacetic acid followed by precipitation of lipoproteins with dextran sulfate. Then clear supernatant after dilution was injected into the HPLC system, and various isoforms of transferrin were separated in an anion exchange column. Detection and quantification of various peaks representing various transferrin isoforms was achieved by UV detection at 470 nm (absorption of iron–transferrin complex), and the relative amount of each isoform was calculated as a percentage of total transferrin (peak areas of all isoform). The authors considered %CDT of 1.7% (representing the ratio of disialotransferrin to total transferrin) as the upper end of the reference range. The mean %CDT determined by this method was 2.4% in male heavy drinkers and 1.77% in female heavy drinkers. At 40, 60, and 80 g of alcohol intake per day, %CDT showed lower test sensitivity in women than in men [74]. Schellenberg *et al.* used the Bio-RAD %CDT method with a Varian HPLC system but used 460-nm wavelength for detection of peaks and 690-nm wavelength for background correction. The method showed good correlation with capillary electrophoresis assay (Sebia) [75]. Daves *et al.* compared capillary zone electrophoresis with HPLC and found that although cutoffs of %CDT ranging from 1.3 to 2.6% have been proposed, 1.6% should be considered as a valid cutoff for forensic investigation for the capillary electrophoresis method of %CDT using Capillarys2 (Sabia). Values between 1.3 and 1.6% should be reevaluated by another method. The cutoff value for HPLC was 1.9%. Although the overall correlation between capillary electrophoresis and the HPLC method (based on analysis of 539 samples) was good with a correlation coefficient of 0.96, discrepancies between values exceeding method-specific cutoff values (1.6% for capillary electrophoresis and 1.9% by HPLC) were observed in a large number of samples (62%), especially when CDT values were between 1.3 and 1.9%. However, with higher values (%CDT $>$2.0%), only 0.6% of specimens showed discrepancies [76]. Del Castillo Busto *et al.* analyzed transferrin isoforms after iron saturation using HPLC coupled with inductively coupled plasma mass spectrometry. Detection was achieved by the presence of iron in each of the separated isoforms of transferrin. After screening of the isoforms containing iron by inductively coupled plasma mass spectrometry, structural characterization of each isoform was carried out using matrix-assisted laser desorption/ionization mass spectrometry with time-of-flight detection (MALDI-TOF). Electrospray mass spectrometry was also used for comparison [77].

6.5 CONCLUSIONS

CDT and %CDT are good alcohol biomarkers for heavy alcohol use due to their good specificity, although sensitivity has varied widely among studies. CDT has been in clinical use for many years and was the first alcohol biomarker approved by the FDA. A number of analytical methods have been developed for analysis of CDT but without any harmonization or calibration to a reference method. As a consequence, there are different cutoff values for the different methods, which is hampering understanding of the diagnostic value of CDT or %CDT in routine clinical use. Standardization of different methods with a reference HPLC method is needed, as is the use of a robust calibration system to achieve standardization of the various CDT methods. Harmonization of methods is possible using human serum-based calibrators [78].

References

[1] Whitehead TP, Clarke CA, Whitfield AG. Biochemical and hematological markers of alcohol intake. Lancet 1978;1(8071):978–81.

[2] Seppa K, Heinila S, Sillanaukee P, Saarni M. Evaluation of macrocytosis by general practitioners. J Stud Alcohol 1996;57:97–100.

[3] Rumsey SE, Hokin B, Magin PJ, Pond D. Macrocytosis—An Australian general practice perspective. Aust Fam Physician 2007;36:571–2.

[4] Wymer A, Becker DM. Recognition and evaluation of red blood cell macrocytosis in the primary care setting. J Gen Intern Med 1980;5:192–7.

[5] Morgan MY, Camilo ME, Luck W, Sherlock S, et al. Macrocytosis in alcohol related liver disease: its value for screening. Clin Lab Hematol 1981;3:35–44.

[6] Koivisto H, Hietala J, Anttila P, Parkkila S, et al. Long-term ethanol consumption and macrocytosis: diagnostic and pathogenic implications. J Lab Clin Med 2006;147:191–6.

[7] Kaferle J, Strzoda CE. Evaluation of macrocytosis. Am Fam Physician 2009;79:203–8.

[8] Eriksson CJ. The role of acetaldehyde in the actions of alcohol (update 2000). Alcohol Clin Exp Res 2001;25:15S–32S.

[9] Niemela O, Parkkila S. Alcoholic macrocytosis—is there a role for acetaldehyde and adducts? Addict Biol 2004;9:3–10.

[10] Yokoyama A, Yokoyama T, Muramatsu T, Omori T, et al. Macrocytosis, a new predictor for esophageal squamous cell carcinoma in Japanese alcoholic men. Carcinogenesis 2003;24: 1773–8.

[11] Yokoyama A, Yokoyama T, Kumagai Y, Kato H, et al. Mean corpuscular volume, alcohol flushing and the predicted risk of squamous cell carcinoma of esophagus in cancer-free Japanese men. Alcohol Clin Exp Res 2005;29:1877–83.

[12] Clemens MR, Kessler W, Schied HW, Schupmann A, et al. Plasma and red cell lipids in alcoholics with macrocytosis. Clin Chim Acta 1986;156:321–8.

[13] Keenan Jr WF. Macrocytosis as an indicator of human disease. J Am Board Fam Pract 1989;2:252–6.

[14] Horstman AL, Serck SL, Go RS. Macrocytosis associated with monoclonal gammopathy. Eur J Hematol 2005;75:146–9.

[15] Younes M, Dagher GA, Dulanton JV, Njeim M, et al. Unexplained macrocytosis. South Med J 2013;106:121–5.

[16] Aslimia F, Mazza JJ, Yale SH. Megaloblastic anemia and other causes of macrocytosis. Clin Med Res 2006;4:236–41.

[17] Riebe D, Thorn W. Influence of carbohydrate moieties of human serum transferrin on the determination of its molecular mass by polyacrylamide gradient gel electrophoresis and staining with periodic acid–Schiff reagent. Electrophoresis 1991;12:287–93.

[18] De Jong G, van Djik JP, van Eijk HG. The biology of transferrin. Clin Chim Acta 1990;190:1–46.

[19] Kamboh MI, Ferrell RE. Human transferrin polymorphism. Hum Hered 1987;37:65–81.

[20] Martensson O, Harlin A, Brandt R, Seppa K, et al. Transferrin isoform distribution: gender and alcohol consumption. Alcohol Clin Exp Res 1997;21:1710–15.

[21] Arndt T. Carbohydrate-deficient transferrin as a marker of chronic alcohol abuse: a critical review of preanalysis, analysis and interpretation. Clin Chem 2001;47:13–27.

[22] Stibler H, Kjellin K. Isoelectric focusing and electrophoresis of the CSF proteins in tremors of different origins. J Neurol Sci 1976;30:269–85.

[23] Liber CS. Carbohydrate deficient transferrin in alcoholic liver disease: mechanism and clinical implications. Alcohol 1999;19:249–54.

[24] Arndt T, Behnken L, Martens B, Hackler R. Evaluation of the cutoff for serum carbohydrate deficient transferrin as a marker of chronic alcohol abuse determination by ChronAlco ID assay. J Lab Med 1999;23:507–10.

[25] Malcom R, Anton RF, Conradi SF, Sutherland S. Carbohydrate deficient transferrin and alcohol use in medical examiner's case. Alcohol 1999;17:7–11.

[26] Sorvajarvi K, Blake JE, Israel Y, Niemela O. Sensitivity and specificity of carbohydrate deficient transferrin as a marker of alcohol abuse are significantly influenced by alterations in serum transferrin: comparison of two methods. Alcohol Clin Exp Res 1996;20:449–54.

[27] Anton RF, Dominick C, Bigelow M, Westby C. Comparison of BioRad %CDT, TIA and CDTect as laboratory biomarkers of heavy alcohol use and their relationship with gamma glutamyltransferase. Clin Chem 2001;47:1769–75.

[28] Fleming M, Mundt M. Carbohydrate deficient transferrin: validity of a new alcohol biomarker in sample patients with diabetes and hypertension. J Am Board Fam Pract 2004;17:245–55.

[29] Hock B, Schwarz M, Domke I, Grunert VP, et al. Validity of carbohydrate deficient transferrin gamma glutamyl transferase (gamma-GT) and mean corpuscular erythrocyte volume (MCV) as biomarkers of chronic alcohol abuse: a study in patients with alcohol dependence and liver disorders of non-alcoholic and alcoholic origin. Addiction 2005;100:1477–86.

[30] Madhubala V, Subhashree AR, Shanthi B. Serum carbohydrate deficient transferrin as a sensitive marker in diagnosing alcohol abuse—A case–control study. J Clin Diagn Res 2013;7:197–200.

[31] Appenzeller BM, Schneider S, Maul A, Wennig R. Relationship between blood alcohol concentration and carbohydrate deficient transferrin among drivers. Drug Alcohol Depend 2005;79:261–5.

[32] Bortolotti F, Trettene M, Gottardo R, Bernini M, et al. Carbohydrate deficient transferrin (CDT): a reliable indicator of the risk of driving under the influence of alcohol when determined by capillary electrophoresis. Forensic Sci Int 2007;170:175–8.

[33] Bortolotti F, Trevisan MT, Micciolo R, Canal L, et al. Re-assessment of the cutoff levels of carbohydrate deficient transferrin (CDT) automated immunoassay and multi-capillary electrophoresis for application in a forensic context. Clin Chim Acta 2013;416:1–4.

[34] Bell H, Tallaksen CM, Try K, Haug E. Carbohydrate deficient transferrin and other markers of high alcohol consumption: a study of 502 patients admitted consecutively to a medical department. Alcohol Clin Exp Res 1994;18:1103–8.

[35] Sillanaukee P, Olsson U. Improved diagnostic classification of alcohol abusers by combining carbohydrate deficient transferrin and γ-glutamyl transferase. Clin Chem 2001; 47:681–5.

[36] Anttila P, Jarvi K, Latvala J, Blake JE, et al. A new modified gamma-%CDT method improves the detection of the problem drinking: studies in alcoholics with or without liver disease. Clin Chim Acta 2003;338:45–51.

[37] Cluver JS, Miller PM, Anton RF. Case studies of utility of serum carbohydrate deficient transferrin (%CDT) in clinical management of alcoholics. J Addict Med 2007;1: 44–7.

[38] Maenhout TM, Baten G, De Buyzere ML, Delanghe JR. Carbohydrate deficient transferrin in a driver's license regranting program. Alcohol Alcohol 2012;47:253–60.

[39] Maenhout TM, Poll A, Vermassen T, De Buyzers ML, et al. Usefulness of indirect alcohol biomarkers for predicting recidivism of drunk-driving among previously convicted drunk driving offenders: results from the Recidivism of Alcohol Impaired Driving (ROAD) study. Addiction 2014;109:71–8.

[40] Popovic V, Atanasijevic T, Nikolic S, Bozic N, et al. Forensic aspects of postmortem serum carbohydrate deficient transferrin analysis as a marker of alcohol abuse. Srp Arh Celok Lek 2013;141:203–6.

[41] Cylwik B, Naklicki M, Chrostek L, Gruszewska E. Congenital disorders of glycosylation: part 1. Defect of protein N-glycosylation. Acta Biochim Pol 2013;60:151–61.

[42] Perez-Cerda C, Quelhas D, Vega AI, Ecay J, et al. Screening using serum percentage of carbohydrate deficient transferrin for congenital disorders of glycosylation in children with suspected metabolic disease. Clin Chem 2008;54:93–100.

[43] Pronicka E, Adamowicz M, Kowalik A, Ploski R, et al. Elevated carbohydrate deficient transferrin (CDT) and its normalization on dietary treatment as a useful biochemical test for hereditary fructose intolerance and galactosemia. Pediatr Res 2007;62:101–5.

[44] Salmela KS, Laitinen K, Nystrom M, Salaspuro M. Carbohydrate-deficient transferrin during 3 weeks' heavy alcohol consumption. Alcohol Clin Exp Res 1994;18:228–30.

[45] Kim SG, Kim JS, Kim SS, Jung JG, et al. Relationship between the level of alcohol consumption and abnormality in biomarkers according to facial flushing in Korean male drinkers. Korean J Fam Med 2013;34:123–30.

[46] Bergstrom J, Helander A. Influence of alcohol use, ethnicity, age, gender, BMI and smoking on the serum transferrin glycoform pattern: implications for use of carbohydrate deficient transferrin (CDT) as alcohol biomarker. Clin Chim Acta 2008;388:59–67.

[47] Fleming MF, Anton RF, Spies CD. A review of genetic, biological, pharmacological and clinical factors that affect carbohydrate deficient transferrin levels. Alcohol Clin Exp Res 2004;28:1347–55.

[48] Fagan KJ, Irvine KM, McWhinney BC, Fletcher LM, et al. BMI but not stage or etiology of nonalcoholic liver disease affects the diagnostic utility of carbohydrate deficient transferrin. Alcohol Clin Exp Res 2013;37:1771–8.

[49] Szabo G, Keller E, Szabo G, Lengyel G, et al. The level of carbohydrate deficient transferrin is highly increased in bodybuilders. Orv Hetil 2008;149:2087–90 [in Hungarian].

[50] Bergstrom J, Helander A. HPLC evaluation of clinical and pharmacological factors reported to cause false positive carbohydrate deficient transferrin (CDT) levels. Clin Chim Acta 2008;389:164–6.

[51] DiMartini A, Day N, Lane T, Beisler AT, et al. Carbohydrate deficient transferrin in abstaining patients with end stage liver disease. Alcohol Clin Exp Res 2001;25:1729–33.

[52] Murawaki Y, Sugisaki H, Yussa I, Kawasaki H. Serum carbohydrate deficient transferrin in patients with nonalcoholic liver disease and with hepatocellular carcinoma. Clin Chim Acta 1997;259:97–108.

[53] Van Pelt J, Azimi H. False positive CDTect values in patients with low ferritin values. Clin Chem 1998;44:2219–20.

[54] Larsson A, Flodin M, Kollberg H. Increased serum concentrations of carbohydrate deficient transferrin (CDT) in patients with cystic fibrosis. Ups J Med Sci 1998;103:231–6.

[55] Nihlen U, Montnemery P, Lindholm LH, Lofdahl CG. Increased serum levels of carbohydrate deficient transferrin in patients with chronic obstructive pulmonary disease. Scand J Clin Lan Invest 2001;61:341–7.

[56] Arndt T, Hackler R, Muller T, Kleine TO, et al. Increased serum concentration of carbohydrate deficient transferrin in patients with combined pancreas and kidney transplantation. Clin Chem 1997;43:344–51.

[57] Assmann B, Hackler R, Peters V, Schaefer JR, et al. Increased carbohydrate deficient transferrin concentration and abnormal protein glycosylation of unknown etiology in a patient with achondroplasia. Clin Chem 2000;46:584–6.

[58] Kenan N, Larsson A, Axelsson O, Helander P. Changes in transferrin glycosylation during pregnancy may lead to false-positive carbohydrate-deficient transferrin (CDT) results in testing for riskful alcohol consumption. Clin Chim Acta 2011;412:129–33.

[59] Reif A, Fallgatter AJ, Schmidtke A. Carbohydrate deficient transferrin parallels disease severity in anorexia nervosa. Psychiatry Res 2005;137:143–6.

[60] Reif A, Keller H, Schneider M, Kamolz S, et al. Carbohydrate deficient transferrin is elevated in catabolic female patients. Alcohol Alcohol 2001;36:603–7.

[61] De Feo TM, Fargion S, Duca L, Mattioli M, et al. Carbohydrate deficient transferrin, a sensitive marker of chronic alcohol abuse, is highly influenced by body iron. Hepatology 1999;29:658–63.

[62] Jensen PD, Peterslund NA, Poulsen JH, Jensen FT, et al. The effect of iron deductive treatment on the serum concentration of carbohydrate deficient transferrin. Br J Haematol 1994;88:56–63.

[63] De Wolf HK, Huijben K, van Wijnen M, de Metz M, et al. A novel C2 transferrin variant interfering with the analysis of carbohydrate deficient transferrin. Clin Chim Acta 2011;412:1683–5.

[64] Helander A, Eriksson G, Stibler H, Jeppsson JO. Interference of transferrin isoform types with carbohydrate-deficient transferrin quantification in the identification of alcohol abuse. Clin Chem 2001;47:1225–33.

[65] Welker MW, Printz H, Hackler R, Rafat M, et al. Identification of elevated carbohydrate deficient transferrin (CDT) serum level as transferrin (Tf) D variant by means of isoelectric focusing. Z Gastroenterol 2004;42:1049–54 [in German].

[66] Helander A, Wielders JP, Jeppsson JO, Weykamp C, et al. Towards standardization of carbohydrate deficient transferrin (CDT) measurements: II. Performance of a laboratory network running the HPLC candidate reference measurement procedures and evaluation of a candidate reference material. Clin Chem Lab Med 2010;48:1585–92.

[67] Stibler H, Jaeken J. Carbohydrate deficient serum transferrin in a new systemic hereditary syndrome. Arch Dis Children 1990;65:107–11.

[68] Delanghe JR, Helander A, Wielders JP, Pekelharing JM, et al. Development and multicenter evaluation of the N-latex CDT direct immunoassay for serum carbohydrate-deficient transferrin. Clin Chem 2007;53:1115–21.

[69] Maenhout TM, Uytterhoeven M, LeCocq E, De Buyzere ML, et al. Immunonephelometric carbohydrate deficient transferrin results and transferrin variants. Clin Chem 2013;59:997–8.

[70] Whitfield JB, Dy V, Madden P, Heath AC, et al. Measuring carbohydrate-deficient transferrin by direct immunoassay: factors affecting diagnostic sensitivity for excess alcohol intake. Clin Chem 2008;54:1158–65.

[71] Joneli J, Wanzenried U, Schiess J, Lanz C, et al. Determination of carbohydrate-deficient transferrin in human serum by capillary zone electrophoresis: evaluation of assay over a 10-year period and quality assurance over a 10-year period in the routine area. Electrophoresis 2013;34:1563–71.

[72] Kenan N, Husand S, Helander A. Importance of HPLC confirmation of problematic carbohydrate-deficient transferrin (CDT) results from a multicapillary electrophoresis routine method. Clin Chim Acta 2010;411:1945–50.

[73] Gonzalo P, Pecquet M, Bon C, Gonzalo S, et al. Clinical performance of the carbohydrate-deficient transferrin (CDT) assay by the Sebia Capillarys2 system in case of cirrhosis: interest of the Bio-Rad %CDT by HPLC test and Siemens N-latex CDT kit as putative confirmatory methods. Clin Chim Acta 2012;413:712–18.

[74] Bergstrom JP, Helander A. Clinical characteristics of carbohydrate-deficient transferrin (% disialotransferrin) measured by HPLC: sensitivity, specificity, gender difference and relationship with other alcohol biomarkers. Alcohol Alcohol 2008;43:436–41.

[75] Schellenberg F, Mennetrey L, Girre C, Nalpas B, et al. Automated measurement of carbohydrate deficient transferrin using Bio-RAD &CDT by HPLC test on a Variant HPLC system: evaluation and comparison with other methods. Alcohol Alcohol 2008;43:569–76.

[76] Daves M, Cemin R, Floreani M, Pusceddu I, et al. Comparative evaluation of capillary zone electrophoresis and HPLC in the determination of carbohydrate-deficient transferrin. Clin Chem Lab Med 2011;49:1677–80.

[77] Del Castillo Busto ME, Montes-Bayon M, Blanco-Gonzalez E, Meija J, et al. Strategies to study human serum transferrin isoforms using integrated liquid chromatography ICPMS, MALDI-TOF and ESI-Q-TOF detection: application to chronic alcohol abuse. Anal Chem 2005;77:5615–21.

[78] Weykamp C, Wielders JP, Helander A, Anton RF, et al. Towards standardization of carbohydrate-deficient transferrin (CDT) measurements: III. Performance of native serum and serum spiked with disialotransferrin proves that harmonization of CDT assay is possible. Clin Chem Lab Med 2013;51:991–6.

CHAPTER 7

β-Hexosaminidase, Acetaldehyde–Protein Adducts, and Dolichol as Alcohol Biomarkers

CONTENTS

7.1 Introduction..163

7.2 β-Hexosaminidase Isoforms..............163

7.3 β-Hexosaminidase as Alcohol Biomarker...........164

7.3.1 Pathophysiological Conditions that Cause Elevated Levels of β-Hexosaminidase.......168

7.4 Laboratory Methods for Measuring β-Hexosaminidase..171

7.5 Acetaldehyde–Protein Adducts as Alcohol Biomarkers 172

7.5.1 Acetaldehyde–Hemoglobin Adducts....173

7.5.2 Acetaldehyde–Erythrocyte Protein Adducts.........................174

7.5.3 IgA Antibody Against Acetaldehyde-Modified Bovine Serum Albumin.........................175

7.1 INTRODUCTION

β-Hexosaminidase, also known as *N*-acetyl-β-D-glucosaminidase, is a lysosomal enzyme found in most body tissues, especially the kidneys, in which concentrations are higher than in other tissues. Total serum β-hexosaminidase activity, particularly the activity of the heat-stable fraction of β-hexosaminidase in serum, as well as total urinary β-hexosaminidase activity, is increased in alcoholics compared to moderate drinkers and nondrinkers. Therefore, β-hexosaminidase is used as an alcohol biomarker. Acetaldehyde, being highly reactive, also rapidly forms stable adducts with a number of compounds, including proteins such as albumin (the most abundant protein in blood) and hemoglobin. These adducts are found mostly in chronic heavy consumers of alcohol because acetaldehyde levels are more significantly elevated in these individuals compared to moderate or social drinkers. The hemoglobin–acetaldehyde adduct has also received attention in the scientific community as a biomarker of alcohol abuse.

7.2 β-HEXOSAMINIDASE ISOFORMS

β-Hexosaminidase is a complex group of glycoprotein lysosomal isoenzymes that releases *N*-acetylglucosamine and *N*-acetylgalactosamine from the nonreducing end of oligosaccharide chains of glycoproteins, glycolipids, and glycosaminoglycans. To prevent accumulation of GM_2 ganglioside in Tay–Sachs disease or gangliosides and oligosaccharides in Sandhoff's disease, higher enzymatic activities of β-hexosaminidase are needed compared to the levels present in these patients. Isoenzymes of β-hexosaminidase are composed of two polypeptide chains designed as α and β encoded by two genes. In humans, the gene encoding the pre-pro-α subunit (*HEX A* gene) is located on chromosome 15 (15q23–q24), whereas the gene encoding the pre-pro-β

A. Dasgupta: Alcohol and Its Biomarkers. DOI: http://dx.doi.org/10.1016/B978-0-12-800339-8.00007-9

7.6 Dolichol as Alcohol Biomarker 175
7.7 Conclusions.. 177
References 177

subunit (*HEX B* gene) is located on chromosome 5 (5q13) [1]. Two major isoenzymes of β-hexosaminidase have been characterized—isoenzyme A (one α and one β chain) and B (two β chains)—whereas isoenzyme S (two α chains) represents only approximately 0.02% of isoenzyme activity [2]. However, there are other isoforms of β-hexosaminidase present in serum that can be separated by isoelectric focusing. Isoform P (the P isoform is elevated in pregnancy, hence the name) and isoforms I_1 and I_2 (intermediate heat-stable forms normally present in serum) all consist of two β subunits. In order of decreasing isoelectric points, the isoenzymes of β-hexosaminidase can be arranged as B, I_1, I_2, P, A, and S. Hexosaminidase C has been purified from human placenta, and this isoform is more active in patients deficient in β-hexosaminidase A and B activity. Hexosaminidase C has a distinct isoelectric point and is found predominantly in the brain [3].

Like other lysosomal enzymes, subunits α and β of β-hexosaminidase are transported as high-molecular-weight precursors through the endoplasmic reticulum and Golgi, where they undergo numerous post-translational modifications in transit or after reaching lysosomes. These modifications include removal of single peptide, N-glycosylation, formation of disulfide bonds, and acquisition of mannose-6-phosphate [4]. The heat-stable isoforms of β-hexosaminidase are hexosaminidase B, I (both I_1 and I_2), and P, whereas hexosaminidase A and S are heat labile. Studies with neuraminidase indicate that β-hexosaminidase A contains more sialic acid residues, and removal of sialic acid from the molecule produces other forms that have similar electrophoretic mobility to those of intermediate and B forms [5]. Mutation of the gene that encodes the α subunit (*HEX A*) leads to a deficiency of β-hexosaminidase isoenzyme A, which causes Tay–Sachs disease, whereas mutation in the gene encoding the β subunit (*HEX B*) leads to deficiency of both β-hexosaminidase isoenzymes A and B in Sandhoff's disease [6].

7.3 β-HEXOSAMINIDASE AS ALCOHOL BIOMARKER

Chronic alcohol consumption results in increased levels of β-hexosaminidase in both serum and urine. In general, more than 90% of alcoholics admitted to hospitals for alcohol intoxication have an increased serum β-hexosaminidase concentration [7]. It has been assumed that consuming greater than 60 g of alcohol (>4.5 standard drinks) per day for 10 days or more results in an increased level of β-hexosaminidase in serum. Karkkainen *et al.* reported that the mean β-hexosaminidase level was 35.0 U/L among drunken men ($n = 25$), whereas it was 16.8 U/L among healthy males ($n = 16$) who were social drinkers. Interestingly, the mean β-hexosaminidase level was 19.8 U/L among teetotalers. The authors further reported that the sensitivity of β-hexosaminidase among heavy drinkers was 85.7%, which was superior to the sensitivity of 47.6% of the alcohol biomarker γ-glutamyl transpeptidase (GGT). The authors further stated that the specificity of β-hexosaminidase as an alcohol biomarker was 97.6% [8]. Therefore, elevated serum β-hexosaminidase concentration is a marker for heavy drinking.

As mentioned previously, β-hexosaminidase in serum exists as heat-labile (A and S) and heat-stable (B, I_1, I_2, and P) isoforms. In serum of alcoholics, the P isoform is often increased more than other isoforms, but isoenzyme separation to detect P isoform is time-consuming using the isoelectric focusing technique [9]. Subsequently, Hultberg *et al.* developed enzyme immunoassays for the determination of β-hexosaminidase isoenzyme A as well as isoenzyme B in human sera using monoclonal antibodies. However, enzyme immunoassay for β-hexosaminidase B also showed similar cross-reactivities to P and intermediate. Therefore, isoenzyme B assay represented the sum of all these isoforms (β-hexosaminidase B, P, and intermediate forms abbreviated as Hex B). Interestingly, Hex B concentration represents the heat-stable fraction of β-hexosaminidase in serum. The authors determined that the upper limit of normal for Hex B was less than 8.2 U/L. However, levels of Hex B were elevated in 38 of 42 patients hospitalized for alcohol detoxification. The mean serum Hex B level was 24.7 U/L in these patients. In comparison, carbohydrate-deficient transferrin (CDT) was elevated in 35 of 42 patients, and the mean value was 41.7 U/L (normal: males, <20 U/L; females, <25 U/L). There was a good correlation between Hex B and CDT levels in these patients. However, neither Hex B nor CDT values correlated with GGT or aspartate aminotransferase (AST) values. The mean half-life of Hex B was 6.5 days, whereas that of CDT was 8.6 days. The authors reported that Hex B had better sensitivity than CDT as an alcohol biomarker to identify heavy alcohol use, and combining both markers would increase the sensitivity further [10]. In another study, the authors observed that mean Hex B concentration was 25.6 U/L among alcoholics ($n = 38$) who were hospitalized for detoxification, but the mean concentration was 4.3 U/L among 20 social drinkers with a median alcohol consumption of 38 g per week. In addition, the mean total β-hexosaminidase levels were 37.9 U/L in alcoholics and 12.5 U/L in social drinkers. Interestingly, previous alcoholics who were abstinent between 6 days and 10 years (median, 17 days) showed a mean Hex B value of 4.3 U/L and a mean total β-hexosaminidase level of 11.6 U/L [11].

Stowell *et al.* measured total β-hexosaminidase level, heat-stable β-hexosaminidase level (after deactivating the heat-labile fraction by heating the specimen at 50°C for 2 hr), and urine β-hexosaminidase level, along with concentrations of various other alcohol biomarkers, including CDT, GGT, alanine aminotransferase (ALT), and AST, in both alcoholic and nonalcoholic subjects. The authors used *p*-nitrophenol-*N*-acetyl-β-D-glucosamine as the substrate for measuring enzymatic activity of β-hexosaminidase (both total activity in serum and activity of the heat-stable fraction) activity in serum and urine. The authors denoted combined activity of the heat-stable fraction of β-hexosaminidase (isoforms B, I, and P) as β-Hex B activity. The authors observed that the mean total β-hexosaminidase level in alcoholics (18 patients admitted to the detoxification center who had consumed 120–300 g of alcohol per day during the preceding 2 weeks before admission) was 2.5 times higher than the mean level observed in moderate drinkers who consumed less than 60 g of alcohol per week (mean value of 49.6 U/L in alcoholics and 19.4 U/L in moderate drinkers). The increase in total β-hexosaminidase activity was mainly due to more than a fivefold increase in heat-stable β-hexosaminidase level (β-Hex B: mean value of 28.4 U/L in alcoholics and 5.7 U/L in moderate drinkers). Interestingly, for nondrinkers, the mean total β-hexosaminidase level was 21.2 U/L, whereas the mean β-Hex B level was 5.9 U/L, indicating that β-hexosaminidase is a marker for heavy alcohol use. For heavy drinkers, who consumed more than 60 g of alcohol per week, the mean total β-hexosaminidase level was 22.4 U/L, whereas the mean heat-stable β-hexosaminidase level was 9.1 U/L. The authors also introduced a parameter called serum β-Hex B%, which can be obtained by dividing β-Hex B activity by total β-hexosaminidase activity. The mean β-Hex B% was 52.4% for alcoholics, 40.2% for heavy drinkers, 29.0% for moderate drinkers, and 27.5% for nondrinkers, indicating that β-Hex B% is a good biomarker for alcohol abuse. This parameter correlated well with serum CDT values. The authors concluded that the cutoff value of serum β-Hex B% was 35% and any value above that indicates heavy alcohol consumption (>60 g per day). At that cutoff value, β-Hex B% had a specificity of 91% and sensitivity of 94%. This marker is more sensitive than GGT, AST, ALT, or mean corpuscular volume (MCV) as an alcohol biomarker. Moreover, serum β-Hex B% is slightly more sensitive than CDT [12]. In a study of β-hexosaminidase using agarose gel isoelectric focusing in sera collected from alcoholics, Maenhout *et al.* observed an additional cathodal band (between pH 6.8 and 7.7). The authors designated this band as Hex-7 and reported that the level of Hex-7 correlated well with %CDT in sera of chronic alcohol abusers [13].

Nystrom *et al.* investigated whether β-hexosaminidase can be used as a marker to identify college students who are heavy drinkers but not alcoholics. The authors measured the enzymatic activity of β-hexosaminidase (total activity) using *p*-nitrophenyl-*N*-acetyl-β-D-glucosamine as the substrate in citrate buffer at pH 4.5. Based on their study of 203 first-year university students, the authors did not find any correlation between serum β-hexosaminidase levels and reported drinking by these students. In addition, mean β-hexosaminidase level in teetotalers did not differ significantly from the mean level observed in the heaviest drinking group. For example, the mean β-hexosaminidase level was 18.9 U/L in teetotalers, and in heavy drinkers it was 20.4 U/L. The mean level was 19.6 U/L in social drinkers. However, β-hexosaminidase levels were elevated in female students who were taking oral contraceptives. The authors concluded that serum β-hexosaminidase is a poor marker of alcohol consumption by young university students, and if an elevated level is observed in female students, use of oral contraceptive must be considered [14].

Binge drinking can significantly increase β-hexosaminidase level. Based on a study of eight nonsmoking men who were irregular binge drinkers and who abstained from drinking for 10 days prior to the study, the authors observed significantly increased serum, urine, and salivary β-hexosaminidase after these subjects consumed 120–160 g of alcohol as vodka (containing 40% alcohol) in a period of 6 hr (between 7 PM and 1 AM). The authors further observed that increased levels of serum and urinary total β-hexosaminidase were mainly due to the β-hexosaminidase A isoenzyme [15]. Interestingly, increased serum total β-hexosaminidase activity in chronic alcohol abusers is mostly attributable to increased activity of heat-stable β-hexosaminidase activity (hexosaminidase B isoenzyme and other heat-stable isoenzymes).

β-Hexosaminidase is a large molecule that is not filtered during glomerular filtration. However, β-hexosaminidase is abundantly present in the cells of proximal tubules, and a small amount is excreted in urine of normal subjects due to the exocytosis process. Elevated total β-hexosaminidase levels (two- or threefold greater than the normal value) in urine have been reported in chronic alcohol abusers. In general, like serum β-hexosaminidase levels, urinary β-hexosaminidase levels are increased following consumption of at least 60 g of alcohol per day for at least 10 consecutive days. According to one study, however, average urinary excretion of β-hexosaminidase was three times higher in alcoholics than in teetotalers, and both serum and urine β-hexosaminidase were more sensitive markers than GGT. Controlled moderate drinking for 10 days was reflected in serum β-hexosaminidase levels only but not in urinary β-hexosaminidase levels. However, in alcoholics, urinary β-hexosaminidase levels remained elevated longer than serum β-hexosaminidase levels after abstinence. The authors further observed that the sensitivity of serum β-hexosaminidase was 69% in identifying alcoholics but 86% among drunkenness arrestees. Urine β-hexosaminidase had a sensitivity of 81% in identifying alcoholics. In addition, the specificity of serum and urine β-hexosaminidase was 98 and 96%, respectively [16]. Serum β-hexosaminidase levels usually return to normal after 7–10 of abstinence, but it may take up to 4 weeks for urinary β-hexosaminidase levels to return to normal after abstinence [17]. Wehr *et al.* suggested that urinary β-hexosaminidase can be used to monitor sobriety in alcohol-dependent individuals [18].

The precise mechanism by which β-hexosaminidase levels in serum and urine are elevated in alcoholics is not clearly established. It has been shown that

BOX 7.1 CHARACTERISTICS OF β-HEXOSAMINIDASE AS ALCOHOL BIOMARKER

- β-Hexosaminidase is an indirect marker of heavy alcohol consumption.
- β-Hexosaminidase is present in serum mostly as A (αβ), and B (ββ), but other minor forms, such as S (αα), I_1, I_2 (intermediate form), and P (elevated in pregnancy), are also present in serum. Isoforms I_1, I_2, and P are composed of two β chains. Total β-hexosaminidase activity in serum represents activities of all these isoenzymes collectively.
- β-Hexosaminidase B, I_1, I_2, and P are heat stable, and these isoenzymes collectively are referred to as heat-stable β-hexosaminidase (abbreviated as β-Hex B or simply Hex B). β-Hexosaminidases A and S are heat labile.
- Both serum and urinary β-hexosaminidase activities are increased significantly after consumption of greater than 60 g of alcohol per day for 10 consecutive days or more.
- Serum β-hexosaminidase levels usually return to normal after 7–10 of abstinence, but it may take up to 4 weeks for urinary β-hexosaminidase levels to return to normal after abstinence.
- At a cutoff value of 35%, serum β-Hex B% (ratio of heat-stable β-hexosaminidase to total β-hexosaminidase activity) has specificity of 91% and sensitivity of 94% to detect heavy alcohol abuse (>60 g per week). However, sensitivity and specificity of serum β-hexosaminidase are 86 and 98%, respectively, among drunkenness arrestees, but sensitivity for identifying alcoholics is only 69%. Sensitivity and specificity of urinary β-hexosaminidase are 81 and 96%, respectively.
- β-Hexosaminidase is a superior alcohol biomarker compared to GGT, AST, and ALT. It may be slightly superior to CDT. However, elevated serum and urine β-hexosaminidase may occur under various pathophysiological conditions (see Table 7.1).
- Although in the past, β-hexosaminidase activity in serum or urine was measured using substrate for β-hexosaminidase, currently immunoassays are available for measuring isoforms of β-hexosaminidase that can be adapted to use in automated analyzers.

increased serum β-hexosaminidase activity in alcoholics is mostly due to elevation of heat-stable β-hexosaminidase isoenzymes. In sera of alcoholics, isoenzyme P is also preferentially elevated. It has been speculated that increased serum β-hexosaminidase levels in alcoholics are related not directly to alcohol but due to alcohol-induced liver dysfunction that results in the release of β-hexosaminidase in the circulation from damaged hepatocytes [11]. Urinary β-hexosaminidase activity can also be used to monitor alcohol consumption during pregnancy [19]. Various characteristics of β-hexosaminidase as an alcohol biomarker are listed in Box 7.1.

7.3.1 Pathophysiological Conditions that Cause Elevated Levels of β-Hexosaminidase

Many conditions and the use of certain medications may cause elevated levels of serum and/or urinary β-hexosaminidase. These causes are listed in Table 7.1. Therefore, elevated levels of β-hexosaminidase due to any of these causes should be considered as false-positive results when evaluating alcohol

Table 7.1 Elevated Serum and Urinary β-Hexosaminidase in Various Conditions

Elevated Serum and Urinary β-Hexosaminidase	Conditions
Elevated serum β-hexosaminidase	Use of oral contraceptives Pregnancy Liver diseases and in liver metastasis Various cancers Viral hepatitis Hypertension Diabetes mellitus Myocardial infarction Cerebral infarction Acute pancreatitis Thyrotoxicosis Inflammatory bowel disease Rheumatoid arthritis Silicosis
Elevated urinary β-hexosaminidase	Renal tubular dysfunctions associated with renal diseases Upper urinary tract infection Kidney rejection after transplant Nephrotoxic drugs Heavy metal toxicity such as with cadmium Diabetes mellitus Hypertension Preeclampsia Chronic heart failure Smokers consuming 10 g of tobacco per day

use in individuals. Many conditions that increase serum β-hexosaminidase levels may also cause elevated urinary β-hexosaminidase levels.

Serum β-hexosaminidase levels are usually increased during pregnancy, possibly due to the impaired capacity of liver nonparenchymal cells to properly clear β-hexosaminidase from blood [20]. Urinary activity of the enzyme is also increased in pregnancy, and, in women with preeclampsia, the level of activity may be increased further [21]. Serum β-hexosaminidase is increased in various liver diseases and can be used as a liver function test [22]. Serum β-hexosaminidase levels are also elevated in patients with liver metastasis, and a variant β-hexosaminidase is observed in sera of these patients [23]. Schmieder *et al.* reported elevated serum activities of *N*-acetyl-β-glucosaminidase (β-hexosaminidase) in patients with essential hypertension, but such elevated levels returned to normal with antihypertensive therapy [24]. The serum level of *N*-acetyl-β-glucosaminidase (β-hexosaminidase) could also be elevated in patients with acute viral hepatitis, acute pancreatitis, and after myocardial infarction [25]. Hultberg *et al.* reported elevated plasma levels of β-hexosaminidase in patients with cerebral infarction, particularly in female patients [26]. Serum β-hexosaminidase levels are elevated in patients with inflammatory bowel disease [27]; they are also elevated in patients with diabetes mellitus, and such levels are further increased in patients who suffer from secondary complications of diabetes [28]. Mungan *et al.* reported elevated urinary *N*-acetyl-β-glucosaminidase (β-hexosaminidase) activities in children and adolescents with type 1 diabetes mellitus [29]. Serum *N*-acetyl-β-glucosaminidase (β-hexosaminidase) levels are also increased in patients with silicosis and workers exposed to silica dust [30]. Elevated levels of β-hexosaminidase have also been reported in patients with thyrotoxicosis [31]. Berenbaum *et al.* reported markedly elevated serum β-hexosaminidase activities in patients with rheumatoid arthritis [32]. Serum levels of β-hexosaminidase may become slightly elevated with advanced age.

Lysosomal enzymes such as β-hexosaminidase are found in renal proximal tubules, and urinary levels of β-hexosaminidase have been associated with renal tubular dysfunctions, various kidney diseases, and drug nephrotoxicity. An elevated level of β-hexosaminidase in urine is also an early warning of kidney rejection after transplant [33]. Increased urinary excretion of *N*-acetyl-β-glucosaminidase (β-hexosaminidase) has been shown in urinary tract infection, and it can also be used as a marker of pyelonephritis and interstitial tubular damage [34]. Bazzi *et al.* commented that *N*-acetyl-β-glucosaminidase (β-hexosaminidase) is a marker of tubular cell dysfunction and a predictor of outcome in primary glomerulonephritis as well as response to therapy [35]. *N*-acetyl-β-glucosaminidase (β-hexosaminidase) levels are also elevated in patients with chronic heart failure but apparently normal kidney function. This elevation may be related to tubular injury caused by chronic heart failure [36]. Elevated urinary excretion of β-hexosaminidase has also been reported in smokers (smoking more than 10 g of tobacco per day) [37].

CASE REPORT 7.1

Urinary *N*-acetyl-β-D-glucosaminidase (β-hexosaminidase) levels were measured in a 5-year-old girl with distal renal tubular acidosis during a 14-month period using cresol sulfonephthalein–glucosaminide as a substrate. Her urinary β-hexosaminidase activities were increased significantly but varied with changes in metabolic status (urinary β-hexosaminidase activities varied from 0.90 to 7.10 U/mmol creatinine; upper limit of normal, 0.6 U/mmol of creatinine), and there was a significant positive correlation of urinary β-hexosaminidase activity with blood pH but no correlation between serum potassium and urinary excretion of calcium. The authors concluded that systematic acidosis, which is a characteristic of distal renal tubular acidosis, was probably causing elevated levels of urinary β-hexosaminidase activities [38].

7.4 LABORATORY METHODS FOR MEASURING β-HEXOSAMINIDASE

The concentration of β-hexosaminidase can be measured in serum or urine either as enzyme activity (activity assay) using a proper substrate for the enzyme or as the amount of the enzyme (mass assay) using a proper immunoassay. The separation of isoenzymes of β-hexosaminidase can be achieved by electrophoresis, isoelectric focusing, capillary zone electrophoresis, or ion exchange chromatography. Although these methods are laborious, one advantage of them is that all isoenzymes of β-hexosaminidase can be separated. This is especially important if an unusual variant of β-hexosaminidase is present in the specimen. Isoelectric focusing can easily separate the two major isoenzymes, hexosaminidase A (isoelectric point, 4.5) and hexosaminidase B (isoelectric point, 7.5), along with other isoforms. Using isoelectric focusing, Plucinsky *et al.* demonstrated that a variant of β-hexosaminidase was present in sera of cancer patients [39]. For capillary zone electrophoresis, a urine sample could be incubated with the synthetic substrate methylumbelliferyl-β-D-glucosaminide, and the reaction mixture could be introduced directly into the instrument. The released reaction product, 4-methylumbelliferone, could be separated at 13.3 kV in a 400-mmol borate buffer at pH 8.1. Detection of the signal could be achieved by either ultraviolet absorption or fluorescence [40].

Several methods, including colorimetry, fluorometry, and chemiluminescent detection, can be utilized for the determination of the activity of β-hexosaminidase in urine as well as serum. The colorimetric method based on stable and soluble *p*-nitrophenyl-*N*-acetyl-β-D-glucosamine has been used widely for this purpose. This method can be used for the determination of total activity as well as activity of heat-stable hexosaminidase (after deactivating the heat-labile fraction by heating the specimen at 50°C for 2 hr) in serum [12]. Yagi *et al.* used 3,4-dinitrophenol-*N*-acetyl-β-D-glucosaminide to measure the enzymatic activity of β-hexosaminidase in urine in aqueous solution at pH 5.0. The authors used 400 nm for measuring the absorbance of 3,4-dinitrophenol released from the substrate by enzymatic activity, and the absorbance was used to determine the activity of β-hexosaminidase in urine [41]. Makise *et al.* developed a kinetic rate assay for urinary *N*-acetyl-β-D-glucosaminidase (β-hexosaminidase) using 2-chloro-4-nitrophenyl-*N*-acetyl-β-D-glucosaminide as substrate. The authors noted that their method can be adopted on various automated analyzers [42]. Yamada and Fujita used 2,4-dinitrophenyl-1-thio-*N*-acetyl-β-D-glucosaminide as the substrate for measuring urinary activity of β-hexosaminidase in patients with renal disease. The analysis is based on the fact that enzyme hydrolyzes the substrate, thus liberating 2,4-dinitrothiophenol, a chromogen that can be measured at

400 nm spectrophotometrically. The authors selected pH 4.6 as the optimum pH for the enzymatic reaction [43]. Fluorometric measurement of the activity of β-hexosaminidase in human urine using 4-methyllumbelliferyl-*N*-acteyl-β-D-glucosaminide as a substrate has also been reported. The author used microtiter plates and an automated reader to measure the fluorescence of 4-methyllumbelliferone liberated from the substrate due to the enzymatic action of β-hexosaminidase present in urine specimens [44]. Chemiluminescent detection has also been reported for measuring urinary activity of β-hexosaminidase using *ortho*-aminophthalylhydrazido-*N*-acetyl-β-D-glucosaminide as the substrate [45].

Isaksson and Hultberg developed an enzyme immunoassay using monoclonal antibody for the determination of β-hexosaminidase isoenzymes A and B in human serum. The enzyme immunoassay for isoenzyme B reacted similarly to those for isoenzyme P and intermediate forms. The authors observed an excellent correlation between total β-hexosaminidase activity in human serum obtained by the conventional enzyme substrate method and total β-hexosaminidase activity obtained as a sum of isoenzymes A and B as determined by enzyme immunoassays. In addition, the proportion of isoenzyme A obtained by the enzyme immunoassay method was similar to values obtained by isoelectric focusing as well as ion exchange chromatography (~60% of total β-hexosaminidase activity). Moreover, enzyme immunoassay specific for B isoenzyme showed that increased total β-hexosaminidase activities in sera of patients with liver cirrhosis and cholestasis were attributable to increased activity of hexosaminidase B isoenzyme, as expected [46]. Numata *et al.* developed an enzyme-linked immunosorbent assay (ELISA) using a monoclonal antibody for analysis of hexosaminidase isoenzyme B in human urine after raising antibody against the isoenzyme obtained from human placenta [47]. Tanaka *et al.* developed a centrifugal microfluidic platform that is also known as Lab-CD (lab on a computer disc) for analysis of urinary activity of *N*-acetyl-β-D-glucosaminidase (β-hexosaminidase) [48]. Various methods used to measure β-hexosaminidase activities in serum or urine are listed in Box 7.2.

7.5 ACETALDEHYDE–PROTEIN ADDUCTS AS ALCOHOL BIOMARKERS

Acetaldehyde is the first and major product of alcohol metabolism, and it is formed in the liver by alcohol dehydrogenase. Although short-lived, acetaldehyde is a toxic metabolite of alcohol and a carcinogen. Acetaldehyde is a highly reactive compound and is a strong electrophilic chemical species that can strongly react with a nucleophilic compound such as protein to form a

BOX 7.2 ANALYTICAL METHODS FOR ANALYSIS OF β-HEXOSAMINIDASE IN SERUM AND URINE

- Electrophoresis
- Isoelectric focusing
- Capillary zone electrophoresis
- Ion exchange chromatography
- Measuring enzymatic activity using 4-nitrophenol-*N*-acetyl-β-D-glucosaminide or a derivative of this substrate—such as 3,4-dinitrophenol-*N*-acetyl-β-D-glucosaminide and 4-methylumbelliferyl-*N*-acteyl-β-D-glucosaminide—using various detection methods, such as ultraviolet detection at 400 nm (which detects 3,4-dinitrophenol), fluorescence detection, or chemiluminescence detection; these can be used depending on the substrate employed. Heat-stable isoenzyme can be measured after deactivating heat-labile isoenzyme (A and S) by heating the specimen at 50°C for 2 hr
- Enzyme immunoassay
- ELISA

Schiff base, thus forming acetaldehyde–protein adducts. As a result, acetaldehyde can form adducts with plasma proteins such as albumin, lipoproteins, erythrocyte membrane proteins, hemoglobin, and also DNA. Acetaldehyde easily forms adduct with human serum albumin in subjects consuming alcohol [49]. Acetaldehyde reacts primarily with lysine residues of proteins to form a stable product, N^{ε}-ethyl lysine (NEL), through formation of unstable intermediate first. Mabuchi *et al.* used gas chromatography–negative ion chemical ionization mass spectrometry for analysis of NEL residue in human plasma after hydrolysis of protein fraction using pronase E in the presence of stable isotope-labeled internal standard followed by derivatization using pentafluorobenzyl bromide. The authors observed significantly elevated levels of NEL in alcoholic patients (mean, 1.17 NEL/1000 lysine) compared to control subjects (mean: 0.26 NEL/1000 lysine), indicating that NEL in plasma may be used as an alcohol biomarker [50]. In general, acetaldehyde adducts to hemoglobin, measurement of IgA antibodies against acetaldehyde-modified erythrocyte proteins, and acetaldehyde–albumin adducts may also be used as potential alcohol biomarkers. Latvala *et al.* developed antibodies against low-density lipoprotein (LDL)–acetaldehyde adduct, very low-density-lipoprotein (VLDL)–acetaldehyde adduct, and bovine serum albumin–acetaldehyde adduct and reported that these antibodies reacted with protein adducts generated at physiologically relevant acetaldehyde concentration *in vitro*. In addition, the antibody prepared against VLDL–acetaldehyde adduct provided the best detection of acetaldehyde–protein adduct *in vivo* in alcoholics compared to control subjects [51].

7.5.1 Acetaldehyde–Hemoglobin Adducts

Stevens *et al.* showed that at physiological concentration, acetaldehyde forms adduct with hemoglobin A, producing a minor variant of hemoglobin [52].

In general, alcoholics show normal glycosylated hemoglobin A_{1c} (HbA_{1c}) but an elevated level of minor hemoglobin that is attributable to hemoglobin–acetaldehyde adduct. Using high-performance liquid chromatography (HPLC), hemoglobin–acetaldehyde adduct can be separated from other hemoglobin variants, such as HbA_{1c}. In general, lysine residues as well as terminal valine residues on hemoglobin molecules are sites of attack by acetaldehyde. Stable Schiff base adducts are formed with N-terminal peptides of hemoglobin β chain as well as with ε-amino groups of lysine due to reaction with acetaldehyde [53]. Hazelett *et al.* used cation exchange HPLC for analysis of hemoglobin–acetaldehyde adduct in red blood cell hemolysates of 182 patients consecutively admitted to the drug and alcohol treatment unit of the authors' institute. The mean hemoglobin–acetaldehyde adduct in patients who reported drinking more than six drinks per day was significantly higher (mean, 0.055% total hemoglobin) than that of those who consumed fewer than six drinks per day (mean, 0.026% total hemoglobin). If a cutoff score of 0.030% was assumed, it had 67% sensitivity and 77% specificity as an alcohol biomarker. A cutoff score of 0.08% produced 100% specificity, but the sensitivity was reduced to 20%. The authors observed no significant gender difference in the levels of hemoglobin–acetaldehyde adduct. The authors concluded that the hemoglobin–acetaldehyde adduct as an alcohol biomarker has better sensitivity and specificity than other alcohol biomarkers, including ALT, AST, and MCV [54].

Lin *et al.* developed an ELISA assay for measuring hemoglobin–acetaldehyde adducts and showed that alcoholics had higher levels of such adducts compared to controls [55]. Hemoglobin–acetaldehyde adduct can also be detected by liquid chromatography combined with time-of-flight mass spectrometry. In one study, it was observed that elevated levels of hemoglobin–acetaldehyde adducts (measured by 21 modified peptide fragments generated by tryptic digestion) indicated recent alcohol consumption because values were reduced during a 5-day period of abstinence [56]. Hemoglobin–acetaldehyde adduct has potential to be an effective alcohol biomarker, but due to technical difficulties, assays are not readily available for routine application in clinical laboratories [57].

7.5.2 Acetaldehyde–Erythrocyte Protein Adducts

Excessive alcohol consumption leads to the formation of circulating antibodies, mostly IgA, capable of recognizing sequential and conformational epitopes generated due to covalent binding between proteins and acetaldehyde. In alcoholic patients with liver disease, serum IgA level is often increased coincident with abnormal IgA tissue deposition. It has been suggested that this anti-adduct IgA can be used as an alcohol biomarker. Hietala *et al.*

developed an ELISA assay to measure specific IgA antibodies against acetaldehyde–erythrocyte protein adducts. For this purpose, microtiter plates were coated with acetaldehyde-modified red cell proteins (or unmodified proteins), and, after incubation with human serum overnight, antigen–antibody complexes were detected using alkaline phosphatase-linked goat anti-human immunoglobulin IgA. The authors used *p*-nitrophenyl phosphate solution for colorimetric detection (absorption measured at 405 nm). The authors observed that mean anti-adduct IgA titer was higher in alcoholics who consumed 40–540 g of alcohol per day (198 U/L) compared to moderate drinkers who consumed 1–40 g of alcohol per day (58 U/L) and nondrinkers (28 U/L). The authors reported that the sensitivity and specificity of this marker were 73 and 94%, respectively. During abstinence, this marker was reduced by approximately 3% for each day of abstinence [58].

7.5.3 IgA Antibody Against Acetaldehyde-Modified Bovine Serum Albumin

Proteins modified *in vitro* by acetaldehyde have been shown to have immunogenic properties, and circulating antibodies against these acetaldehyde–protein adducts have been demonstrated in mice and rats fed alcohol as well as in human alcoholics. Worrall *et al.* developed an ELISA by coating microtiter plates with acetaldehyde-modified bovine serum albumin or unmodified bovine serum albumin to detect circulating immunoglobulin against such acetaldehyde–albumin adducts. For this purpose, after incubating with sera from alcoholics or controls, microtiter plates were further incubated with biotinylated antibodies for total human immunoglobulin or class-specific immunoglobulin (IgA, IgG, or IgM). Social drinkers demonstrated elevated IgM reactivity with acetaldehyde-modified bovine serum albumin, whereas alcoholics and heavy drinkers showed elevated IgA as well as elevated IgM immunoreactivities against such modified protein. No elevation of IgG was observed in any subjects (alcoholics or social drinkers). The authors concluded that elevated IgA reactivity with acetaldehyde-modified epitopes in heavy drinkers (alcohol intake >130 g per week for females and >150 g per week for males) is a potential alcohol biomarker [59]. Various acetaldehyde–protein adducts that may be used as alcohol biomarkers are listed in Table 7.2.

7.6 DOLICHOL AS ALCOHOL BIOMARKER

Dolichol, a homologous series of α-saturated polyisoprenoid alcohols containing 14–24 isoprene units, is synthesized by a common isoprenoid pathway from acetate and accumulates in tissues during aging [60]. Roine *et al.*

Table 7.2 Commonly Used Acetaldehyde–Protein Adducts as Alcohol Biomarkers

Acetaldehyde–Protein Adducts	Comments
Hemoglobin–acetaldehyde adduct	This can be measured by cation exchange chromatography, liquid chromatography combined with time-of-flight mass spectrometry, immunoassay, and carbon-13 nuclear magnetic resonance (^{13}C-NMR). Cutoff level determines sensitivity and specificity of the assay. Hemoglobin–acetaldehyde adduct has the greatest potential as an alcohol biomarker among various acetaldehyde–protein adducts, but due to technical difficulty, assays are not readily available in clinical laboratories for routine application.
Acetaldehyde–erythrocyte protein adducts	In general, these adducts are measured indirectly by measuring circulating IgA in serum directed against them. In one study, the authors observed a sensitivity of 73% and specificity of 94%. However, this alcohol biomarker is not routinely tested in clinical laboratories and is more or less a research tool.
IgA antibody against acetaldehyde-modified bovine serum albumin	Albumin is a major protein found in serum. Circulating IgA antibodies that recognize acetaldehyde-modified bovine serum albumin can also be used as an alcohol biomarker. However, this biomarker is not routinely tested in clinical laboratories.

reported that urinary dolichol normalized to urinary creatinine was significantly higher in alcohols than in controls. The authors also commented that sensitivity of urinary dolichol in identifying alcoholics was 68% compared to 44% sensitivity for GGT. The control group had only 3.9% false-positive results. The authors concluded that urinary dolichol is a potential alcohol biomarker [61]. In another study, the authors measured serum dolichol levels using HPLC and observed that serum dolichol levels were significantly higher among 95 alcoholics (mean, 182.7 ng/mL) compared to 41 social drinkers (mean, 142.1 ng/mL). However, in alcoholics, the higher serum dolichol levels did not decrease as rapidly as urine dolichol levels when these individuals practiced abstinence [62]. Stetter *et al.* also studied dolichol as a potential biomarker using 21 alcohol-dependent subjects and 21 healthy controls and employing HPLC for analysis of dolichol. The authors observed that urinary dolichol was only slightly elevated in alcoholics compared to controls, and when dolichol was normalized to urinary creatinine (dolichol/creatinine; micrograms/millimoles), no difference between alcoholic subjects and control subjects was seen. In addition, the authors determined that the sensitivity of urinary dolichol as an alcohol biomarker was only 9–19%, which was significantly inferior to that of GGT. They concluded that the utility of urinary dolichol as an alcohol biomarker was doubtful [63].

7.7 CONCLUSIONS

Serum and urinary total β-hexosaminidase levels as well as serum β-hexosaminidase heat-stable isoenzyme fraction (B, P, and intermediate forms) are useful as alcohol biomarkers and superior to GGT but comparable to CDT. Compared to β-hexosaminidase, the acetaldehyde–protein adduct is less commonly used as an alcohol biomarker, and only hemoglobin–acetaldehyde adduct has potential as an alcohol biomarker. However, such tests are not routinely available in clinical laboratories. In contrast, measurement of β-hexosaminidase levels in serum is readily available as a test in many large medical centers and reference laboratories because β-hexosaminidase levels are significantly reduced in patients with Tay–Sachs disease and this test is used as a diagnostic test along with molecular testing. Use of dolichol as an alcohol biomarker is questionable. Moreover, only chromatographic methods are available for analysis of dolichol in serum or urine.

Salsolinol, a condensation product between dopamine and acetaldehyde, may play a role in alcoholism, but the use of salsolinol as an alcohol biomarker is controversial because salsolinol levels in humans are increased significantly after banana intake. Moreover, acute ethanol intake may not change plasma salsolinol levels in humans [64].

References

[1] Zwierz K, Zalewska A, Zoch-Zwierz A. Isoenzymes of N-acetyl-β-hexosaminidase. Acta Biochim Pol 1999;46:739–51.

[2] Borzym-Kluczyk M, Radziejewska I, Olszewska E, Szajda S, et al. Statistical evaluation of isoform pattern of N-acetyl-β-hexosaminidase from human renal cancer tissue separated by isoelectrofocusing. Clin Biochem 2007;40:403–6.

[3] Beutler E, Kuhl W. The tissue distribution of hexosaminidase S and hexosaminidase C. Ann Human Genet 1977;41:163–7.

[4] Hultberg B, Isaksson A, Nordstrom M, Kjellstrom T. Release of β-hexosaminidase isoenzymes in cultured human fibroblast. Clin Chim Acta 1993;216:73–9.

[5] Banerjee DK, Basu D. Purification of normal urinary N-acetyl-β-hexosaminidase A by affinity chromatography. Biochem J 1975;145:113–18.

[6] Casal JA, Cano E, Tutor JC. β-Hexosaminidase isoenzyme profiles in serum, plasma, platelets and mononuclear, polymorphonuclear and unfractionated total leukocytes. Clin Biochem 2005;38:938–42.

[7] Hultberg B, Isakasson A, Tiderstrom G. β-Hexosaminidase, leucine, aminopeptidase, hepatic enzymes and bilirubin in serum of chronic alcoholics with acute ethanol intoxication. Clin Chim Acta 1980;105:317–23.

[8] Karkkainen P, Poikolainen K, Salaspuro M. Serum β-hexosaminidase as a marker of heavy drinking. Alcohol Clin Exp Res 1990;14:187–90.

[9] Hultberg B, Isaksson A. Isoenzyme pattern of serum β-hexosaminidase in liver disease, alcohol intoxication and pregnancy. Enzyme 1983;30:166–71.

[10] Hultberg B, Isaksson A, Berglund M, Alling C. Increases and time course variations in β-hexosaminidase isoenzyme B and carbohydrate-deficient transferrin in serum of alcoholics are similar. Alcohol Clin Exp Res 1995;19:452–6.

[11] Hultberg B, Isaksson A, Berglund M, Moberg AL. Serum β-hexosaminidase: a sensitive marker for alcohol abuse. Alcohol Clin Exp Res 1991;15:549–52.

[12] Stowell L, Stowell A, Garrett N, Robinson G. Comparison of serum β-hexosaminidase isoenzyme B activity with serum carbohydrate deficient transferrin and other markers of alcohol abuse. Alcohol Alcohol 1997;32:713–14.

[13] Maenhout T, Poll A, Wuyts B, Lecocq E, et al. Microheterogeneity of serum β-hexosaminidase in chronic alcohol abusers in a driver's license regranting program. Alcohol Clin Exp Res 2013;37:1264–70.

[14] Nystrom M, Perasalo J, Salaspuro M. Serum β-hexosaminidase in young university students. Alcohol Clin Exp Res 1991;15:877–80.

[15] Waszkiewicz N, Szajda SD, Jankowska A, Kepka A, et al. The effect of the binge drinking session on the activity of salivary, serum and urinary β-hexosaminidase: preliminary data. Alcohol Alcohol 2008;43:446–50.

[16] Karkkainen P, Salaspuro M. Beta-hexosaminidase in the detection of alcoholism and heavy drinking. Alcohol Alcohol Suppl 1991;1:459–64.

[17] Martines D, Morris AL, Gilmore IT, Ansari MA, et al. Urinary enzyme output during detoxification of chronic alcoholic patients. Alcohol Alcohol 1989;24:113–20.

[18] Wehr H, Habrat B, Czartoryska B, Gorska D, et al. Urinary β-hexosaminidase activity as a marker for the monitoring of sobriety. Psychiatr Pol 1995;29:689–96 [in Polish].

[19] Niemiec KT, Raczynski P, Laskowska-Klita T, Czerwinska B, et al. Determination of β-hexosaminidase and γ-glutamyltranspeptidase activities in urine of pregnant women to monitor alcohol consumption. Med Wieku Rozwoj 2003;7:629–38 [in Polish].

[20] Hultberg B, Isaksson A. A possible explanation for the occurrence of increased β-hexosaminidase activity in pregnancy serum. Clin Chim Acta 1981;113:135–40.

[21] Perez-Blanco FJ, Sanabria MC, Huertas JM, Cantero J, et al. Urinary N-acetyl-β-glucosaminidase in the prediction of preeclampsia. Clin Nephrol 1998;50:169–71.

[22] Severini G, Aliberti LM, Koch M, Capurso L, et al. Clinical evaluation of serum N-acetyl-β-D-glucosaminidase as a liver function test. Biochem Med Metab Biol 1990;44:247–51.

[23] Alhandeff JA, Prorok JJ, Dura PA, Plucinsky MC, et al. Atypical β-hexosaminidase in sera of cancer patients with liver metastases. Cancer Res 1984;44:5422–6.

[24] Schmieder RE, Rockstroh JK, Munch HG, Ruddel H, et al. Elevated serum activity of N-acetyl-β-glucosaminidase in essential hypertension: diagnostic value and reversal to normal after antihypertensive therapy. Am J Kidney Dis 1991;18:638–48.

[25] Calvo P, Barba JL, Cabezas JA. Serum β-*N*-acetylglucosaminidase, β-D-glucosidase, α-D-glucosidase, β-D-fucoside, α-L-fucosidase and β-D-galactosidase levels in acute viral hepatitis, pancreatitis, myocardial infarction and breast cancer. Clin Chim Acta 1982;119:15–19.

[26] Hultberg B, Isaksson A, Lindgren A, Israelsson B, et al. Plasma β-hexosaminidase isoenzymes A and B in patients with cerebral infarction. Clin Chim Acta 1996;244:35–44.

[27] Scapa E, Neuman M, Eshchar J. Inflammatory bowel disease and serum β-N-acetyl hexosaminidase. Enzyme 1990;43:146–50.

[28] Rao GM, Morghom LO, Abukhris AA. Serum β-glycosidases in diabetes mellitus. Clin Physiol Biochem 1989;7:161–4.

[29] Mungan N, Yuksel B, Bakman M, Topaloglu AK, et al. Urinary N-acetyl-β-D-glucosaminidase activity in type 1 diabetes mellitus. Indian Pediatr 2003;40:410–14.

[30] Koskinen H, Jarvisalo J, Pitkanen E, Mutanen P, et al. Serum β-N-acetylglucosaminidase and β-glucuronidase activities in patients and in workers exposed to silica dust. Br J Dis Chest 1984;78:217–24.

[31] Oberkotter LV, Tenore A, Palmieri MJ, Koldovsky O. Relationship of thyroid status and serum N-acetyl-β-glucosaminidase isoenzyme activities in human. Clin Chim Acta 1979;94:281–6.

[32] Berenbaum F, Le Gars L, Toussirot E, Sanon A, et al. Marked elevation of serum N-acetyl-β-D-hexosaminidase activity in rheumatoid arthritis. Clin Exp Rheumatol 2000;18:63–6.

[33] Tassi C, Mancuso F, Feligioni L, Marangi M, et al. Expression modes of urinary N-acetyl-β-D-glucosaminidase in patients with chronic renal diseases. Clin Chim Acta 2004;346: 129–33.

[34] Mohkam M, Karimi A, Habibian S, Sharifian M. Urinary N-acetyl-β-D-glucosaminidase as a diagnostic marker of acute pyelonephritis in children. Iran J Kidney Dis 2008;2:24–8.

[35] Bazzi C, Petrini C, Rizza V, Arrigo G, et al. Urinary N acetyl β glucosaminidase excretion is a marker of tubular cell dysfunction and a predictor of outcome in primary glomerulonephritis. Nephrol Dial Transplant 2002;17:1890–6.

[36] Jungbauer CG, Birner C, Jung B, Buchner S, et al. Kidney injury molecule-1 and N-acetyl-β-D-glucosaminidase in chronic heart failure: possible biomarkers of cardiorenal syndrome. Eur J Heart Fail 2011;13:1104–10.

[37] Hultberg B, Isaksson A, Brattstrom L, Israelsson B. Elevated urinary excretion of β-hexosaminidase in smokers. Eur J Clin Chem Clin Biochem 1992;30:131–3.

[38] Ring E, Erwa W, Haim-Kuttnig M. Urinary N-acetyl-β-D-glucosaminidase activity in a girl with distal renal tubular acidosis. Eur J Pediatr 1992;151:314.

[39] Plucinsky MC, Prorok JJ, Alhadeff JA. Variant serum β-hexosaminidase as a biochemical marker of cancer patients. Cancer 1986;58:1484–7

[40] Friedberg M, Shihabi ZK. Analysis of urinary N-acetyl-β-glucosaminidase by capillary zone electrophoresis. J Chromatogr B Biomed Sci Appl 1997;695:187–91.

[41] Yagi T, Hisada R, Shibata H. 3,4-Dinitrophenyl-N-acetyl-β-D-glucosaminide, a synthetic substrate for direct spectrophotometric assay on N-acetyl-β-D-glucosaminidase or N-acetyl-β-D-hexosaminidase. Anal Biochem 1989;183:245–9.

[42] Makise J, Saito E, Obuchi M, Kanayama M, et al. Kinetic rate assay of urinary N-acetyl-β-D-glucosaminidase with 2-chloro-4-nitrophenyl-N-acetyl-β-D-glucosaminide as substrate. Clin Chem 1988;34:2140–3.

[43] Yamada M, Fujita T. Analysis of urinary N-acetyl-β-D-glucosaminidase using 2,4-dinitrophenyl-1-thio-N-acetyl-β-d-glucosaminide as substrate. J Clin Lab Anal 2003;17:127–31.

[44] Linko-Lopponen S. Fluorometric measurement of urinary N-acetyl-β-D-glucosaminidase and its correlation with uremia. Clin Chim Acta 1986;160:123–7.

[45] Sasamoto K, Zenko R, Ueno K, Ohkura Y. Chemiluminescence assay of N-acetyl-β-D-glucosaminidase using O-aminophthalylhydrazido-N-acetyl-β-D-glucosaminide. Chem Pharm Bull (Tokyo) 1991;39:1317–19.

[46] Isaksson A, Hultberg B. Immunoassay of β-hexosaminidase isoenzymes in serum in patients with raised total activities. Clin Chim Acta 1989;183:155–62.

[47] Numata Y, Morita A, Kosugi Y, Shibata K, et al. New sandwich ELISA for human urinary N-acetyl-β-D-glucosaminidase isoenzyme B as a useful clinical test. Clin Chem 1997;43:569–74.

[48] Tanaka Y, Okuda S, Sawai A, Suzuki S. Development of a N-acetyl-β-D-glucosaminidase (NAG) assay on a centrifugal lab on a computer disk (Lab-CD) platform. Anal Sci 2012;28:33–8.

[49] Romanazzi V, Schiliro T, Carraro E, Gilli G. Immune response to acetaldehyde–human serum albumin adducts among healthy subjects related to alcohol intake. Environ Toxicol Pharmacol 2013;36:378–83.

[50] Mabuchi R, Kurita A, Miyoshi N, Yokoyama A, et al. Analysis of N(ε)-ethyl lysine in human plasma by gas chromatography–negative ion chemical ionization/mass spectrometry as a biomarker for exposure to acetaldehyde and alcohol. Alcohol Clin Exp Res 2012;36:1013–20.

[51] Latvala J, Melkko J, Parkkila S, Jarvi K, et al. Assays for acetaldehyde derived adducts in blood of patients based on antibodies against acetaldehyde/lipoprotein condensates. Alcohol Clin Exp Res 2001;25:1648–53.

[52] Stevens VJ, Fantl WJ, Newman CB, Sims RV, et al. Acetaldehyde adducts with hemoglobin. J Clin Invest 1981;67:361–9.

[53] Braun KP, Pavlovich JG, Jones DR, Peterson CM. Stable acetaldehyde adducts: structural characterization of acetaldehyde adducts on human hemoglobin N-terminal β-globin chain peptides. Alcohol Clin Exp Res 1997;21:40–3.

[54] Hazelett SE, Liebelt RA, Brown WJ, Androulakakis D, et al. Evaluation of acetaldehyde-modified hemoglobin and other markers of chronic heavy alcohol use: effects of gender and hemoglobin concentrations. Alcohol Clin Exp Res 1998;22:1813–19.

[55] Lin RC, Shahidi S, Kelly TJ, Lumeng C, et al. Measurement of hemoglobin–acetaldehyde adducts in alcoholic patients. Alcohol Clin Exp Res 1993;17:669–74.

[56] Toennes SW, Wagner MG, Kauert GF. Application of LC-TOF MS to analysis of hemoglobin acetaldehyde adducts in alcohol detoxification patients. Anal Bioanal Chem 2010;398:769–77.

[57] Ingall GB. Alcohol biomarkers. Clin Lab Med 2012;32:391–406.

[58] Hietala J, Koivisto H, Latvala J, Anttila P, et al. IgAs against acetaldehyde-modified red cell protein as a marker of ethanol consumption in male alcoholic subjects, moderate drinkers and abstainers. Alcohol Clin Exp Res 2006;30:1693–8.

[59] Worrall S, de Jersey J, Wilce PA, Seppa K, et al. Relationship between alcohol intake and immunoglobulin A immunoreactivity with acetaldehyde-modified bovine serum albumin. Alcohol Clin Exp Res 1996;20:836–40.

[60] Carroll KK, Guthrie N, Ravi K. Dolichol: function, metabolism and accumulation in human tissues. Biochem Cell Biol 1992;70:382–4.

[61] Roine RP, Turpeinen U, Ylikahri R, Salaspuro M. Urinary dolichol—A new marker of alcoholism. Alcohol Clin Exp Res 1987;11:525–7.

[62] Roine RP, Nykanen I, Ylikahri R, Heikkila J, et al. Effect of alcohol on blood dolichol concentration. Alcohol Clin Exp Res 1989;13:519–22.

[63] Stetter F, Gaertner HJ, Wiatr G, Mann K, et al. Urinary dolichol—A doubtful marker of alcoholism. Alcohol Clin Exp Res 1991;15:938–41.

[64] Lee J, Ramchandani VA, Hamazaki K, Engleman EA, et al. A critical evaluation of influence of ethanol and diet on salsolinol enantiomers in human and rats. Alcohol Clin Exp Res 2010;34:242–50.

CHAPTER 8

Direct Alcohol Biomarkers Ethyl Glucuronide, Ethyl Sulfate, Fatty Acid Ethyl Esters, and Phosphatidylethanol

CONTENTS

8.1 Introduction 181

8.2 Ethyl Glucuronide and Ethyl Sulfate 182

8.3 Ethyl Glucuronide and Ethyl Sulfate as Alcohol Biomarkers 186

8.3.1 Ethyl Glucuronide and Ethyl Sulfate Observed Due to Incidental Exposure to Alcohol.................. 188

8.3.2 Ethyl Glucuronide and Ethyl Sulfate Cutoff Concentrations in Urine...................... 191

8.3.3 Ethyl Glucuronide and Ethyl Sulfate Cutoff Concentrations in Hair, Meconium, and other Matrices..................... 192

8.3.4 False-Positive/False-Negative Results with Ethyl Glucuronide............... 194

8.3.5 Application of Ethyl Glucuronide and Ethyl Sulfate as Alcohol Biomarkers................ 196

8.1 INTRODUCTION

Biomarkers for alcohol can be broadly classified as "indirect markers" and "direct markers" of alcohol use. Common markers such as γ-glutamyl transferase (GGT; also known as γ-glutamyl transpeptidase), mean corpuscular volume (MCV), carbohydrate-deficient transferrin (CDT), and β-hexosaminidase are indirect markers of alcohol consumption because these markers are elevated as a consequence of alcohol abuse but not directly related to alcohol. Although acetaldehyde–protein adducts such as acetaldehyde–hemoglobin and adducts of acetaldehyde with erythrocyte membrane proteins are direct markers of alcohol consumption, due to the unavailability of commercially available assays, these markers are not routinely used in clinical practice. However, ethyl glucuronide and ethyl sulfate are direct alcohol biomarkers because these are minor metabolites of alcohol. Similarly, fatty acid ethyl esters and phosphatidylethanol are also direct alcohol biomarkers because these products are formed due to the reaction of alcohol with fatty acids and phosphatidylcholine, respectively.

Ethyl glucuronide and ethyl sulfate are minor metabolites of alcohol that are found in various body fluids and also in human hair. Ethyl glucuronide is formed by the direct conjugation of ethanol and glucuronic acid through the action of a liver enzyme. Ethyl sulfate is formed directly by the conjugation of ethanol with a sulfate group. These compounds are water soluble. Fatty acid ethyl esters are also direct markers of alcohol abuse because they are formed due to the chemical reaction between fatty acids and alcohol. Fatty acid ethyl esters are formed primarily in the liver and pancreas and then are released into the circulation. These compounds are also incorporated into hair follicles through sebum. Phosphatidylethanol is formed via the action of phospholipase D in the presence of ethanol. Various characteristics of ethyl glucuronide, ethyl sulfate, fatty acid ethyl esters, and phosphatidylethanol as

A. Dasgupta: Alcohol and Its Biomarkers. DOI: http://dx.doi.org/10.1016/B978-0-12-800339-8.00008-0

8.3.6 Laboratory Methods for Determination of Ethyl Glucuronide and Ethyl Sulfate 199
8.4 Fatty Acid Ethyl Esters as Alcohol Biomarkers 201
8.4.1 Fatty Acid Ethyl Esters in Hair 202
8.4.2 Fatty Acid Ethyl Esters in Meconium .. 204
8.4.3 Laboratory Analysis of Fatty Acid Ethyl Esters 206
8.5 Phosphatidyl-ethanol as Alcohol Biomarker 207
8.5.1 Cutoff Concentration of Phosphatidylethanol . 210
8.5.2 Laboratory Analysis of Phosphatidylethanol . 211
8.6 Sensitivity and Specificity of Direct Alcohol Biomarkers 213
8.7 Conclusions 215
References 215

alcohol biomarkers are listed in Table 8.1. Pathways for forming these biomarkers are shown in Figure 8.1.

One advantage of ethyl glucuronide and ethyl sulfate compared to many alcohol biomarkers (most commonly indirect markers) is that any consumption of alcohol can cause detectable ethyl glucuronide and ethyl sulfate in blood or urine. Therefore, unlike markers such as MCV, GGT, or CDT, which are markers for heavy alcohol consumption, ethyl glucuronide or ethyl sulfate values in blood or urine reflect any amount of alcohol consumption and not just heavy alcohol consumption. Moreover, ethyl glucuronide and ethyl sulfate can be detected in blood and urine within hours of alcohol consumption, whereas conventional markers such as CDT and %CDT require 1 week before a significant change in values can be observed after repeated heavy drinking for at least 1 week.

8.2 ETHYL GLUCURONIDE AND ETHYL SULFATE

Ethyl glucuronide and ethyl sulfate are minor metabolites of alcohol that are formed through the non-oxidative pathway, whereas the major metabolic pathway of alcohol is the oxidative pathway involving alcohol dehydrogenase. Alcohol dehydrogenase oxidizes alcohol into acetaldehyde and then aldehyde dehydrogenase, which further oxidizes acetaldehyde into acetate. In alcoholics, alcohol may also be oxidized by CYP2E1 in the liver (see Chapter 2).

Formation of ethyl glucuronide is characterized by the addition of glucuronic acid to ethanol, and this metabolic pathway is catalyzed by the uridine 5′-diphospho (UDP)-glucuronosyltransferase superfamily of enzymes. Multiple isoenzymes of UDP-glucuronosyltransferase (UGT) are involved in catalyzing the formation of ethyl glucuronide, but isoforms UGT1A1 and UGT2B7 play a major role in such transformation. Approximately 0.02–0.06% of the total amount of ethanol consumed is eliminated as ethyl glucuronide in urine. However, clearance of ethyl glucuronide takes place at a much slower rate than clearance of alcohol [1]. It is usually assumed that ethyl glucuronide may be detected for up to 2 or 3 days in urine after alcohol consumption, but after heavy consumption it may be detected for as long as 5 days.

Like ethyl glucuronide, ethyl sulfate is a minor metabolite of ethanol. Experiments on laboratory animals such as rabbits and rats have demonstrated that ethanol undergoes sulfate conjugation with 3′-phosphoadenosine-5′-phosphosulfate through the action of the cytosolic enzyme sulfotransferase, yielding ethyl sulfate. Using mass spectrometry, ethyl sulfate has been characterized in humans after alcohol consumption [2]. The sulfoconjugation of small molecules such as ethyl alcohol is mediated through soluble

Table 8.1 Characteristics of Ethyl Glucuronide, Ethyl Sulfate, Fatty Acid Ethyl Esters, and Phosphatidylethanol as Alcohol Biomarkers

Alcohol Biomarker	Cutoff Level	Detection Window	Comments
Ethyl Glucuronide Can be measured in serum but is commonly measured in urine; also may be measured in hair and meconium	■ Although cutoff levels of 500 ng/mL for urinary ethyl glucuronide and 200 ng/mL for urinary ethyl sulfate have been used by some laboratories, in order to show total abstinence, 100 ng/mL cutoff is most appropriate for ethyl glucuronide and 25 ng/mL for ethyl sulfate. ■ 4 to 30 pg/mg cutoffs in hair have been proposed for ethyl glucuronide. A value greater than 30 pg/mg indicates chronic heavy consumption of alcohol (60 g or more per day). ■ German guidelines for re-granting a driver's license recommend a cutoff level of 100 ng/mL for urinary ethyl glucuronide and 7 pg/mg for hair ethyl glucuronide to document complete abstinence. ■ For meconium, cutoff values of 50 or 120 ng/g have been suggested.	■ Blood up to 8 hr. ■ Urine up to 130 hr but commonly 2 or 3 days after moderate consumption (0.5 g/kg of body weight). Urinary glucuronide is sometimes referred to as the 80-hr test for alcohol, but unless a person consumes a large amount of alcohol, this detection window is unrealistic. ■ Patients with renal insufficiency may have a longer window of detection of ethyl glucuronide and ethyl sulfate in urine compared to healthy people. ■ In hair, the detection window may be up to 3 months.	■ Direct alcohol biomarker formed via a minor metabolic pathway involving UDP-glucuronosyltransferase enzyme and appears in the urine within 1 hr of consuming alcohol. ■ Ethyl glucuronide is found in hair. The level of ethyl glucuronide in pubic hair may be slightly higher than that in scalp hair.
Ethyl Sulfate Commonly measured in urine	■ 25 ng/mL of ethyl sulfate in urine as a cutoff has been suggested to establish complete abstinence. However, a cutoff value as high as 200 ng/mL has also been suggested.	■ Up to 2 days after moderate consumption (0.5 g/kg of bodyweight). ■ Concentration may be less than that of ethyl glucuronide.	■ Direct alcohol biomarker formed via a minor metabolic pathway involving sulfotransferase and appears in the urine within 1 hr of alcohol consumption. It is usually measured in urine.

Continued...

Table 8.1 Characteristics of Ethyl Glucuronide, Ethyl Sulfate, Fatty Acid Ethyl Esters, and Phosphatidylethanol as Alcohol Biomarkers *Continued*

Alcohol Biomarker	Cutoff Level	Detection Window	Comments
Fatty Acid Ethyl Esters Can be measured in serum but are commonly measured in hair and meconium.	■ 0.5 ng/mg of hair is the cutoff for living people and 0.8 ng/mg is the cutoff in postmortem cases, with sensitivity and specificity of approximately 90% to differentiate alcohol abuse (>60 g/day) from social drinking. ■ For meconium, 500 ng/g or 2 nmol/g (600 ng/g) has been used.	■ Up to 24 hr in serum, but a much longer window of detection in hair (~3 months). Therefore, it is most commonly measured in hair.	■ Fatty acid ethyl esters are formed due to esterification of fatty acids by ethanol. Usually, four to six fatty acids that are found in appreciable concentrations are used for calculation of total fatty acid ethyl ester levels.
Phosphatidylethanol Commonly measured in whole blood.	■ Not established, but levels from 0.2 to 0.7 μmol (140–492 ng/mL) have been suggested in whole blood. In Sweden, 0.7 μmol/L has been adopted as the clinical cutoff.	■ Up to 14–21 days after abstinence in heavy alcohol users.	■ Phosphatidylethanol is measured in whole blood (membrane phosphatidylethanol formed on erythrocyte membrane by the action of phospholipase D on phosphatidylcholine). ■ Two major molecular species are 16:0/18:1 and 16:0/18:2. However, 48 different molecular species have been characterized. Specificity is almost 100%.

sulfotransferases, which are members of a common gene/enzyme family termed *SULT*. A total of 11 *SULT* forms have been identified in humans, and dietary and physiological factors (hormonal status, menstrual cycle, blood pressure, obesity, and selenium deficiency) affect the expression of *SULT*. Members of the *SULT2* family are often termed alcohol sulfotransferases and members of the *SULT1* family, "phenol sulfotransferases," the highest activity being observed with *SULT1B1*, followed by *SULT1A2*, *SULT1A3*, and *SULT1C2*. If the expression levels in tissues are also taken into account, then *SULT1A3* may be the predominant form for the sulfonation of ethanol *in vitro* [3]. A very small fraction of ethanol ingested is excreted as ethyl sulfate (0.010–0.016% of the total ethanol dosage).

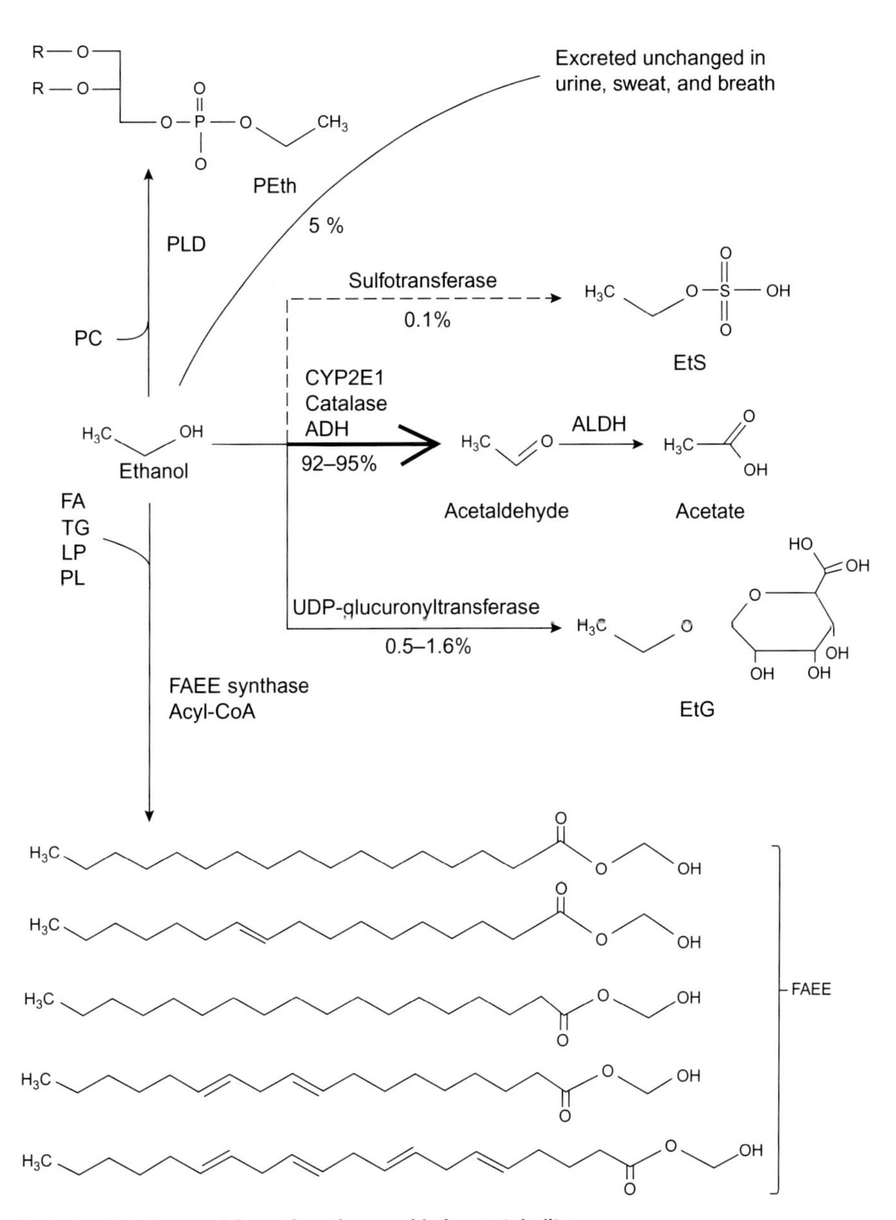

FIGURE 8.1 Alcohol metabolism and formation of non-oxidative metabolites.
ADH, alcohol dehydrogenase; ALDH, acetaldehyde dehydrogenase; EtG, ethyl glucuronide; EtS, ethyl sulfate; FA, fatty acids; FAEE, fatty acid ethyl esters; LP, lipoproteins; PC, phosphatidylcholine; PEth, phosphatidylethanol; PL, phospholipid; PLD, phospholipase D; TG, triglycerides; UDP, uridine 5′-diphospho. Source*: From Maenhout et al.* [4]. *© Elsevier. Reprinted with permission.*

8.3 ETHYL GLUCURONIDE AND ETHYL SULFATE AS ALCOHOL BIOMARKERS

In general, ethyl glucuronide can be detected in blood, urine, or hair specimens. Ethyl sulfate is measured either in urine or in blood. Most investigators have used blood and/or urine to study kinetics and other aspects of ethyl glucuronide or ethyl sulfate. Ethyl glucuronide and ethyl sulfate may also be determined in meconium and maternal hair.

Both ethyl glucuronide and ethyl sulfate are excellent direct biomarkers of alcohol. After variable dosages of ethanol, ethyl glucuronide appears in serum approximately 45 min after the appearance of alcohol in blood. The maximum ethyl glucuronide concentration in blood is reached at between 2.0 and 3.5 hr, and there are significant interindividual variations. Hoiseth *et al.* studied the pharmacokinetics of ethyl glucuronide in blood and urine using 10 male volunteers who consumed alcohol at a fixed dosage of 0.5 g/kg of body weight in a fasting stage. The authors collected blood specimens for 14 hr and urine specimens for 45–50 hr following consumption of alcohol, and they observed that median time for maximum blood ethyl glucuronide concentration was 4 hr (range, 3.5–5 hr), whereas peak alcohol concentration was observed at between 0.5 and 2.0 hr (median, 1.0 hr). The median half-life of elimination of ethyl glucuronide from blood was 2.2 hr (range, 1.7–3.1 hr). Blood alcohol was detected up to 5.0–7.0 hr after consumption of alcohol, whereas ethyl glucuronide in blood was detected up to 10–14 hr after consumption (median, 10 hr). The urinary concentration of ethyl glucuronide was generally much higher than the serum concentration. For example, the maximum median blood ethyl glucuronide concentration was 0.32 mg/L (320 ng/mL), whereas the median urinary maximum ethyl glucuronide concentration was 46.5 mg/L (46.5 μg/mL or 46,500 ng/mL). The median time of detection of ethyl glucuronide in urine was 30 hr. The total amount of ethyl glucuronide excreted in urine (21.5–39.7 mg) represented only 0.017% of total ethanol consumed on a molar basis [5].

Halter *et al.* studied serum and urinary excretion of ethyl glucuronide and ethyl sulfate after consumption of a moderate amount of alcohol by 13 volunteers. The peak ethyl glucuronide concentration was observed 4 hr (mean value) after initiation of drinking, whereas peak ethyl sulfate level was reached 3.0 hr (mean value) after initiation of drinking. The mean time difference between reaching maximum blood alcohol level and blood ethyl glucuronide level was 2.3 hr, whereas for ethyl sulfate the difference was 1.2 hr. Maximum serum ethyl glucuronide and ethyl sulfate levels showed wide interindividual variations and did not correlate with serum ethanol level. Mean time for peak urinary ethyl glucuronide concentration after initiation of drinking was 6.2 hr, whereas for ethyl sulfate the mean peak time for maximum urinary concentration was 5.3 hr [6]. In a study of 32 alcohol-dependent patients who had a blood alcohol level of 100–340 mg/dL on admission to the hospital, Helander *et al.* observed that detection time of ethyl glucuronide in urine was weakly correlated with initial alcohol concentration. Ethyl glucuronide in urine was detected (at a cutoff level <0.5 μg/mL (500 ng/mL)) for up to 130 hr (median window of detection, 78 hr; range, 40–130 hr), with a similar time course observed for ethyl sulfate. The authors concluded that during alcohol detoxification, ethyl glucuronide and ethyl sulfate remained detectable in urine for several days, and the detection window also showed wide interindividual variations [7]. In contrast, Wurst *et al.* reported that both ethyl glucuronide and ethyl sulfate could be detected in urine for up to 36 hr after consumption of ethanol [8]. Such wide variations in the detection window reported by different investigators may be related to the amount of alcohol consumed by subjects as well as the wide interindividual variation in metabolism of alcohol through minor metabolic pathways producing ethyl glucuronide or ethyl sulfate. However, renal impairment may prolong detection time of ethyl glucuronide and ethyl sulfate in patients with decreased renal function [9].

Only a very small amount of ethyl glucuronide (picograms per milligram of hair) is deposited in hair, and until recently, detecting such a small amount of ethyl glucuronide was a major analytical challenge. However, with the development of liquid chromatography combined with tandem mass spectrometry (LC–MS/MS), detecting ethyl glucuronide at 0.7 pg/mg of hair can be achieved, making possible the determination of ethyl glucuronide in hair at a very low concentration [10]. Therefore, ethyl glucuronide in hair has relatively recently begun to be used as an alcohol biomarker. One advantage of analysis of ethyl glucuronide in hair is that this biomarker can be detected months after alcohol consumption and it has the longest window of detection compared to blood or even urine. Morini *et al.* described the determination of ethyl glucuronide and ethyl sulfate in maternal hair and meconium and observed that of 99 specimens analyzed, ethyl glucuronide was detected in 82 meconium specimens and ethyl sulfate in only 19 specimens. However, neither maternal nor neonatal hair appears to be a good predictor of gestational alcohol consumption and subsequent fetal exposure in mother–infant dyads [11]. Because urine may be diluted or concentrated, some authors express urinary ethyl glucuronide or ethyl sulfate concentration as micrograms per gram of creatinine. Alternatively, the ethyl glucuronide (EtG) level can be normalized to 100 mg/dL of creatinine, and this is often referred as EtG_{100}. Individuals may attempt to beat the ethyl glucuronide test in urine by diluting urine specimens, and some laboratories may reject a specimen if the creatinine concentration in urine is less than 20 mg/dL.

8.3.1 Ethyl Glucuronide and Ethyl Sulfate Observed Due to Incidental Exposure to Alcohol

A very low level of ethyl glucuronide or ethyl sulfate may be observed in blood or urine due to incidental exposure to ethyl alcohol-containing products such as mouthwash and handwash or from the consumption of nonalcoholic beer and wine. Moreover, eating certain foods, such as ripe bananas, may results in detectable amounts of ethyl glucuronide in urine. Therefore, a cutoff concentration must be established to differentiate between incidental exposures to alcohol and the direct consumption of alcohol. Various causes of positive ethyl glucuronide in urine other than the consumption of alcohol are listed in Table 8.2.

Thierauf *et al.* observed that after consuming nonalcoholic beer, the urinary concentrations of ethyl glucuronide and ethyl sulfate were 0.30–0.87 μg/mL (300–870 ng/mL) and 0.04–0.07 μg/mL (40–70 ng/mL), respectively, which should be considered as positive if a cutoff concentration of 0.1 μg/mL (100 ng/mL) for ethyl glucuronide was selected. However, these volunteers each consumed 2.5 L of nonalcoholic beer, which could be considered an excessive amount for one session. The positive ethyl glucuronide and ethyl sulfate in urine may be due to the presence of a very small amount of ethanol (up to 0.5% by volume) present in nonalcoholic beers [12]. Hoiseth *et al.*

Table 8.2 Causes of Detectable Ethyl Glucuronide and Ethyl Sulfate in Urine

Source of Exposure	Comments
Nonalcoholic beer	Ethyl glucuronide in the range of 300–870 ng/mL and ethyl sulfate in the range of 40–70 ng/mL have been reported in urine of volunteers who consumed a very large amount (2.5 L) of nonalcoholic beer.
Nonalcoholic wine	Drinking a large amount of nonalcoholic wine (750 mL) may result in the presence of ethyl sulfate in urine but no ethyl glucuronide.
Ethanol-containing mouthwash	After extensive gargling with mouthwash containing alcohol, the highest ethyl glucuronide and ethyl sulfate levels were 173 and 104 ng/mL, respectively, in urine in one study.
Ethanol-containing hand sanitizer	After very extensive use of high-ethanol-containing hand sanitizer, a maximum ethyl glucuronide concentration of 2001 ng/mL in urine was observed in one subject. The maximum ethyl sulfate level was 84 ng/mL.
Yeast in combination with sugar	Eating bakers' yeast and sugar resulted in the highest urinary ethyl glucuronide and ethyl sulfate levels of 500 and 1050 ng/mL, respectively. Peak concentration was observed 4.5 hr after consuming bakers' yeast and sugar. However, consuming brewers' yeast tablets with sugar did not produce any detectable ethyl glucuronide or sulfate in urine.
Fruit juices	Drinking a large amount of apple juice is not associated with detectable ethyl glucuronide or ethyl sulfate in urine, but drinking grape juice may result in a detectable amount of ethyl sulfate in urine (but no ethyl glucuronide).
Ripe bananas	Eating ripe bananas may cause detectable amounts of ethyl glucuronide and ethyl sulfate in urine.

reported that no ethyl glucuronide was observed in any urine specimens of 12 subjects after the latter had consumed a large amount of nonalcoholic wine (750 mL). However, small amounts of ethyl sulfate were detected in urine specimens, and the maximum observed concentration was 2.15 μg/mL (a small amount of ethyl sulfate may be present in nonalcoholic wine). Moreover, no ethyl glucuronide or ethyl sulfate was detected in any urine specimens when subjects used mouthwash containing 0.2 g/L of alcohol [13].

Reisfield *et al.* studied the effect of intensive exposure to high ethanol content mouthwash on urinary excretion of ethyl glucuronide and ethyl sulfate using 10 volunteers who gargled with Listerine antiseptic four times daily for 3 days and 6 hr. This mouthwash product contains 26.9% alcohol. Prior to the experiment, no subjects had any urinary ethyl glucuronide, but 2 subjects showed very small amounts of ethyl sulfate (6 and 82 ng/mL). The authors observed no urinary ethanol in any subjects during the experiment, and only 1 subject showed a detectable amount of ethyl glucuronide in urine (173 ng/mL) 2 hr post-gargling on the final day of the experiment. Although ethyl sulfate was detected in urine specimens from 7 of 10 subjects (maximum concentration, 104 ng/mL), in no specimen were both ethyl glucuronide and ethyl sulfate detected. The authors concluded that the currently accepted cutoff concentration of 500 ng/mL for ethyl glucuronide is adequate to distinguish between ethanol consumption and four times daily use of a high ethanol content mouthwash [14].

Use of alcohol-containing hand sanitizers may also result in detectable amounts of ethyl glucuronide and/or ethyl sulfate in urine. Rohrig *et al.* reported that after moderate use of the hand sanitizer Germ-X (every 60 min or every 30 min), no ethyl glucuronide or ethyl sulfate was detectable in urine of any subjects (measured using LC MS/MS with a limit of detection of 50 ng/mL). However, when subjects used Germ-X (62% ethanol) more extensively (every 15 min throughout the workday), an ethyl glucuronide concentration of 62 ng/mL was detected in 1 subject. These results indicate that use of hand sanitizer containing alcohol should not cause a positive ethyl glucuronide result even at a cutoff of 100 ng/mL [15]. In another study, 11 volunteers used Purell hand sanitizer containing 62% alcohol every 5 min for 10 hr on 3 consecutive days. The maximum ethyl glucuronide and ethyl sulfate in urine during the course of study were 2001 and 84 ng/mL, respectively, indicating that extensive dermal exposure to alcohol may produce a significant amount of ethyl glucuronide and ethyl sulfate in urine [16]. However, in a study of 5 subjects who intensively used high alcohol-containing hand sanitizer and 2 subjects who were merely exposed to sanitizer vapor but had no dermal contact, Arndt *et al.* demonstrated that the presence of ethyl glucuronide in urine in these subjects was predominantly due to inhalation of ethyl alcohol but not due to transdermal resorption of alcohol. Ethyl glucuronide levels of up to 0.6 μg/mL (600 ng/mL) were observed in 2 subjects who were exposed to vapor only and had no dermal contact with hand sanitizer [17].

Rosano and Lin reported that urine ethyl glucuronide levels in 39 adult abstainers ranged from less than 10 μg/L to 62 μg/L (<10–62 ng/mL), and such levels in 13 children ranged from less than 10 μg/L to 80 μg/L (<10–80 ng/mL). Repeated use of hand sanitizer (60% ethanol) by 9 adults resulted in ethyl glucuronide concentrations of less than 10 ng/L to 114 ng/mL in 88 first-morning void specimens. In contrast, ethyl glucuronide concentration in urine ranged from 12,200 to 83,200 ng/mL after 3–8 hr of consumption of 24 g of ethanol by 4 adults. Ethyl glucuronide was detected in urine for 25–39 hr after alcohol consumption. The authors concluded that low levels of ethyl glucuronide may be present even in urine of people who are abstainers. Moreover, use of hand sanitizer containing high amounts of alcohol may produce urine ethyl glucuronide levels greater than 100 ng/mL [18].

Thierauf *et al.* investigated whether consuming barker's yeast or brewer's yeast with sugar caused any *in vivo* fermentation that may result in detectable amounts of ethyl glucuronide or ethyl sulfate in urine. After 5 days of abstinence, volunteers consumed bakers' yeast and 50 g of sugar on a slice of bread and also consumed 300 mL of water. Eating bakers' yeast and sugar resulted in the highest urinary ethyl glucuronide and ethyl sulfate levels of 500 and 1050 ng/mL, respectively. Peak concentration was observed 4.5 hr after consuming bakers' yeast and sugar. However, consuming brewers' yeast tablets with sugar did not produce any detectable ethyl glucuronide or sulfate in urine [19].

Consuming commercially available fruit juices may result in detectable amounts of ethyl glucuronide and/or ethyl sulfate in urine. Musshoff *et al.* reported that freshly opened commercial apple juice contained 0.1–0.4 g/L of ethanol, whereas freshly opened commercially available grape juice contained 0.3–1.8 g/L of ethanol. Ethanol content of grape juice also increased slightly during storage. According to German regulations, alcohol concentrations of up to 3 g/L are allowed in fruit juices. When volunteers consumed large amounts of apple juice (1.1–2.0 L), no ethyl glucuronide or ethyl sulfate was detected in any urine specimens. After drinking 1.5–2 L of grape juice, no ethyl glucuronide was detected in any urine specimen. However, ethyl sulfate was detected, with peak concentrations between 107 and 648 ng/mL. Peak concentration was observed between 4.5 and 12.5 hr after drinking fruit juice, and ethyl sulfate in urine was detectable for up to 35 hr. After consumption of 670–690 g of mature peeled bananas, urinary peak concentrations of ethyl glucuronide in volunteers ranged from 40 to 120 ng/mL, and peak ethyl sulfate concentrations ranged from less than 15 ng/mL to 55 ng/mL.

Peak concentrations were observed between 3 and 8 hr after ingestion of bananas. Ethyl glucuronide and ethyl sulfate were detectable up to 24 h and 20 h respectively. In the same investigation, the authors reported that peak urinary ethyl glucuronide levels ranged from 211 to 512 ng/mL and peak ethyl sulfate levels ranged from 134 to 169 ng/mL after volunteers consumed 2 or 3 L of nonalcoholic beer. The peak concentrations were reached between 5 and 7.5 hr after consumption of nonalcoholic beer [20]. Such values of ethyl glucuronide were comparable to values reported by Thierauf *et al.* when volunteers consumed 2.5 L of nonalcoholic beer [12].

8.3.2 Ethyl Glucuronide and Ethyl Sulfate Cutoff Concentrations in Urine

Both ethyl glucuronide and ethyl sulfate appear in urine within 1 hr of consuming alcohol, and both are markers of any alcohol consumption—not just heavy alcohol consumption. Because small amounts of ethyl glucuronide may be present even in teetotalers and after incidental exposure to alcohol such as that present in alcohol-containing hand sanitizer or mouthwash, various cutoff concentrations have been proposed to differentiate incidental exposure to ethanol from direct consumption of alcohol. Ethyl glucuronide and ethyl sulfate are often used as alcohol biomarkers in drug and alcohol rehabilitation programs to document abstinence from alcohol by patients. Therefore, a reliable cutoff value for ethyl glucuronide and ethyl sulfate in urine is essential. Moreover, ethyl glucuronide and ethyl sulfate are often used in forensic investigations, and cutoff levels may differ significantly in clinical versus forensic settings.

A 100 ng/mL (0.1 mg/L) cutoff value for ethyl glucuronide and ethyl sulfate in urine has been criticized because in some reports incidental exposure to alcohol resulted in a urine ethyl glucuronide level that exceeded 100 ng/mL. A cutoff value of 500 ng/mL for ethyl glucuronide is widely used by many laboratories, although higher cutoff levels have also been proposed, especially in forensic situations. The proposed cutoff value for ethyl sulfate ranges from 50 to 200 ng/mL. Although the majority of ethyl glucuronide/ethyl sulfate is eliminated within 24 hr of alcohol consumption, these compounds can be detected up to 48 hr after consumption of ethanol dosages of 0.25–0.5 g/kg body weight. Moreover, ethyl glucuronide may be detected up to 5 days after heavy alcohol consumption. In a study of 13 patients who were heavily intoxicated with alcohol, Albermann *et al.* observed that after 36 hr, only 1 of 13 urine specimens tested negative for ethyl glucuronide using the 100 ng/mL cutoff, but 8 of 13 specimens could be considered negative if the cutoff concentration of 500 ng/mL was used. Moreover, an ethyl sulfate concentration of less than 100 ng/mL was observed in 4 of 13 specimens after 36 hr. The authors also studied the effect of moderate drinking on ethyl glucuronide and ethyl sulfate levels using 12 healthy volunteers who consumed a moderate amount of alcohol (one or two glasses of white wine, each of which contained 100 mL of wine, or one or two bottles of beer, each of which contained 330 mL of beer). Following consumption, an ethyl glucuronide level greater than 500 ng/mL was observed in 2 urine specimens 24–27 hr post consumption in individuals who consumed the highest amount of alcohol (200 mL of wine or 660 mL of beer). Ethyl sulfate concentration was detectable for up to 28.5 hr. The authors concluded that an ethyl glucuronide cutoff of 100 ng/mL seems to be appropriate for determining repeated consumption of alcohol, and a false-positive result due to unintentional alcohol consumption is limited to a few hours. However, 500 ng/mL may be too high to show abstinence. For ethyl sulfate, the authors proposed a cutoff of 50 ng/mL [21].

Although ethyl glucuronide cutoff concentrations of between 100 and 1100 ng/mL (1.1 μg/mL) have been proposed, Kummer *et al.* proposed a cutoff of 0.1 μg/mL (100 ng/mL) for both ethyl glucuronide and ethyl sulfate because urine analysis of teetotalers shows no ethyl glucuronide or ethyl sulfate above that level [22]. In general, a lower cutoff concentration such as 100 ng/mL is usually used to demonstrate complete abstinence. Jatlow *et al.* noted that any alcohol consumption at night should be detectable the following morning with an ethyl glucuronide cutoff of 100 or even 200 ng/mL. The sensitivity was poor after 48 hr regardless of dosage of alcohol (light to moderate drinking) and regardless of cutoff. The average ratio of ethyl sulfate to ethyl glucuronide was 0.47 in moderate drinkers and 0.52 in subjects who practiced abstinence, thus justifying a cutoff level approximately 50% lower for ethyl sulfate compared to ethyl glucuronide. However, the authors observed few specimens in which ethyl glucuronide was negative (<100 ng/mL cutoff) but ethyl sulfate was positive (>50 ng/mL). The authors also commented that testing for both ethyl glucuronide and ethyl sulfate is unnecessary; monitoring only ethyl glucuronide is sufficient. Moreover, an ethyl glucuronide cutoff at 100 ng of ethyl glucuronide per mg of creatinine is suitable for monitoring abstinence during the 12- to 24-hr period before testing [23].

8.3.3 Ethyl Glucuronide and Ethyl Sulfate Cutoff Concentrations in Hair, Meconium, and other Matrices

In recent years, a great deal of attention has been focused on determining the concentration of ethyl glucuronide in hair. Usually, a scalp hair specimen is collected by cutting hair using scissors as closely as possible to the skin, and an approximately 3-cm proximal segment is used for analysis. A very small amount of ethyl glucuronide is present in teetotalers. Pirro *et al.* reported that ethyl glucuronide level in hair was less than 1.0 pg/mg in 95% of abstainers whom they studied ($n = 44$) [24]. Various cutoff concentrations have been proposed for analysis of ethyl glucuronide in hair, where the value is expressed as picograms per milligram of hair. A cutoff concentration of 4 pg/mg of hair could differentiate alcohol intake higher than 30 g/day, whereas a cutoff value of 23 pg/mg could differentiate higher alcohol consumption of 60 g/day. However, use of a cutoff of 27 pg/mg for daily consumption of alcohol greater than 60 g/day for a period of 3 months was proposed by Morini *et al.*, who further stated that 27 pg/mg could provide the best compromise between sensitivity (92%) and specificity (96%). The authors also commented that hair color, gender, age, body mass index, smoking, and cosmetic treatment of hair did not influence hair analysis for ethyl glucuronide [25]. Based on a meta-analysis of 15 records, Boscolo-Berto *et al.* calculated that the mean ethyl glucuronide concentration in hair of social drinkers was 7.5 pg/mg (95% confidence interval (CI), 4.7–10.2 pg/mg) and in hair of heavy drinkers, 142.7 pg/mg (95% CI, 99.9–185.5 pg/mg); and in deceased subjects with a known history of chronic alcohol abuse, the mean hair glucuronide level was 586.1 pg/mg (95% CI, 177.2–995.0 pg/mg). The authors commented that a cutoff of 30 pg/mg limits the false-negative effect in differentiating heavy drinking from social drinking, whereas the 7 pg/mg cutoff may be used only when active alcohol use is suspected and not to prove complete abstinence [26].

Berger *et al.* analyzed hair and fingernails from 447 participants (undergraduate college students) and observed that ethyl glucuronide levels in hair ranged from 0 to 180.5 pg/mg and in fingernails, from 0 to 397.08 pg/mg. For ethyl glucuronide in hair, sensitivity was highest (93%) at an ethyl glucuronide cutoff of 8 pg/mg, but the corresponding specificity was 71% for the high-risk drinking group (>30 standard drinks per week). However, at a 30 pg/mg cutoff, sensitivity and specificity were 43 and 92%, respectively, for identifying high-risk alcohol consumption. For fingernails, at an 8 pg/mg cutoff, sensitivity and specificity were 100 and 64%, respectively, for identifying high-risk alcohol consumption. The authors concluded that ethyl glucuronide in nails may be a useful alcohol biomarker [27].

In Germany, drivers' licenses may be revoked for driving while intoxicated, and the driver must prove abstinence for 1 year in order to regain a license. German drivers' license re-granting program guidelines use a cutoff of 100 ng/mL (0.1 mg/L) for urinary ethyl glucuronide and a cutoff of 7 pg/mg for ethyl glucuronide in hair to demonstrate abstinence (using a 3-cm long hair sample for analysis). Agius *et al.* reported that hair testing was more adequate than was urine testing for ethyl glucuronide to monitor long-term abstinence from alcohol [28]. In contrast, a cutoff of 30 pg/mg of ethyl glucuronide in hair could be useful to demonstrate chronic excessive drinking (daily alcohol consumption >60 g) [29].

Ethyl glucuronide in meconium is also measured to investigate possible exposure of the fetus to maternal alcohol use. In a study of 557 women singleton births and available data including meconium specimens, Goecke *et al.* reported that only ethyl glucuronide in meconium showed association with alcohol consumption history (204 of 557 women admitted drinking during pregnancy). A value above the cutoff (120 ng ethyl glucuronide per gram of meconium) indicated significant alcohol consumption during pregnancy, even if women denied alcohol consumption [30]. Bana *et al.* used a cutoff of 50 ng/g of meconium for ethyl glucuronide and 1000 ng/g of meconium for fatty acid ethyl esters and reported that 34.65% of women in their study consumed alcohol during pregnancy, and 17% of women showed positive results with both alcohol biomarkers. When fatty acid ethyl esters values exceeded 5000 ng/g of meconium, 50% of these women had children with low birth weight [31].

8.3.4 False-Positive/False-Negative Results with Ethyl Glucuronide

Although ethyl glucuronide and ethyl sulfate are widely used as direct biomarkers of alcohol consumption, false-positive results are encountered, probably more often with ethyl glucuronide analysis. False-positive results may be due to interference in the analytical method used for the determination of ethyl glucuronide in a biological matrix; such false-positive results are encountered mostly when an immunoassay is used for measuring ethyl glucuronide levels. However, false-positive results may be analytically true positive, and a common example is positive ethyl glucuronide in hair due to use of a hair product containing ethyl glucuronide.

Propyl alcohol/isopropyl alcohol-based hand sanitizers are widely used, and false-positive results of up to 4 mg/L (4000 ng/mL) may be observed using DRI enzyme immunoassay for ethyl glucuronide in urine after normal use of such hand sanitizer. Even passive inhalation of the sanitizer vapor may result in the detectability of up to 0.89 mg/L (890 ng/mL) of ethyl glucuronide, which is higher than the 0.5 mg/L cutoff recommended by the manufacturer of the assay (Microgenics). A positive ethyl glucuronide level in urine of up to a concentration of 0.63 mg/L (630 ng/mL) was observed in a subject even 6 hr after use of hand sanitizer. Analysis of these positive specimens using LC–MS/MS showed no ethyl glucuronide in these specimens but confirmed the presence of 1-propyl glucuronide and 2-propyl glucuronide (isopropyl glucuronide). The authors concluded that false-positive ethyl glucuronide in these specimens as determined by the DRI ethyl glucuronide enzyme immunoassay was due to the presence of propyl glucuronide [32]. Arndt *et al.* also reported false-positive ethyl glucuronide test results with DRI ethyl glucuronide enzyme immunoassay due to intake of chloral hydrate. Ethyl glucuronide concentrations in urine of up to 8.0 mg/L (8000 ng/mL) were observed in ethanol-abstaining women under buprenorphine treatment (medications: levetiracetam, gabapentin, clomethiazole, and chloral hydrate). When control subjects consumed 500 mg of chloral hydrate in a single dosage, ethyl glucuronide in urine up to a concentration of 0.28 mg/L (280 ng/mL) was observed. However, LC–MS/MS analysis did not show the presence of ethyl glucuronide in these specimens. The authors concluded that false-positive enzyme immunoassay test results were due to cross-reactivity with trichloroethyl glucuronide, an important chloral hydrate metabolite with antibody used in the DRI ethyl glucuronide enzyme immunoassay [33].

Urinary tract infection may cause a false-negative urinary ethyl glucuronide level, but the ethyl sulfate level is not affected. One of the common causes of urinary tract infection is *Escherichia coli*, which contains the enzyme β-glucuronidase that can hydrolyze ethyl glucuronide. When urine specimens containing *E. coli* were supplemented with 1 mg/L each of ethyl glucuronide and ethyl sulfate, the majority of the specimens showed marked decreases in ethyl glucuronide concentration over time when they were stored at 22°C, but no decline was observed in specimens stored at 4 or −20°C. In three specimens, complete hydrolysis of ethyl glucuronide was observed within 24 hr when they were stored at 22°C. However, no decrease in ethyl sulfate levels was observed in any specimen, even when stored at 22°C. The authors concluded that *E. coli* hydrolyzes only ethyl glucuronide but not ethyl sulfate if present in urine. Helander and Dahl recommended measuring ethyl sulfate in urine to circumvent this problem. However, if sodium fluoride is used as a preservative, hydrolysis of ethyl glucuronide can be prevented [34]. Drinking a large amount of water (~1 L) may significantly lower the concentration of urinary ethyl glucuronide due to water-induced diuresis. However, normalizing ethyl glucuronide values with urine creatinine concentration can circumvent this problem. Goll *et al.* recommended measuring urinary creatinine along with ethyl glucuronide to establish the integrity of the specimen. The authors recommended a creatinine cutoff value of 25 mg/dL to identify diluted urine specimens [35]. However, in general, a creatinine concentration cutoff of 20 mg/dL is commonly used by many laboratories to identify diluted urine during drugs of abuse testing, and such cutoff may also be used to identify diluted urine submitted for ethyl glucuronide testing.

Ethyl glucuronide may be present in commercial hair tonics, especially herbal hair tonics. In one report, the authors observed ethyl glucuronide concentrations of between 0.07 and 1.06 mg/L in hair tonics in seven products from four manufacturers (11 tonics from eight manufacturers were analyzed). Ethyl sulfate was found in only three products. Therefore, using such hair tonics may cause positive ethyl glucuronide test results in hair (true analytical-positive but false-positive result) [36]. Sporkert *et al.* applied a herbal hair lotion containing 2.7 μg/mL of ethyl glucuronide and 35 mg/mL of ethyl alcohol to the hair of a volunteer over a period of 6 weeks, ethyl glucuronide at a level of 72 pg/mg subsequently being detected in the hair. When a negative hair-strand was incubated with such lotion overnight, an ethyl glucuronide concentration of 140 pg/mg was observed [37]. Some commercial hair tonics may contain alcohol, but in one study, seven volunteers used a commercial hair tonic containing 44% ethanol by volume over a period of 1 or 2 months and no increase in ethyl glucuronide levels in hair specimens collected from these volunteers was observed, indicating that use of ethanol-containing hair tonic should not cause a false-positive ethyl glucuronide level in hair [38]. *In vitro* study has shown that although hair coloring has no significant effect on ethyl glucuronide concentration in hair, bleaching and permanent wave treatment may cause significant decreases in ethyl glucuronide levels in hair due to the leaching out effect (during bleaching) and chemical degradation [39].

8.3.5 Application of Ethyl Glucuronide and Ethyl Sulfate as Alcohol Biomarkers

The Substance Abuse and Mental Health Services Administration (SAMHSA) issued a black box warning to alert clinicians that using ethyl glucuronide testing to determine whether or not a patient has been consuming alcohol is not foolproof. Furthermore, SAMHSA cautioned drug courts, licensing agencies, and other organizations that a positive ethyl glucuronide test result is not definitive proof that a person has been consuming alcohol [40]. Therefore, when ethyl glucuronide is used alone or in combination with other alcohol biomarkers in either a clinical or a legal situation, it is important to rule out alternative explanations for positive ethyl glucuronide test results, such as incidental exposure to alcohol as discussed previously in this chapter. Moreover, if an immunoassay is used to measure ethyl glucuronide, use of chloral hydrate or exposure to isopropyl alcohol-containing hand sanitizer must also be ruled out.

Ethyl glucuronide and ethyl sulfate are widely used as alcohol biomarkers for detecting recent alcohol consumption. Ethyl glucuronide is used for documenting abstinence in alcohol and drug rehabilitation programs, and this is sometimes referred to as the 80-hr test; however, such a long window may not be suitable if a person has not consumed a large amount of alcohol. In one study involving 19 healthy participants who had consumed a low to high amount of alcohol (1 to 6.4 standard drinks), urinary ethyl glucuronide levels were less than the 100 ng/mL cutoff after 48 hr of consumption, even in individuals who had consumed relatively large amounts of alcohol (5 or more standard drinks). For low alcohol consumption, tests were negative after 24 hr. The authors commented that ethyl glucuronide testing to document abstinence is not a practical test because it must be performed within 24 hr of when an individual is suspected of consuming a low amount of alcohol [41]. In general, it is assumed that the window of detection is 24 hr or less if a person consumes only 1 standard drink; a longer window, such as 72 hr, is appropriate after heavy alcohol consumption. Although both 500 and 100 ng/mL cutoffs are used in the clinical setting, to show complete abstinence a cutoff of 100 ng/mL is more appropriate. Lande *et al.* reported that ethyl glucuronide testing may be adopted in military substance abuse programs [42].

Testing for ethyl glucuronide and ethyl sulfate is very useful in medical legal cases in which a positive alcohol test result has been challenged. The production of urinary ethanol after sample collection due to the presence of glucose and microorganisms in urine has been reported [43]. If glucose is present in high concentration along with *Candida albicans* (due to urinary tract infection) in the urine specimen, then ethanol may be produced *in vitro*, causing a false-positive result. This is especially important if sodium fluoride is not used as a preservative. However, ethyl glucuronide or ethyl sulfate are not produced under such conditions. Therefore, a positive ethanol test in urine but negative ethyl glucuronide and/or ethyl sulfate is indicative of a false-positive ethanol result due to post-collection production of ethanol in urine.

Ethyl glucuronide and ethyl sulfate are widely used as alcohol biomarkers in postmortem investigations in order to demonstrate antemortem alcohol ingestion. Hoiseth *et al.* reported their findings in 36 forensic death investigations in which postmortem formation of ethanol was suspected. The authors measured ethyl glucuronide and ethyl sulfate in blood and urine and observed that in 19 cases, both ethyl glucuronide and ethyl sulfate were positive. The median concentration of ethyl glucuronide in urine was 35.9 mg/L (range, 1.0–182 mg/L), and that of ethyl sulfate, 8.5 mg/L (range, 0.3–99 mg/L). In another 16 cases, no trace of ethyl glucuronide or ethyl sulfate was detected in any body fluid.

CASE REPORT 8.1

A commercial international maritime drug testing program utilized urine specimen for testing of illicit drugs and alcohol. A subject showed a very high urinary alcohol level (108 g/dL), but the concentration of ethyl glucuronide in urine was only 4.8 mg/L, and that of ethyl sulfate was 1.2 mg/L. Such ethyl glucuronide and ethyl sulfate levels did not correlate with high ethanol in urine. Yeast (*C. albicans*) was present in the urine, indicating that the ethanol level was falsely elevated due to some *in vitro* production of ethanol by yeast. In another case, the urine ethanol level was 80.5 mg/dL, but no ethyl glucuronide or ethyl sulfate was detected in urine. This urine specimen was also positive for *C. albicans*, and the authors concluded that the positive ethanol level in urine of this subject was indeed a false-positive test result due to post-collection production of ethanol in the urine specimen [44].

The authors concluded that antemortem consumption of alcohol was likely in 19 cases in which ethyl glucuronide and ethyl sulfate were identified in both blood and urine, whereas antemortem consumption of alcohol was unlikely in 16 cases in which no trace of ethyl glucuronide or ethyl sulfate was detected. In 1 case, ethyl glucuronide and ethyl sulfate in urine were positive, but only ethyl sulfate was detected in blood. This may be due to postmortem degradation of ethyl glucuronide in blood [45]. The disappearance of ethyl glucuronide during heavy putrefaction has also been reported [46].

Ethyl glucuronide level in hair is also routinely measured and used for both clinical and forensic purposes because hair analysis can provide a longer period for detection of any alcohol consumption. Lees *et al.* reported that ethyl glucuronide was detected in 29 of 100 hair samples collected from 100 study participants ranging from teetotalers to high-risk drinkers (>50 standard drinks per week). Using a 30 pg/mg cutoff, the authors reported that 57.9% of high-risk drinkers, 45.5% of increasing risk drinkers (22–50 drinks per week), but only 9.8% of low-risk drinkers (1–21 standard drinks per week) were identified by hair ethyl glucuronide testing. The authors concluded that although sensitivity was relatively low (52% to identify high-risk drinkers at 30 pg/mg cutoff), any positive result is highly likely to indicate alcohol consumption during the past 3 months [47]. If scalp hair is not available, chest, arm, or leg hair can be used for ethyl glucuronide testing [48]. A major advantage of testing hair for ethyl glucuronide compared to urine is that hair testing provides evidence of long-term abstinence from alcohol [28]. However, patients with decreased kidney function may have higher levels of ethyl glucuronide in hair. In one study, ethyl glucuronide levels in hair of 12 patients with decreased kidney function ranged from none detected to 134 pg/mg, and these patients consumed between 0.1 and 12 g of alcohol per day (up to 1 standard drink). These levels were significantly higher for the same level of alcohol consumption compared to those of healthy volunteers. The authors concluded that ethyl glucuronide levels in hair should be interpreted with caution if a patient has decreased kidney function [49].

Determination of ethyl glucuronide in hair improves the evaluation of long-term alcohol abstention in liver transplant candidates because prior to listing patients for orthotopic liver transplant, a 6-month period of abstention may be required, especially if the patient has a history of alcohol dependence. Patients with alcoholic liver cirrhosis may require liver transplant, and it is important to ensure that these patients do not consume any alcohol. Sterneck *et al.* evaluated 63 transplant candidates with alcoholic liver cirrhosis and 25 control patients who had cirrhosis not related to alcohol abuse. They measured hair and urine ethyl glucuronide as well as blood alcohol, methanol, and CDT. For interpreting hair ethyl glucuronide level, the authors considered 7 pg/mg as an indication of abstinence or rare drinking, 7–30 pg/mg as an indication of alcohol consumption of more than 10 g per day, and values greater than 30 pg/mg as an indication of heavy alcohol consumption of 60 g of alcohol or more per day. The urinary ethyl glucuronide cutoff was 500 ng/mL. The authors reported that during interviews, only 19 of 63 patients admitted alcohol consumption during the past 6 months, but positive alcohol markers were observed in 39 of 63 patients. In 18 of these 39 patients, only hair ethyl glucuronide value was positive, indicating alcohol consumption during the past 6 months. Of 44 patients who denied alcohol use, 23 tested positive for at least one alcohol biomarker. In 9 patients who denied alcohol consumption, the only biomarker that tested positive was hair ethyl glucuronide. Moreover, 20 of 63 patients tested showed hair ethyl glucuronide concentrations greater than 30 pg/mg; the highest concentration observed was 409.7 ng/mg (0- to 3-cm hair segment tested). In contrast, no ethyl glucuronide was detected in hair in 25 patients who were used as controls (liver cirrhosis not related to alcohol). The authors concluded that hair ethyl glucuronide is a promising alcohol biomarker to determine abstinence in liver transplant patients and has a long detection period [50]. Ethyl glucuronide and ethyl sulfate can also be detected in vitreous humor and can be used as a postmortem evidence marker for ethanol consumption prior to death [51]. Various causes of false-positive or false-negative results in ethyl glucuronide testing are listed in Table 8.3.

CASE REPORT 8.2

A 5-year-old boy participated in a party held by his parents. The child may have consumed vodka containing 40% alcohol (estimated consumption of 120–150 g vodka representing 50–60 g of pure alcohol), and he showed clear signs indicating a state of inebriation. The symptoms did not alert his parents because according to testimony of witnesses, such episodes had occurred in the past. However, this time the child was found dead in his bed the next morning. During autopsy, blood alcohol level was 0.4 g/dL (400 mg/dL) and urine alcohol level was 0.5 g/dL (500 mg/dL), indicating that death occurred during the phase of high blood alcohol elimination. The blood alcohol level of 400 mg/dL justified the ruling of sudden death as a consequence of alcohol inebriation. In addition, hair analysis revealed 46 pg/mg of ethyl glucurone in the first 2-cm segment of hair and 54 pg/mg in the next 2-cm segment (total length of hair analyzed was 4 cm), indicating that the boy may have occasionally imbibed alcohol prior to death; pathological lesions of the liver observed in histopathology supported this hypothesis [52].

8.3.6 Laboratory Methods for Determination of Ethyl Glucuronide and Ethyl Sulfate

For determination of ethyl glucuronide in urine, enzyme immunoassay (DRI assay; Microgenics) is commercially available, but false-positive test results may occur due to therapy with chloral hydrate or the use of propanol/isopropyl alcohol-containing hand sanitizers. However, chromatographic methods based on gas chromatography combined with mass spectrometry (GC–MS) or on LC–MS or LC–MS/MS are superior to immunoassay because these methods

Table 8.3 Sources of False-Positive/False-Negative Results in Ethyl Glucuronide Testing Using Urine or Hair Specimens

Source of False-Positive/ False-Negative Result	Comment
Therapy with chloral hydrate	False-positive ethyl glucuronide in urine using DRI enzyme immunoassay (Microgenics) due to cross-reactivity with trichloroethyl glucuronide, a major metabolite of chloral hydrate. However, LC–MS/MS-based assays are not affected.
Use of propanol/isopropyl alcohol-containing hand sanitizer	False-positive ethyl glucuronide in urine using DRI enzyme immunoassay (Microgenics) due to cross-reactivity with propyl glucuronide or isopropyl glucuronide. However, LC–MS/MS-based assays are not affected.
Other sources of positive urinary ethyl glucuronide	Incidental exposure to ethanol or factors listed in Table 8.2 may cause false-positive urinary ethyl glucuronide test results.
Urinary tract infection with *E. coli*	*E. coli* contains β-glucuronidase enzyme, which may hydrolyze ethyl glucuronide if present in urine, but ethyl sulfate is not affected. This may cause false-negative results. Testing for ethyl sulfate may be more appropriate in such situations. Alternatively, use of sodium fluoride as a preservative can prevent hydrolysis of ethyl glucuronide by *E. coli*.
Hair tonic containing ethanol or herbal hair tonic containing ethanol and ethyl glucuronide	Herbal hair tonic containing ethyl glucuronide may be incorporated into hair and may cause a false-positive result, but using herbal tonic containing only ethanol should not cause a false-positive result.
Bleaching and perming of hair	Although hair coloring has no significant effect on ethyl glucuronide levels in hair, bleaching and perming may reduce these levels due to chemical degradation of ethyl glucuronide.

offer more sensitivity as well as specificity. One of the limitations of GC–MS for analysis of ethyl glucuronide in a biological matrix is that derivatization is needed prior to analysis. However, for LC–MS- or LC–MS/MS-based methods, ethyl glucuronide can be analyzed directly without the need for derivatization. Krivankova *et al.* described a capillary electrophoresis method for analysis of ethyl glucuronide in human serum [53].

Wurst *et al.* used both GC–MS and LC–MS/MS for determination of serum and urine ethyl glucuronide in various patients. For determination of ethyl glucuronide, the authors used deuterated ethyl glucuronide (d_5-ethyl glucuronide) as the internal standard. Ethyl glucuronide, along with the internal standard, was extracted from serum or urine using methanol. Then, for GC–MS analysis, the organic layer was evaporated followed by derivatization using pyridine and *N,O*-bis(trimethylsilyl)trifluoroacetamide (BSTFA). After evaporating excessive derivatizing reagent, dry residue was reconstituted with ethyl acetate followed by analysis using GC–MS. However, for LC–MS/MS analysis (triple-quadrupole mass spectrometer API 365 equipped with a nebulizer-assisted electrospray source for detection), no derivatization was necessary. The MS/MS ion transitions monitored were m/z 221→75 transition for ethyl glucuronide and m/z 226→75 transition for the internal standard. The detection limit was 0.1 mg/L for both GC–MS and LC–MS/MS [54].

Al-Asmari *et al.* described a hydrophilic interaction liquid chromatography combined with electrospray ionization tandem mass spectrometric method (LC/ESI–MS/MS) for direct determination of urinary ethyl glucuronide and ethyl sulfate in postmortem urine specimens [55]. Helander *et al.* compared liquid chromatography combined with electrospray ionization mass spectrometry (LC/ESI–MS) with LC/ESI–MS/MS for analysis of ethyl glucuronide in urine using 482 specimens; they commented that LC–MS/MS should be considered as the reference method [56].

LC–MS/MS can also be successfully used for determination of ethyl glucuronide in hair, but proper washing of hair and extraction of ethyl glucuronide prior to analysis are necessary. Dichloromethane and/or methanol can be used for extracting ethyl glucuronide from hair. Yaldiz *et al.* described a hydrophilic interaction LC–MS/MS for determination of ethyl glucuronide in human hair. The authors washed the hair specimen using water, and after supplementing with internal standard in methanol (d_5-ethyl glucuronide), the specimen was vortex-mixed, followed by the addition of acetonitrile/water and ultrasonication to extract ethyl glucuronide from the hair. After centrifugation, supernatant was analyzed by LC–MS/MS. The assay was linear up to an ethyl glucuronide concentration of 200 pg/mg, and the limit of detection was 0.05 pg/mg. The limit of quantitation was 0.18 pg/mg [57].

8.4 FATTY ACID ETHYL ESTERS AS ALCOHOL BIOMARKERS

Fatty acid ethyl esters are minor metabolites of ethanol that are formed after alcohol consumption in virtually all tissues due to interaction of ethanol with free fatty acids as well as triglycerides, lipoproteins, and phospholipids. This pathway is an enzyme-mediated esterification of fatty acid or fatty acetyl-CoA by ethanol. Hydrolysis of a fatty acid from a phospholipid or a triglyceride molecule in the presence of ethanol can also lead to the formation of fatty acid ethyl esters. It has been speculated that fatty acid ethyl esters are toxic and are responsible for alcohol-induced organ damage. High amounts of fatty acid ethyl esters have been observed in the liver and pancreas during autopsy of subjects who were intoxicated at death. The fatty acid ethyl ester synthase may play a role in the formation of fatty acid ethyl esters because the liver and pancreas often have a high activity of this enzyme. Carboxyl ester lipase, which has the ability to liberate fatty acids from complex lipids, also has fatty acid ethyl ester synthase capability. In serum, fatty acid ethyl esters appear after alcohol consumption and are bound to albumin and also found in the core of lipoproteins along with other neutral lipids. Fatty acid ethyl esters can be used as a marker to determine both acute and chronic ingestion of alcohol. The concentrations of fatty acid ethyl esters in serum parallel ethanol levels in serum after acute ingestion of alcohol. If serum alcohol is negative but the fatty acid ethyl ester test is positive, then it can be assumed that alcohol consumption has occurred in the past 24 hr [58].

8.4.1 Fatty Acid Ethyl Esters in Hair

Although serum can be used to determine the concentrations of fatty acid ethyl esters, recently the focus has been on the determination of hair concentration of fatty acid ethyl esters in both clinical and forensic investigations. Fatty acid ethyl esters are incorporated in hair mainly from sebum glands attached to each hair root. This steady deposition into the hair shaft leads to the accumulation of fatty acid ethyl esters in hair, increasing from the proximal end to approximately 6–10 cm of hair, beyond which the concentration may decrease due to degradation or washout effects. Fatty acid ethyl esters are also stored in fatty tissues (e.g., adipose tissue) after long-term heavy drinking, and delayed incorporation in hair from such tissues may occur after beginning abstinence due to the release of fatty acid ethyl esters into the circulation from fatty tissues. Moreover, hair from a strictly abstinent person may show a very small amount of fatty acid ethyl ester, which may be related to a trace amount of endogenous ethanol production, nutrition, or use of hair cosmetics. Hair specimens collected from the pubic region, armpit, chest, arm, or thigh show comparable fatty acid ethyl esters to a specimen collected from scalp hair. Although approximately 15–20 fatty acid ethyl esters can be detected in a specimen, the most common ones used for calculation in varying amounts include ethyl laurate, ethyl myristate, ethyl palmitate, ethyl palmitoleate, ethyl stearate, ethyl oleate, ethyl linoleate, ethyl linolenate, ethyl arachidonate, and ethyl docosahexaenoate. Sometimes only the sum of the concentrations of four fatty acid ethyl esters (ethyl myristate, ethyl oleate, ethyl palmitate, and ethyl stearate) is used to determine fatty acid ethyl ester concentration in a specimen. The sum of these fatty acid ethyl esters in hair specimens varies from less than 0.2 ng/mg of hair in strict teetotalers to more than 30 ng/mg of hair (usually expressed as 30 ng/mg) in samples from alcoholic death. Usually for hair analysis, a 0.5 ng/mg cutoff (proximal scalp hair 0–3 cm used for analysis) is used in living persons and 0.8 ng/mg is used in postmortem cases to show abstinence with sensitivity and specificity of approximately 90%. In one study, analysis of 644 hair specimens revealed the presence of fatty acid ethyl esters in all specimens with a range of concentration between 0.11 and 31 ng/mg (mean, 1.77 ng/mg; median, 0.82 ng/mg). Comparison of fatty acid ethyl ester data to serum CDT and GGT readings revealed that a large portion of negative CDT and GGT results showed positive fatty acid ethyl ester, indicating that fatty acid ethyl ester has a much longer window of detection than traditional markers such as CDT and GGT [59]. Although the sum of four fatty acid ethyl esters is used to determine the total amount of fatty acid ethyl esters, some investigators have used a combination of six fatty acid ethyl esters for their calculation (ethyl palmitate, ethyl palmitoleate, ethyl stearate, ethyl oleate, ethyl linoleate, and ethyl arachidonate).

Bertol *et al.* proposed a cutoff of 0.5 ng/mg for fatty acid ethyl ester concentration in hair (3-cm segment analyzed) to differentiate between social drinking and excessive alcohol consumption (>60 g/day). In a study of 160 healthy volunteers ranging from teetotalers to individuals who consumed an excessive amount of alcohol, the authors observed that fatty acid ethyl ester concentrations in hair ranged from 0.01 to 10.78 ng/mg (expressed as the sum of the four fatty acid ethyl esters), with an average value of 1.16 ng/mg and a median value of 0.60 ng/mg. The estimated daily consumption of alcohol ranged from 0 to 246 g of ethanol. Using the 0.5 ng/mg cutoff to differentiate between excessive drinking and social drinking, the authors observed a specificity of 87% [60]. Although the Society of Hair Testing also recommends 0.5 ng/mg cutoff (scalp hair 0- to 3-cm proximal segment) for determination of chronic alcohol abuse, Hastedt *et al.* found that, based on a study of 1057 autopsy cases, median fatty acid ethyl ester concentration was 0.302 ng/mg (range, 0.008–14.3 ng/mg) in social drinkers and 1.346 ng/mg (range, 0.010–83.7 ng/mg) in the group of alcohol abusers. The authors proposed a cutoff of 1 ng/mg to demonstrate excessive alcohol consumption [61].

Combining fatty acid ethyl ester and ethyl glucuronide in hair to differentiate social drinking from heavy drinking has also been proposed. Pragst *et al.* commented that the cutoff value of 0.5 ng/mg for fatty acid ethyl ester in hair and the cutoff value of 30 pg/mg for ethyl glucuronide in proximal hair segments 0–3 cm long seem to be an optimal compromise for differentiating heavy consumption of alcohol from social drinking [62]. In a study to determine if pregnant women were consuming alcohol, the Pragst and Yegles found that strict abstinence is excluded or improbable at a fatty acid ethyl ester concentration in hair greater than 0.2 ng/mg and ethyl glucuronide concentration greater than 7 pg/mg. Moderate social drinkers should have a fatty acid ethyl ester concentration less than 0.5 ng/mg and an ethyl glucuronide concentration less than 25 pg/mg [63].

Hartwig *et al.* evaluated the effect of hair care products on fatty acid ethyl ester concentration in hair. In general, shampooing and cosmetic treatment (permanent wave, dyeing, or bleaching or shading) at the usual frequency have no significant effect on the use of fatty acid ethyl esters as an alcohol biomarker in hair, and such treatments do not lead to false-negative results. Some hair wax products also contain fatty acid ethyl esters, but use of such hair wax should not increase the concentration of fatty acid ethyl ester in hair. In doubtful cases, pubic hair should be analyzed for comparison [64].

Gareri *et al.* studied the effect of hair care products on fatty acid ethyl ester concentrations in hair. Nine individuals who were identified, based on fatty acid ethyl ester results in hair, to be chronic alcohol abusers denied any alcohol abuse. Specimens were further tested for ethyl glucuronide. Although fatty acid ethyl ester levels in these individuals ranged from 0.5 to 4.9 ng/mg, all ethyl glucuronide values were less than the 30 pg/mg cutoff level. The hair products used by these individuals contained as low as 10% alcohol by volume to as high as 95% by volume. The authors concluded that regular use of hair products containing as little as 10% alcohol can impact fatty acid ethyl ester values in hair. In this case, ethyl glucuronide should be tested because ethyl glucuronide values in hair seem to be unaffected by use of hair products containing alcohol [65].

8.4.2 Fatty Acid Ethyl Esters in Meconium

Meconium analysis of fatty acid ethyl esters is a valid method for identifying heavy prenatal ethanol exposure. However, small amounts of fatty acid ethyl esters are found in the meconium of neonates without any maternal use of alcohol. This may originate from endogenous ethanol or from trace amounts of alcohol in food. Depending on the combination of fatty acid ethyl esters used for calculation, various cutoff values have been proposed. Because the median molecular weight of fatty acid ethyl esters is approximately 300, the concentration expressed as nanomoles per gram should be multiplied by 300 to derive the concentration in nanograms per gram (2 nmol/g is approximately 600 ng/g), which is widely used for the cutoff by many investigators. Some investigators have used a combination of six fatty acid ethyl esters (ethyl palmitate, ethyl palmitoleate, ethyl stearate, ethyl oleate, ethyl linoleate, and ethyl arachidonate) and a cutoff of 600 ng/g, whereas others have used a combination of four fatty acid ethyl esters (ethyl palmitate, ethyl stearate, ethyl oleate, and ethyl linoleate) but the same cutoff concentration of 600 ng/g. Bakdash *et al.* used a combination of ethyl palmitate, ethyl linoleate, ethyl oleate, and ethyl stearate with a cutoff concentration of 500 ng/g of meconium to study the detection of alcohol abuse by mothers during pregnancy. The authors used LC–MS/MS for analysis of meconium. A concentration of fatty acid ethyl esters greater than the cutoff was observed in 43 of 602 meconium specimens analyzed (7.1% of specimens); the concentration ranged between 507 and 22,580 ng/g, except for one outlier with a concentration of 150,000 ng/g (no ethyl glucuronide was detected in the specimen). Ethyl glucuronide was detected in 97 cases. The authors noted that by using a cutoff of 274 ng/g for ethyl glucuronide, optimal agreement between two markers was obtained. The authors concluded that combining fatty acid ethyl glucuronide with fatty acid ethyl ester in meconium should be useful in studying potential alcohol consumption by mothers during pregnancy [66]. Moore *et al.* used a combination of ethyl palmitate, ethyl palmitoleate, ethyl stearate, ethyl oleate, ethyl linoleate, and ethyl arachidonate for the determination of fatty acid ethyl esters in meconium and reported that total fatty acid ethyl ester concentration exceeding 10,000 ng/g of meconium may indicate that a newborn was exposed to significant amounts of alcohol during gestation. The authors also commented that fatty acid ethyl esters are unstable in meconium if stored at room temperature. Therefore, specimens should be stored and transported frozen to the laboratory prior to analysis [67]. False-positive meconium test results for fatty acid ethyl esters may be observed secondary to delayed sample collection. In one study, the authors collected 136 specimens of meconium from 30 neonates during their first few days of life. Although the first collected meconium specimens tested negative at a 2 nmol/g cutoff in all infants, later specimens tested positive above the cutoff in 19 of 30 neonates. Median time to appearance of fatty acid ethyl esters in meconium was 59.2 hr postpartum. The authors concluded that samples collected later in the postpartum period may lead to false-positive test results due to dietary components in postnatally produced stool and ethanol-producing microorganisms [68]. Critical issues regarding the interpretation of hair and meconium fatty acid ethyl ester concentrations are summarized in Box 8.1.

BOX 8.1 CRITICAL ISSUES REGARDING INTERPRETATION OF HAIR AND MECONIUM FATTY ACID ETHYL ESTER CONCENTRATIONS

- Although the serum concentration of fatty acid ethyl esters was initially measured as an alcohol biomarker (window of detection ~24 hr after consumption of alcohol), fatty acid ethyl esters are currently measured in hair or meconium specimens.
- Although approximately 15–20 fatty acid ethyl esters can be detected in a specimen, the most common ones used for calculation in varying amounts include ethyl laurate, ethyl myristate, ethyl palmitate, ethyl palmitoleate, ethyl stearate, ethyl oleate, ethyl linoleate, ethyl linolenate, ethyl arachidonate, and ethyl docosahexaenoate.
- Sometimes only the sum of the concentrations of four fatty acid ethyl esters (ethyl myristate, ethyl oleate, ethyl palmitate, and ethyl stearate) is used to determine fatty acid ethyl ester concentration in a specimen, but six (e.g., ethyl palmitate, ethyl palmitoleate, ethyl stearate, ethyl oleate, ethyl linoleate, and ethyl arachidonate) or eight fatty acid combinations may also be used for calculating total fatty acid ethyl ester concentrations. Therefore, cutoff concentrations may vary from report to report due to the use of various combinations of fatty acid ethyl esters for calculation.
- Although a cutoff concentration of 0.5 ng/mg for fatty acid ethyl ester in hair is usually used, it is better to use this value in combination with a cutoff value of 30 pg/mg for ethyl glucuronide in proximal hair segments (0–3 cm used for analysis) for interpretation of results because hair fatty acid ethyl values may be falsely elevated due to the use of ethanol-containing hair products, but the ethyl glucuronide value in hair is not affected.
- Usually, a cutoff of 2 nmol/g or 600 ng/g is used for meconium fatty acid ethyl esters using six (ethyl palmitate, ethyl palmitoleate, ethyl stearate, ethyl oleate, ethyl linoleate, and ethyl arachidonate) or four (ethyl palmitate, ethyl stearate, ethyl oleate, and ethyl linoleate) fatty acid combinations, but another cutoff has been proposed.
- For analysis of meconium, it is important to collect the specimen as soon as possible after delivery because delayed specimen collection may cause a false-positive test result.

8.4.3 Laboratory Analysis of Fatty Acid Ethyl Esters

Chromatographic methods such as GC–MS, LC–MS, and LC–MS/MS are used for the determination of fatty acid ethyl esters in hair and meconium, although these methods can be used for analysis of serum if needed. Many protocols are available for analysis of fatty acid ethyl esters for both hair and meconium specimens. For hair analysis, after washing the specimen with water or an organic solvent, fatty acid ethyl esters are extracted using a variety of organic solvents. Auwarter *et al.* used a combination of dimethyl sulfoxide and heptane for extraction followed by GC–MS for analysis. The authors used ethyl myristate, ethyl palmitate, ethyl oleate, and ethyl stearate for quantification of total fatty acid ethyl esters in hair and used corresponding deuterated fatty acid ethyl esters as internal standards [69].

Hutson *et al.* developed a GC–MS method for analysis of fatty acid ethyl esters in meconium. The authors quantified four fatty acid ethyl esters (ethyl palmitate, ethyl linoleate, ethyl oleate, and ethyl stearate) using 0.5 mg of meconium, and corresponding deuterated (d_5) ethyl esters were used as internal standards. Fatty acid ethyl esters were extracted from meconium using liquid–liquid extraction with heptane and acetone. For GC–MS analysis, the authors selected m/z values of 284, 308, 310, and 312 for ethyl palmitate, ethyl linoleate, ethyl oleate, and ethyl stearate, respectively. For the corresponding internal standards, m/z values were 289 (d_5-ethyl palmitate), 313 (d_5-ethyl linoleate), 315 (d_5-ethyl oleate), and 317 (d_5-ethyl stearate). The detection limits of four fatty acid ethyl esters ranged from 0.020 to 0.042 nmol/g [70]. Roehsig *et al.* described a protocol for the determination of the following eight fatty acid ethyl esters in meconium samples by solid phase microextraction and GC–MS: ethyl laurate, ethyl myristate, ethyl palmitate, ethyl palmitoleate, ethyl stearate, ethyl oleate, ethyl linoleate, and ethyl arachidonate. The authors used corresponding deuterated (d_5) fatty acid ethyl esters as internal standards. The limit of detection was less than 100 ng/g for fatty acid ethyl esters. The authors concluded that their method was suitable for analysis of fatty acid ethyl esters in meconium to detect alcohol-exposed newborns at a cumulative cutoff concentration of 600 ng/g [71].

Pichini *et al.* used LC–MS/MS for analysis of fatty acid ethyl esters in meconium. The authors quantified nine fatty acid ethyl esters (ethyl laurate, ethyl myristate, ethyl palmitate, ethyl palmitoleate, ethyl stearate, ethyl oleate, ethyl linoleate, ethyl linolenate, and ethyl arachidonate) and used ethyl heptadecanoate as the internal standard. Analytes were initially extracted from meconium with hexane followed by further solid phase extraction using an aminopropyl-silica column. Chromatographic analysis was performed using a C-8 reverse-phase column and a mobile phase consisting of water/isopropyl alcohol/acetonitrile (20:40:40 by volume). Mass spectrometric analysis was achieved by a triple quadrupole mass spectrometer that monitored the ion transitions in multiple reactions monitoring mode. The limit of quantification varied from 0.12 to 0.20 nmol/g [72].

8.5 PHOSPHATIDYLETHANOL AS ALCOHOL BIOMARKER

Phosphatidylethanol is a unique phospholipid that is formed only in the presence of ethanol via the action of enzyme phospholipase D. Like ethyl glucuronide, ethyl sulfate, and fatty acid ethyl esters, phosphatidylethanol is a direct alcohol biomarker. Normally, phospholipase D catalyzes hydrolysis of phosphatidylcholine into phosphatidic acid. However, phospholipase D has a high affinity for short-chain alcohol (100- to 1000-fold higher) compared to water. Therefore, in the presence of ethanol, this enzyme promotes a transphosphatidylation producing phosphatidylethanol. In humans, two isoforms of phospholipase D have been characterized. Phospholipase D_1 is distributed perinuclearly and requires phosphokinase C activation, whereas phospholipase D_2 is localized in the cellular membrane and is constitutively active. However, both isoforms of phospholipase D catalyze the formation of phosphatidylethanol on human erythrocyte membranes [73].

Phosphatidylethanol is not a single molecule but, rather, a group of glycerophospholipid homologs with a common phosphoethanol head group and two long fatty acid chains attached to a glycerol backbone (*sn*-1 and *sn*-2 positions). Because phosphatidylethanol is derived from phosphatidylcholine, there are similarities between molecular species (fatty acid composition in *sn*-1 and *sn*-2 positions) in phosphatidylcholine and phosphatidylethanol. There are many combinations of fatty acid chain length and the number of double bonds in the fatty acids that are attached to *sn*-1 and *sn*-2 positions of glycerol backbone, but phosphatidylethanol 16:0/18:1 (palmitic acid/oleic acid) and phosphatidylethanol 16:0/18:2 (palmitic acid/linoleic acid) are the two major molecular species extracted from human erythrocytes, representing 37 and 25%, respectively, of all phosphatidylethanol molecular species. Phosphatidylethanol 16:0/20:4 (palmitic acid/arachidonic acid) represents approximately 13% of all molecular species [74]. Gnann *et al.* also reported that phosphatidylethanol 16:0/18:1 and 16:0/18:2 are the two most abundant molecular species present in blood. However, using liquid chromatography combined with time-of-flight tandem mass spectrometry, the authors characterized 48 different molecular species of phosphatidylethanol [75]. Although a combination of phosphatidylethanol molecular species 16:0/18:1 and 16:0/18:2 may be sufficient to represent total phosphatidylethanol level in blood, sometimes a combination of more molecular species, such as 16:0/18:1, 16:0/18:2, 16:0/20:4, 18:1/18:1, and 18:1/18:2, is also used. However, some investigators prefer to use only the 16:0/18:1 molecular species, which is the most abundant molecular species for quantification because the reference standard is commercially available.

Phosphatidylethanol is a good marker for excessive alcohol consumption, and blood phosphatidylethanol is found almost exclusively in the erythrocyte fraction. Consumption of one alcoholic drink is not sufficient to produce a detectable level of phosphatidylethanol in blood, and consumption of 50 g or more alcohol per day for several weeks is necessary to produce a detectable amount of phosphatidylethanol. However, once positive, it remains positive for 2 or 3 weeks, whereas blood alcohol may be not be detectable several hours after consumption of the last drink. The half-life of phosphatidylethanol in blood is 4 days, and the amount of alcohol consumed correlates with blood phosphatidylethanol level. This marker is almost 100% specific because its formation is totally dependent on the presence of ethanol. In addition, this marker is more sensitive than other traditional alcohol biomarkers, including CDT, GGT, and MCV [76]. In one study, 5 healthy volunteers were given 32–47 g of alcohol in a single dose (approximately two or three standard drinks), and no phosphatidylethanol was detected in blood of any volunteers. However, when 12 volunteers consumed various amounts of alcohol for 3 weeks, no phosphatidylethanol was detected in blood of volunteers who consumed lower amounts of alcohol (mean, 742 g); but with higher intake of alcohol (mean, 1630 g), detectable levels of phosphatidylethanol (1.0–2.1 μmol/L) were observed in blood specimens. The authors concluded that an individual must consume a substantial amount of alcohol before phosphatidylethanol can be detected in blood. The authors commented that phosphatidylethanol is more sensitive than serum CDT as an alcohol biomarker [77]. In another study, after 3 weeks of alcohol abstinence, 11 volunteers drank sufficient amounts of alcohol to achieve a blood alcohol concentration of 0.1% for 5 successive days. The phosphatidylethanol 16:0/18:1 concentration in blood reached a maximum of 74–237 ng/mL after 3–6 days [78]. Marques *et al.* commented that phosphatidylethanol levels in alcohol-dependent outpatients may have a mean value between 2.47 and 3.4 μmol/L, but the mean value of alcohol-dependent hospitalized patients may be as high as 7.7 μmol/L. However, values were much lower in drivers charged with driving while intoxicated (mean, 1.45 μmol/L) [79]. Wurst *et al.* reported that in an alcohol detoxification center, phosphatidylethanol levels on day 1 ranged from 0.63 to 26.95 μmol/L (mean, 6.22 μmol/L), and there were no false negatives, indicating that sensitivity was 100%. In contrast, MCV and CDT showed a sensitivity of 40.4 and and 69.2%, respectively. No gender difference was observed in phosphatidylethanol level in alcohol-dependent patients [80]. Gunnarsson *et al.* observed no phosphatidylethanol in blood of male controls, but concentrations between 5 and 13 μmol/L were observed in three alcoholics. However, phosphatidylethanol values decreased during 3 or 4 weeks of abstinence [81]. Because some authors have expressed phosphatidylethanol in micromoles per liter (μmol/L) and others have used a traditional unit of nanograms per milliliter (ng/mL), to avoid confusion, values in micromoles per liter can be multiplied by 703 to obtain the concentration in nanograms per milliliter (the molecular weight of phosphatidylethanol 16:0/18:1, the most abundant molecular species, is 702.98). Alternatively, values reported in nangrams per milliliter can be divided by 703 to obtain the value of phosphatidylethanol in micromoles per liter [79].

Kwak *et al.* reported that phosphatidylethanol was not detected in pregnant women who abstained from alcohol during pregnancy. In contrast, phosphatidylethanol 16:0/18:1 was above the cutoff concentration in women who consumed alcohol. This molecular species was detected for up to 4 weeks after cessation of alcohol exposure [82]. In another study, the authors observed that 4.8% of abstaining pregnant women had positive phosphatidylethanol, but these women had a positive history of alcohol use before pregnancy and this marker may be positive up to 3 or 4 weeks after the last drink. Phosphatidylethanol correlated with the amount of alcohol and the number of days of alcohol consumption per week [83]. Phosphatidylethanol is also superior to CDT for investigating alcohol consumption during pregnancy because CDT values increase with gestational age, whereas phosphatidylethanol values are not affected. In a study in which pregnant women were followed from prenatal care visit to term, it was observed that during recruitment (mean gestational age, 22.6 weeks), mean %CDT was 1.49%, but at term it had increased to 1.67%. Using a conventional cutoff of 1.7% for %CDT, 22.9% of specimens during the prenatal visit and 45.7% of specimens at term should be classified as positive. In contrast, no phosphatidylethanol in blood was detected, and ethyl glucuronide levels were also negative, indicating that %CDT values were false-positive values because these pregnant women were abstinent. The authors proposed that the %CDT cutoff should be increased to 2% in assessing pregnant women [84].

Phosphatidylethanol may be a superior marker to ethyl glucuronide and ethyl sulfate. In one study involving 252 participants, all 10 subjects who tested positive for ethyl glucuronide or ethyl sulfate but denied consuming alcohol were told about the possibility of further testing for phosphatidylethanol. After the test procedure was explained, 3 of the 10 subjects admitted drinking, but 7 subjects who still denied drinking were tested for phosphatidylethanol. With this testing, 5 subjects showed negative phosphatidylethanol levels, supporting their claim of not consuming alcohol.

However, for the 2 other subjects, phosphatidylethanol levels were positive (220 and 320 ng/mL), and it was determined that they had consumed alcohol and were sent for further treatment. The authors concluded that phosphatidylethanol results along with with previous low positive ethyl glucuronide or ethyl sulfate results allow differentiating between no consumption of alcohol and consumption. Negative phosphatidylethanol testing after a low positive ethyl glucuronide result supports a patient's claim of recent abstinence [85]. Phosphatidylethanol testing in postmortem blood is useful to investigate previous heavy drinking. Hansson *et al.* measured phosphatidylethanol in femoral blood in 85 consecutive forensic autopsies and observed phosphatidylethanol concentrations between 0.8 and 22.0 μmol/L in 35 cases. Of these cases, no ethanol in blood was detected in 12 cases despite positive phosphatidylethanol levels. However, in 2 cases, ethanol levels were positive but no phosphatidylethanol was present in blood. Interestingly, in 1 case of fatal methanol poisoning, a chromatographic peak was observed at a position for phosphatidylethanol but it was characterized as phosphatidylmethanol because phospholipase D also accepts other alcohols, such as methanol, and substrate-forming phosphatidylmethanol is likely in a subject who consumes methanol. The authors concluded that phosphatidylethanol testing is useful in postmortem investigations to show heavy alcohol consumption in cases in which no blood alcohol may be detected or alcohol consumption is not otherwise evident [86].

8.5.1 Cutoff Concentration of Phosphatidylethanol

No cutoff concentration of has been firmly established for the clinical application of phosphatidylethanol, but values of between 0.2 and 0.7 μmol/L have been proposed based on the limit of quantitation of the high-performance liquid chromatography (HPLC) method. In Sweden, 0.7 μmol/L (492 ng/mL) has been proposed, which can identify alcohol consumption of 50 g or more per day, but at a cutoff of 0.2 μmol/L (140 ng/mL), alcohol consumption of 40 g or less may be detected [74]. Stewart *et al.* studied phosphatidylethanol levels in 80 nonpregnant women ages 18–35 years and concluded that 93% of subjects who consumed an average of 2 or more drinks per day had detectable phosphatidylethanol levels in blood, whereas phosphatidylethanol could be quantified in 53% of subjects who consumed 1 drink per day during the preceding 14 days prior to blood analysis. In the entire group of subjects, median alcohol consumption was 23 drinks per week or 1.6 drinks per day on average, whereas the median blood phosphatidylethanol level was 45 ng/mL (range, 0–565 ng/mL). The authors concluded that a phosphatidylethanol level above the limit of quantitation was highly sensitive for alcohol consumption averaging at least 2 drinks per day, and values over 127 ng/mL were highly specific for drinking in excess of 2 drinks per day [87]. Based on the receiver operating curve, Hartmann *et al.* observed that phosphatidylethanol at a cutoff of 0.36 μmol/L (253 ng/mL) has a sensitivity of 94.5% and specificity of 100% for differentiating between drinkers and sober patients. Sensitivity and specificity of phosphatidylethanol were superior to those of CDT, MCV, and GGT [88].

However, phosphatidylethanol may also be formed *in vitro* if ethanol is present. Varga and Alling reported that the formation of phosphatidylethanol in human red blood cells is concentration dependent. Incubation of red cells with 50 mmol ethanol yielded detectable phosphatidylethanol after 12 hr, and a maximum value was achieved after 60 hr of incubation. *In vitro* formation of phosphatidylethanol was also higher when red blood cells were collected from alcoholics (mean, 5.2 μmol/L) compared to controls (mean, 2.4 μmol/L). The authors concluded that phosphatidylethanol is formed in red blood cells *in vitro* at physiological alcohol concentrations [89].

Interestingly, antibodies specific to phosphatidylethanol may be present in sera of individuals who consume alcohol. These antibodies may be IgG, IgA, and IgM and bind specifically to phosphatidylethanol. Nissinen *et al.* analyzed antibodies to phosphatidylethanol in sera of 20 heavy drinkers, 58 patients with alcoholic pancreatitis, and 24 control subjects using a chemiluminescence immunoassay, and phosphatidylethanol level in blood was measured using LC–MS. The authors observed lower concentrations of these antibodies (IgG, IgA, and IgM) in sera of individuals who were heavy drinkers or patients with alcoholic pancreatitis, compared to controls (social drinkers). The plasma IgG titers correlated with whole blood phosphatidylethanol in heavy drinkers. The authors concluded that subjects with heavy alcohol consumption showed markedly lower levels of plasma antibodies to phosphatidylethanol, potentially making them useful as a biomarker to distinguish heavy drinking from moderate alcohol use [90]. Critical issues regarding the interpretation of results relating to phosphatidylethanol are summarized in Box 8.2.

8.5.2 Laboratory Analysis of Phosphatidylethanol

Various methods can be used in clinical laboratories for the determination of phosphatidylethanol, including thin-layer chromatography, HPLC coupled with use of a evaporative light scattering detector, GC–MS, nonaqueous capillary electrophoresis, immunoassay using phosphatidylethanol-specific monoclonal antibody, and LC–MS/MS [91]. Phosphatidylethanol-specific antibody can be generated using the traditional hybridoma technique, and such monoclonal antibody can be incorporated to develop an immunoassay for detection of phosphatidylethanol in whole blood [92]. Yon and Han described a GC–MS protocol for the determination of trimethylsilyl

BOX 8.2 CRITICAL ISSUES REGARDING INTERPRETATION OF RESULTS RELATING TO PHOSPHATIDYLETHANOL

- Phosphatidylethanol is almost exclusively found in erythrocyte membranes, and it is analyzed in whole blood only, preferably by using liquid chromatography combined with tandem mass spectrometry.
- In general, phosphatidylethanol 16:0/18:1 (palmitic acid/oleic acid) and phosphatidylethanol 16:0/18:2 (palmitic acid/linoleic acid) are two major molecular species which are usually extracted from human erythrocytes. Combining these two most abundant molecular species is sufficient to calculate the total phosphatidylethanol level in blood. However, some investigators prefer to use 16:0/18:1 molecular species, which is the most abundant molecular species for quantification, because the reference standard is commercially available. Moreover, a combination of more molecular species, such as 16:0/18:1, 16:0/18:2, 16:0/20:4, 18:1/18:1, and 18:1/18:2, may also be used.
- Phosphatidylethanol has almost 100% specificity and high sensitivity to individuals who consume moderate to heavy amounts of alcohol (50 g of alcohol or more per day for 2 or 3 weeks). After abstinence, phosphatidylethanol may be detected for up to 2 or 3 weeks in blood due to its long half-life. Phosphatidylethanol is a superior biomarker compared to traditional alcohol biomarkers such as GGT, MCV, and %CDT.
- Phosphatidylethanol is useful for determining whether a subject is practicing abstinence, especially if ethyl glucuronide and/or ethyl sulfate levels are low. In addition, %CDT values increase in pregnancy, but phosphatidylethanol levels are unaffected. Therefore, phosphatidylethanol is a good marker to assess alcohol use during pregnancy. Phosphatidylethanol is also a good marker to determine antemortem heavy alcohol consumption during forensic autopsy.
- In Sweden, a cutoff of 0.7 μmol/L (492 ng/mL) has been proposed for phosphatidylethanol, although a firm cutoff value is not yet established internationally (multiplying micromoles per liter (μmol/L) by 703 yields a value in nanograms per microliter (ng/mL)).
- Phosphatidylethanol may be formed *in vitro* in the presence of alcohol.
- Phosphatidylethanol is not stable in whole blood at room temperature or if frozen, but it is stable in whole blood for at least 30 days if stored at −80°C. However, phosphatidylethanol is stable in dried blood spots even if stored at room temperature.
- Phospholipase D is capable of producing phosphatidylmethanol in a person who has overdosed with methanol. Such peak usually occurs near the peak of phosphatidylethanol in the chromatogram during LC–MS/MS analysis.

derivatization products of phosphatidylethanol. The two major products formed during derivatization with BSTFA are ethyl *bis* (trimethylsilyl)-phosphate and *tris* (trimethylsilyl)-phosphate; *bis* (trimethylsilyl)-phosphate, with a molecular weight of 270.09, can be used as a marker of phosphatidylethanol [93]. However, analysis of intact molecules using HPLC or LC–MS/MS is the superior and most accepted method for analysis of phosphatidylethanol in clinical laboratories.

Several methods have been reported for analysis of phosphatidylethanol in whole blood using LC–MS/MS. Whole blood may be collected using an anticoagulant such as heparin. Cabarcos *et al.* described LC–MS/MS analysis of phosphatidylethanol 16:0/18:1, the main molecular species in human whole blood, using phosphatidylbutanol 16:0/18:1 as the internal standard after liquid–liquid microextraction. Chromatographic separation was achieved using a reverse-phase C-8 column, and mass spectrometric analysis was conducted using negative ion mode electrospray ionization. Data were acquired in multiple reaction monitoring mode with phosphatidylethanol 16:0/18:1 (m/z 701.4 → 255.2 and 281.1) and the internal standard (m/z 729.6 to 209.2, 155.3, and 465.2). The linearity of the assay was from limit of quantification to up to 10 μg/mL. The limit of detection was 0.01 μg/mL. A set of 50 blood samples were analyzed, and the range of phosphatidylethanol detected was from the lower limit of detection to 1.71 μg/mL [94]. However, it is important to note that phosphatidylethanol in whole blood may degrade even if stored frozen; if stored at −80°C, it is stable for at least 30 days. Interestingly, phosphatidylethanol in dried blood spot is stable even if stored at room temperature [95]. Using phosphatidylpropanol as the internal standard, Faller *et al.* analyzed phosphatidylethanol molecular species in both whole blood and dried blood and demonstrated that whole blood values match dried blood values [96]. Bakhireva *et al.* studied the validity of phosphatidylethanol in dried blood spots of newborns for the identification of prenatal alcohol exposure and concluded that newborn phosphatidylethanol analyzed in dried blood spot is a highly specific biomarker and can facilitate accurate detection of prenatal alcohol exposure in conjunction with other biomarkers [97].

8.6 SENSITIVITY AND SPECIFICITY OF DIRECT ALCOHOL BIOMARKERS

As expected, the sensitivity and specificity of ethyl glucuronide, ethyl sulfate, fatty acid ethyl esters, and phosphatidylethanol vary with cutoff concentration as well as the specimen in which the analyte is analyzed. However, phosphatidylethanol in blood has the highest specificity of 100% at a cutoff value of 0.36 μmol/L. The sensitivity at this cutoff is 94.5% to differentiate between drinkers and sober patients [88]. Bakhireva *et al.* reported that sensitivity of phosphatidylethanol in dried blood spot of newborn to identify prenatal alcohol exposure was only 32.1%, although specificity was 100% at a cutoff level of 8 ng/mL. However, a battery consisting of maternal direct ethanol metabolites (urinary ethyl glucuronide, urinary ethyl sulfate, and blood phosphatidylethanol) increased sensitivity to 50% without a substantial compromise of specificity of 93.8% [97].

For hair, ethyl glucuronide sensitivity of 96% and specificity of 99% have been reported at a cutoff concentration of 30 pg/mg of hair to identify individuals who consume alcohol chronically at an amount exceeding 60 g/day [29].

However, sensitivity and specificity of hair ethyl glucuronide at a cutoff of 8 pg/mg of hair were 92 and 87%, respectively, to identify individuals who consumed on average at least 28 g of alcohol each day [98]. Berger *et al.* reported that at 30 pg/mg cutoff level, sensitivity and specificity of ethyl glucuronide in hair were 43 and 92%, respectively, for identifying high-risk drinkers [27].

Stewart *et al.* reported that for urinary glucuronide at a cutoff of 100 ng/mL, sensitivity and specificity were 76 and 93%, respectively. The authors also determined that sensitivity and specificity of urinary ethyl sulfate at 25 ng/mL cutoff were 82 and 86%, respectively, for identifying alcohol consumption 3–7 days prior to clinic visits [99]. Wurst *et al.* reported that at a cutoff of 435 ng/mL for urinary ethyl glucuronide, the sensitivity and specificity were 90.8 and 76.5%, respectively, for determining sobriety for less than 24 hr [100]. Ethyl glucuronide at a positive cutoff of 2.0 nmol/g of meconium (444.2 ng/g because the molecular weight of ethyl glucuronide is 222.1) showed both sensitivity and specificity of 100% to discriminate true prenatal exposure of ethanol [101]. Interestingly, in one report, although specimens were collected from the same socioeconomic groups in two cities, the authors observed that the median value of ethyl glucuronide in meconium was 101.5 ng/g ($n = 81$) in specimens collected from Barcelona, Spain, but was 15.6 ng/g ($n = 96$) in samples collected from Reggio Emilia, Italy. The authors commented that in the Barcelona cohort, ethyl glucuronide values allowed differentiation between specimens with fatty acid ethyl ester values below or above the cutoff concentration (2 nmol/g) [102]. Chan *et al.* reported that at a cutoff of 2 nmol of total fatty acid ethyl esters per gram of meconium, the sensitivity and specificity were 100 and 98.4%, respectively, for identifying maternal alcohol consumption [103]. Sensitivity and specificity of direct alcohol biomarkers are summarized in Table 8.4.

Table 8.4 Sensitivity and Specificity of Direct Alcohol Biomarkers

Alcohol Biomarker	Specimen	Cutoff Level	Sensitivity (%)	Specificity (%)
Ethyl glucuronide	Urine	100 ng/mL	76	93
		435 ng/mL	90.8	76.5
Ethyl glucuronide	Hair	30 pg/mg	96	99
		8 pg/mg	92	87
Ethyl sulfate	Urine	25 ng/mL	82	86
FAEE	Hair	0.29 ng/mg	90	100
FAEE	Meconium	2 nmol/g/600 ng/mL	100	98.4
Phosphatidylethanol	Blood	0.36 μmol/L	94.5	100

8.7 CONCLUSIONS

Direct biomarkers of alcohol, including ethyl glucuronide, ethyl sulfate, fatty acid ethyl esters, and phosphatidylethanol, are used in various clinical settings, such as workplace drug and alcohol testing, monitoring alcohol consumption if suspected in pregnant women, and military drug and alcohol testing programs, as well as drug and rehabilitation programs. These alcohol biomarkers are also routinely used in forensic investigations. Although ethyl glucuronide and ethyl sulfate represent single-molecule, fatty acid ethyl esters and phosphatidylethanol, each represents a class of molecules. In general, phosphatidylethanol has specificity close to 100% and is an emerging alcohol biomarker. Although immunoassay is commercially available for monitoring ethyl glucuronide, chromatographic methods are more applicable for the determination of fatty acid ethyl esters in hair and phosphatidylethanol in blood.

References

[1] Fort RS, Fisher MB. Assessment of UDP-glucuronosyltransferase catalyzed formation of ethyl glucuronide in human liver microsomes and recombinant UGTs. Forensic Sci Int 2005;153:109–16.

[2] Helander A, Beck O. Mass spectrometric identification of ethyl sulfate as an ethanol metabolite in humans. Clin Chem 2004;50:936–7.

[3] Schneider H, Glatt H. Sulpho-conjugation of ethanol in humans in vivo and by individual sulphotransferase forms in vitro. Biochem J 2004;383:543–9.

[4] Maenhout TM, De Buyzere ML, Delanghe JR. Non-oxidative ethanol metabolites as a measure of alcohol intake. Clin Chim Acta 2013;415:322–9.

[5] Hoiseth G, Bernard JP, Karinen R, Johnsen L, et al. A pharmacokinetic study of ethyl glucuronide in blood and urine: application to forensic toxicology. Forensic Sci Int 2007;172:119–24.

[6] Halter CC, Dresen S, Auwaerter V, Wurst FM, et al. Kinetics in serum and urinary excretion of ethyl sulfate and ethyl glucuronide after medium dose ethanol intake. Int J Legal Med 2008;122:123–8.

[7] Helander A, Bottcher M, Fehr C, Dahmen N, et al. Detection times of urinary ethyl glucuronide and ethyl sulfate in heavy drinkers during alcohol detoxification. Alcohol Alcohol 2009;44:55–61.

[8] Wurst FM, Dresen S, Allen JP, Wiesbeck G, et al. Ethyl sulfate: a direct ethanol metabolite reflecting recent alcohol consumption. Addiction 2006;101:204–11.

[9] Hoiseth G, Nordal K, Pettersen E, Morland J. Prolonged urinary detection times of EtG and EtS in patients with decreased renal function. Alcohol Clin Exp Res 2012;36:1148–51.

[10] Crunelle CL, Yegles M, van Nuijs AL, Covaci A, et al. Hair glucuronide levels as a marker for alcohol use and abuse: a review of the current state of the art. Drug Alcohol Depend 2014;134:1–11.

[11] Morini L, Marchei E, Vagnarelli F, Garcia Algar O, et al. Ethyl glucuronide and ethyl sulfate in meconium and hair-potential biomarkers of intrauterine exposure to ethanol. Forensic Sci Int 2010;196:74–7.

[12] Thierauf A, Gnann H, Wohlfarth A, Auwarter V, et al. Urine tested positive for ethyl glucuronide and ethyl sulfate after the consumption of "non-alcoholic" beer. Forensic Sci Int 2010;202:82–5.

[13] Hoiseth G, Yttredal B, Karinen R, Gjwerde H, et al. Levels of ethyl glucuronide and ethyl sulfate in oral fluid, blood and urine after use of mouthwash, and ingestion of nonalcoholic wine. J Anal Toxicol 2010;34:84–8.

[14] Reisfield G, Goldberger BA, Pesce AJ, Crews BO, et al. Ethyl glucuronide, ethyl sulfate, and ethanol in urine after intensive exposure to high ethanol content mouthwash. J Anal Toxicol 2011;35:264–8.

[15] Rohrig TP, Huber C, Goodson L, Ross W. Detection of ethyl glucuronide in urine following the application of GermX. J Anal Toxicol 2006;30:703–4.

[16] Reisfield G, Goldberger BA, Crews BO, Pesce AJ, et al. Ethyl glucuronide, ethyl sulfate, and ethanol in urine after sustained exposure to an ethanol-based hand sanitizer. J Anal Toxicol 2011;35:85–91.

[17] Arndt T, Schrofel S, Gussregen B, Stemmerich K. Inhalation but not transdermal resorption of hand sanitizer ethanol causes positive ethyl glucuronide findings in urine. Forensic Sci Int 2014;237:126–30.

[18] Rosano TG, Lin J. Ethyl glucuronide excretion in humans following oral administration of and dermal exposure to ethanol. J Anal Toxicol 2008;32:594–600.

[19] Thierauf A, Wohlfarth A, Auwarter V, Perdekamp MG, et al. Urine tested positive for ethyl glucuronide and ethyl sulfate after the consumption of yeast and sugar. Forensic Sci Int 2010;202:e45–7.

[20] Musshoff F, Albermann E, Madea B. Ethyl glucuronide and ethyl sulfate after consumption of various beverages and food: misleading results? Int J Legal Med 2010;124:623–30.

[21] Albermann ME, Musshoff F, Doberentz E, Heese P, et al. Preliminary investigations on ethyl glucuronide and ethyl sulfate cut-offs for detecting alcohol consumption on the basis of an ingestion experiment and on data from withdrawal treatment. Int J Legal Med 2012;126:757–64.

[22] Kummer N, Wille S, Di Fazio V, Lambert W, et al. A fully validated method for the quantification of ethyl glucuronide and ethyl sulfate in urine by UPLC–ESI–MS/MS applied in a prospective alcohol self-monitoring program. J Chromatogr B 2013;929:149–54.

[23] Jatlow PI, Agro A, Wu R, Nadim H, et al. Ethyl glucuronide and ethyl sulfate assays in clinical trials, interpretation and limitations: results of a dose ranging alcohol challenge study and 2 clinical trials. Alcohol Clin Exp Res 2014;38:2056–65.

[24] Pirro V, Di Corcia D, Seganti F, Salomone A, et al. Determination of ethyl glucuronide levels in hair for assessment of alcohol abstinence. Forensic Sci Int 2013;232:229–36.

[25] Morini L, Politi L, Polettini A. Ethyl glucuronide in hair: a sensitive and specific marker of chronic heavy drinking. Addiction 2009;104:8915–20.

[26] Boscolo-Berto R, Viel G, Montisci M, Terranova C, et al. Ethyl glucuronide concentration in hair for detecting heavy drinking and/or abstinence: a meta-analysis. Int J Legal Med 2013;127:611–19.

[27] Berger L, Fendrich M, Jones J, Fuhrmann D, et al. Ethyl glucuronide in hair and fingernails as a long-term alcohol biomarker. Addiction 2013;109:425–31.

[28] Agius R, Nadulski T, Kahl HG, Dufaux B. Ethyl glucuronide in hair—A highly effective test for the monitoring of alcohol consumption. Forensic Sci Int 2012;218:10–14.

[29] Boscolo-Berto R, Favretto D, Cecchetto G, Vincenti M, et al. Sensitivity and specificity of EtG in hair as a marker of chronic excessive drinking: pooled analysis of raw data and meta-analysis of diagnostic accuracy studies. Ther Drug Monit 2014. [E-pub ahead of print]. PMID: 24577122.

[30] Goecke TW, Burger P, Fasching PA, Bakdash A, et al. Meconium indicators of maternal alcohol abuse during pregnancy and association with patients' characteristics. Biomed Res Int 2014;2014:702848.

[31] Bana A, Tabernero MJ, Perez-Munuzuri A, Lopez-Suarez O, et al. Prenatal exposure and its repercussion on newborns. J Neonatal Perinatal Med 2014;7:47–54.

[32] Arndt T, Gruner J, Schrofel S, Stemmerich K. False positive ethyl glucuronide immunoassay screening caused by a propyl alcohol based hand sanitizer. Forensic Sci Int 2012;223: 359–63.

[33] Arndt T, Gierten B, Gussregen B, Werle A, et al. False positive ethyl glucuronide immunoassay screening associated with chloral hydrate medication as confirmed by LC–MS/MS and self-medication. Forensic Sci Int 2009;184:e27–9.

[34] Helander A, Dahl H. Urinary tract infection: a risk factor for false negative urinary ethyl glucuronide but not ethyl sulfate in the detection of recent alcohol consumption. Clin Chem 2005;51:1728–30.

[35] Goll M, Schmitt G, Ganssmann B, Aderjan RE. Excretion profiles of ethyl glucuronide in human after internal dilution. J Anal Toxicol 2002;26:262–6.

[36] Arndt T, Schrofel S, Stemmerich K. Ethyl glucuronide identified in commercial hair tonics. Forensic Sci Int 2013;231:195–8.

[37] Sporkert F, Kharbouche H, Augsburger MP, Klemm C, et al. Positive EtG findings in hair as a result of cosmetic treatment. Forensic Sci Int 2012;218:97–100.

[38] Martins Ferreira L, Binz T, Yegles M. The influence of ethanol containing cosmetics on ethyl glucuronide concentration in hair. Forensic Sci Int 2012;218:123–5.

[39] Kerekes I, Yegles M. Coloring, bleaching and perming: influence on EtG content in hair. Ther Drug Monit 2013;35:527–9.

[40] Center for Substance Abuse Treatment. The role of biomarkers in the treatment of alcohol use disorders. Substance Abuse Treatment Advisory 2006;5:4.

[41] Wojcik MH, Hawthorne JS. Sensitivity of commercial ethyl glucuronide (RTG) testing in screening for alcohol abstinence. Alcohol Alcohol 2007;42:317–20.

[42] Lande RG, Marin B, Chang AS. Clinical application of ethyl glucuronide testing in U.S. Army. J Addict Dis 2011;30:39–44.

[43] Saady JJ, Poklis A, Dalton HP. Production of urinary ethanol after sample collection. J Forensic Sci 1993;38:1467–71.

[44] Helander A, Hagelberg CA, Beck O, Petrini C. Unreliable alcohol testing in a shipping safety programme. Forensic Sci Int 2009;189:e45–7.

[45] Hoiseth G, Karinen R, Christophersen A, Morland J. Practical use of ethyl glucuronide and ethyl sulfate in postmortem cases as markers of antemortem alcohol ingestion. Int J Legal Med 2010;124:143–8.

[46] Hoiseth G, Karinen R, Johnsen L, Normann PT, et al. Disappearance of ethyl glucuronide during heavy putrefaction. Forensic Sci Int 2008;176:147–51.

[47] Lees R, Kingston R, Williams TM, Henderson G, et al. Comparison of ethyl glucuronide in hair with self-reported alcohol consumption. Alcohol Alcohol 2012;47:267–72.

[48] Pianta A, Liniger B, Baumgartner MR. Ethyl glucuronide in scalp and non-head hair: an intra-individual comparison. Alcohol Alcohol 2013;48:295–302.

[49] Hoiseth G, Morini L, Ganss R, Nordal K, et al. Higher levels of hair ethyl glucuronide in patients with decreased kidney function. Alcohol Clin Exp Res 2013;37(Suppl. 1): E14–16.

[50] Sterneck M, Yegles M, von Rothkirch G, Staufer K, et al. Determination of ethyl glucuronide in hair improves evaluating long-term alcohol abstention in liver transplant candidates. Liver Int 2014;34:469–76.

[51] Thierauf A, Kempf J, Perdekamp MG, Auwarter V, et al. Ethyl sulfate and ethyl glucuronide in vitreous humor as postmortem evidence marker for ethanol consumption prior to death. Forensic Sci Int 2011;210:63–8.

[52] Klys M, Wozniak K, Rojek S, Rzepecka-Wozniak E, et al. Ethanol-related death in a child: an unusual case report. Forensic Sci Int 2008;179:e1–4.

[53] Krivankova L, Caslavska J, Malaskova H, Gebauer P, et al. Analysis of ethyl glucuronide in human serum by capillary electrophoresis with sample self-stacking and indirect detection. J Chromatogr A 2005;1081:2–8.

[54] Wurst FM, Kempter C, Seidl S, Alt S. Ethyl glucuronide—a marker of alcohol consumption and relapse marker with clinical and forensic implications. Alcohol Alcohol 1999;34:71–7.

[55] Al-Asmari AI, Anderson RA, Appelblad P. Direct determination of ethyl glucuronide and ethyl sulfate in postmortem urine specimens using hydrophilic interaction liquid chromatography–electrospray ionization tandem mass spectrometry. J Anal Toxicol 2010;34:261–72.

[56] Helander A, Kenan N, Beck O. Comparison of analytical approaches for liquid chromatography/mass spectrometry determination of the alcohol biomarker ethyl glucuronide in urine. Rapid Commun Mass Spectrom 2010;24:1737–43.

[57] Yaldiz F, Daglioglu N, Hilal A, Keten A, et al. Determination of ethyl glucuronide in human hair by hydrophilic interaction liquid chromatography–tandem mass spectrometry. J Forensic Leg Med 2013;20(7):799–802.

[58] Laposata M. Fatty acid ethyl esters: short-term and long-term serum markers of ethanol intake. Clin Chem 1997;43:1527–34.

[59] Sube S, Selavka CM, Mieczkowski T, Pragst F. Fatty acid ethyl ester concentrations in hair and self-reported alcohol consumption in 644 cases from different origins. Forensic Sci Int 2010;196:111–17.

[60] Bertol E, Bravo ED, Vaiano F, Mari F, et al. Fatty acid ethyl esters in hair: correlation with self-reported ethanol intake in 160 subjects and influence of estroprogestin therapy. Drug Test Anal 2014;6(9):930–5.

[61] Hastedt M, Bossers L, Krumbiegel F, Herre S, et al. Fatty acid ethyl esters as alcohol markers: estimating a reliable cut-off point by evaluation of 1,057 autopsy cases. Forensic Sci Med Pathol 2013;9:184–93.

[62] Pragst F, Rothe M, Moench B, Hastedt M, et al. Combined use of fatty acid ethyl esters and ethyl glucuronide in hair for diagnosis of alcohol abuse: interpretation and advantages. Forensic Sci Int 2010;196:101–10.

[63] Pragst F, Yegles M. Determination of fatty acid ethyl esters (FAEE) and ethyl glucuronide (EtG) in hair: a promising way for retrospective detection of alcohol abuse during pregnancy. Ther Drug Monit 2008;30:255–63.

[64] Hartwig S, Auwarter V, Pragst F. Effect of hair care cosmetics on the concentrations of fatty acid ethyl esters in hair as markers of chronically elevated alcohol consumption. Forensic Sci Int 2003;131:90–7.

[65] Gareri J, Appenzeller B, Walasek P, Koren G. Impact of hair care products on FAEE hair concentrations in substance abuse program. Anal Bioanal Chem 2011;400:183–8.

[66] Bakdash A, Burger P, Goecke TW, Fasching PA, et al. Quantification of fatty acid ethyl esters (FARR) and ethyl glucuronide (EtG) in meconium from newborns for detection of alcohol abuse in a maternal health evaluation study. Anal Bioanal Chem 2010;396:2469–77.

[67] Moore C, Jones J, Lewis D, Buchi K. Prevalence of fatty acid ethyl esters in meconium specimens. Clin Chem 2003;49:133–6.

[68] Zelner L, Hutson JR, Kapur BM, Feig DS, et al. False positive meconium test results for fatty acid ethyl esters secondary to delayed sample collections. Alcohol Clin Exp Res 2012;36:1497–506.

[69] Auwarter V, Sporkert F, Hartwig S, Pragst F, et al. Fatty acid ethyl esters in hair as markers of alcohol consumption: segmental hair analysis of alcoholics, social drinkers and teetotalers. Clin Chem 2001;47:2114–23.

[70] Hutson JR, Rao C, Fulga N, Aleksa K, et al. An improved method for rapidly quantifying fatty acid ethyl esters in meconium suitable for prenatal alcohol screening. Alcohol 2011;45:193–9.

[71] Roehsig M, de Paula DM, Moura S, Diniz EM, et al. Determination of eight fatty acid esters in meconium samples by headspace solid phase microextraction and gas chromatography–mass spectrometry. J Sep Sci 2010;33:2115–22.

[72] Pichini S, Pellegrini M, Gareri J, Koren G, et al. Liquid chromatography–tandem mass spectrometry for fatty acid ethyl esters in meconium: assessment of prenatal exposure of alcohol in two European cohorts. J Pharm Biomed Anal 2008;48:927–33.

[73] Colley WC, Sung TC, Roll R, Jenco J, et al. Phospholipase D2, a distinct phospholipase D isoform with novel regulatory properties that provokes cytoskeletal reorganization. Curr Biol 1997;7:191–201.

[74] Helander A, Zheng Y. Molecular species of the alcohol biomarker phosphatidylethanol in human blood measured by LC–MS. Clin Chem 2009;55:1395–405.

[75] Gnann H, Engelmann C, Skopp G, Winkler M, et al. Identification of 48 homologues of phosphatidylethanol in blood by LC–ESI–MS/MS. Anal Bioanal Chem 2010;396:2415–23.

[76] Isaksson A, Walther L, Hansoon T, Andersson A, et al. Phosphatidylethanol in blood (B-PEth): a marker for alcohol use and abuse. Drug Test Anal 2011;4:195–200.

[77] Varga A, Hansson P, Lundqvist C, Alling C. Phosphatidylethanol in blood as a marker of ethanol consumption in healthy volunteers: comparison with other markers. Alcohol Clin Exp Res 1998;22:1832–7.

[78] Gnann H, Weinmann W, Thierauf A. Formation of phosphatidylethanol and its subsequent elimination during and extensive drinking experiment over 5 days. Alcohol Clin Exp Res 2012;36:1507–11.

[79] Marques P, Hansson T, Isaksson A, Walther L, et al. Detection of phosphatidylethanol (PEth) in the blood of drivers in an alcohol ignition interlock program. Traffic Inj Prev 2011;12:136–41.

[80] Wurst FM, Thon N, Aradottir S, Hartman S, et al. Phosphatidylethanol: normalization during detoxification, gender aspects and correlation with other biomarkers and self-report. Addict Biol 2010;15:88–95.

[81] Gunnarsson T, Karlsson A, Hansson P, Johnson G, et al. Determination of phosphatidylethanol in blood from alcoholic male using high-performance liquid chromatography and evaporative light or electrospray mass spectrometric detection. J Chromatogr B Biomed Appl 1998;705: 243–9.

[82] Kwak HS, Han JY, Kim MH, Ryu HM, et al. Blood levels of phosphatidylethanol in pregnant women reporting positive alcohol ingestion, measured by an improved LC–MS/MS analytical method. Clin Toxicol (Phila) 2012;50:886–91.

[83] Kwak HS, Han JY, Choi JS, Ahn HK, et al. Characterization of phosphatidylethanol blood concentrations for screening alcohol consumption in early pregnancy. Clin Toxicol (Phila) 2014;52:25–31.

[84] Bakhireva LN, Cano S, Rayburn WF, Savich RD, et al. Advanced gestational age increases serum carbohydrate-deficient transferrin levels in abstinent pregnant women. Alcohol Alcohol 2012;47:683–7.

[85] Skipper GE, Thon N, Dupont RL, Baxter L, et al. Phosphatidylethanol: the potential role in further evaluating low positive urinary ethyl glucuronide and ethyl sulfate results. Alcohol Clin Exp Res 2013;37:1582–6.

[86] Hansson P, Varga A, Krantz P. Phosphatidylethanol in post-mortem blood as a marker of previous heavy drinking. Int J Legal Med 2001;115:158–61.

[87] Stewart SH, Law TL, Randall PK, Newman RR. Phosphatidylethanol and alcohol consumption in reproductive age woman. Alcohol Clin Exp Res 2010;34:488–92.

[88] Hartmann S, Aradottir S, Graf M, Wiesbeck M, et al. Phosphatidylethanol as a sensitive and specific biomarker: comparison with gamma-glutamyl transpeptidase, mean corpuscular volume, and carbohydrate-deficient transferrin. Addict Biol 2007;12:81–4.

[89] Varga A, Alling C. Formation of phosphatidylethanol in vitro in red blood cells from healthy volunteers and chronic alcoholics. J Lab Clin Med 2002;140:79–83.

[90] Nissinen AE, Laitinen LM, Kakko S, Helander A, et al. Low plasma antibodies specific for phosphatidylethanol in alcohol abusers and patients with alcoholic pancreatitis. Addict Biol 2012;17:1057–67.

[91] Viel G, Boscolo-Berto R, Cecchetto G, Faris P, et al. Phosphatidylethanol in blood as a marker of chronic alcohol use: a systematic review and meta-analysis. Int J Mol Sci 2012;13:14788–812.

[92] Nissinen AE, Makala SM, Varistor JT, Liisanantti MK, et al. Immunochemical detection of in vitro phosphatidylethanol—A biomarker with monoclonal antibodies. Alcohol Clin Exp Res 2008;32:921–8.

[93] Yon C, Han JS. Analysis of trimethylsilyl derivatization products of phosphatidylethanol by gas chromatography–mass spectrometry. Exp Mol Med 2000;32:243–5.

[94] Cabarcos P, Angel Cocho J, Moreda A, Miguez M, et al. Application of dispersive liquid–liquid microextraction for the determination of phosphatidylethanol in blood by liquid chromatography tandem mass spectrometry. Talanta 2013;111:189–95.

[95] Faller A, Richter B, Kluge M, Koenig P, et al. Stability of phosphatidylethanol in spiked and authentic whole blood and matching dried blood spot. Int J Legal Med 2013;127:603–10.

[96] Faller A, Richter B, Kluge M, Koenig P, et al. LC–MS/MS analysis of phosphatidylethanol in dried blood spots versus conventional blood specimens. Anal Bioanal Chem 2011;410:1163–6.

[97] Bakhireva LN, Leeman L, Savich RD, Cano S, et al. The validity of phosphatidylethanol in dried blood spots of newborns for the identification of prenatal alcohol exposure. Alcohol Clin Exp Res 2014;38:1078–85.

[98] Stewart SH, Koch DG, Willner IR, Randall PK, et al. Hair ethyl glucuronide is highly sensitive and specific for detecting moderate to heavy drinking in patients with liver disease. Alcohol Alcohol 2013;48:83–7.

[99] Stewart SH, Koch DG, Burgess DM, Willner IR, et al. Sensitivity and specificity of urinary glucuronide and ethyl sulfate in liver disease patients. Alcohol Clin Exp Res 2013;37:150–5.

[100] Wurst FM, Wiesbeck GA, Metzger JM, Weinmann W. On sensitivity, specificity and the influence of various parameters on ethyl glucuronide levels in urine: results from the WHO/ISBRA study. Alcohol Clin Exp Res 2004;28:1220–8.

[101] Morini L, Groppi A, Marchei E, Vagnarelli F, et al. Population baseline of meconium ethyl glucuronide and ethyl sulfate concentrations in newborns of nondrinking women in 2 Mediterranean cohorts. Ther Drug Monit 2010;32:359–63.

[102] Pichini S, Morini L, Marchei E, Palmi I, et al. Ethyl glucuronide and ethyl sulfate in meconium to assess gestational ethanol exposure: preliminary results in two Mediterranean cohorts. Can J Clin Pharmacol 2009;16:e370–5.

[103] Chan D, Bar-Oz B, Pellerin B, Paciorek C, et al. Population baseline of meconium fatty acid ethyl esters among infants of nondrinking women in Jerusalem and Toronto. Ther Drug Monit 2003;25:271–8.

CHAPTER 9

Less Commonly Used Alcohol Biomarkers and Proteomics in Alcohol Biomarker Discovery

CONTENTS

9.1 Introduction 221
9.2 Total Sialic Acid in Serum as Alcohol Biomarker 221
9.2.1 Other Causes of Elevated Plasma Sialic Acid Concentrations.. 224
9.2.2 Laboratory Determination of Total Sialic Acid 225
9.3 Sialic Acid Index of Apolipoprotein J as Alcohol Biomarker 227
9.3.1 Laboratory Methods for the Determination of the Sialic Acid Index of Plasma Apolipoprotein J 229
9.4 5-Hydroxytryptophol as Alcohol Biomarker 230
9.4.1 Laboratory Methods for Determining 5-HTOL and 5-HIAA 235
9.5 Other Alcohol Biomarkers 236

9.1 INTRODUCTION

There are several alcohol biomarkers that are used less often than common indirect alcohol state biomarkers such as γ-glutamyl transferase (GGT), alanine aminotransferase (ALT), aspartate aminotransferase (AST), mean corpuscular volume (MCV), carbohydrate deficient transferrin (CDT), and β-hexosaminidase. Direct alcohol biomarkers such as ethyl glucuronide, ethyl sulfate, and fatty acid ethyl esters are also widely used in both clinical and forensic settings. Acetaldehyde, the first metabolite of alcohol, is a very reactive molecule that forms adducts with proteins. Hemoglobin–acetaldehyde adduct is a direct alcohol biomarker but is less commonly used than ethyl glucuronide, ethyl sulfate, and fatty acid ethyl esters due to technical difficulties associated with its measurement. Another direct alcohol biomarker, phosphatidylethanol, has almost 100% specificity and is gaining more applications in both clinical and forensic investigations. There are other alcohol biomarkers, such as total sialic acid in serum, plasma sialic acid index of apolipoprotein J (Apo J), 5-hydroxytryptophol, cholesteryl ester transfer protein, homocysteine, and circulating cytokines, but these are used less often in clinical and legal investigations compared to biomarkers such as ethyl glucuronide. Various characteristics of three biomarkers—total sialic acid in serum, plasma sialic acid index of Apo J, and 5-hydroxytryptophol—are listed in Table 9.1.

9.2 TOTAL SIALIC ACID IN SERUM AS ALCOHOL BIOMARKER

Sialic acids comprise a family of more than 50 naturally occurring carbohydrates that are derivatives of the 9-carbon monosaccharide neuraminic acid (5-amino-3,5-dideoxy-D-glycero-D-galactononulsonic acid). One branch of the

A. Dasgupta: Alcohol and Its Biomarkers. DOI: http://dx.doi.org/10.1016/B978-0-12-800339-8.00009-2

9.6 Proteomics in Alcohol Biomarker Discovery 237
9.6.1 Specific Proteins Identified as Alcohol Biomarkers Using the Proteomics Approach.................... 238
9.7 Conclusions 241
References 241

Table 9.1 Characteristics of Plasma Total Sialic Acid, Sialic Acid Index of Apo J, and 5-Hydroxytryptophol as Alcohol Biomarkers

Alcohol Biomarker	Cutoff Level	Comments
Plasma (serum) total sialic acid	Measured in serum or plasma with cutoff value of 77.8 mg/dL in females and 80 mg/dL in males [6].	Total sialic acid value in serum or plasma is increased following excessive drinking. After 3 weeks of abstinence, values are reduced more significantly in women. However, various disease states may increase total plasma sialic acid levels. Plasma total sialic acid is an indirect alcohol biomarker for heavy alcohol consumption. Sensitivity is 57.7% in females and 47.8% in males. Specificity is 95.5% in females and 81.3% in males.
Sialic acid index of plasma Apo J	Measured in HDL fraction of serum or plasma, but cutoff value is not yet established.	Alcohol interferes with incorporation of sialic acid moiety in the Apo J molecule. Usually, with consumption of 50–60 g of alcohol per day for 30 days, sialic acid index of plasma Apo J may be reduced by 50%, but with abstinence, values may return to normal in 8 weeks (half-life, 4–5 weeks). Specificity is approximately 100%, and sensitivity is 90–92%. Sialic acid index of plasma Apo J is an indirect alcohol biomarker for heavy alcohol consumption.
5-Hydroxytryptophol	Measured in urine and usually expressed as the ratio of picomoles of 5-hydroxytryptophol to nanomoles of 5-hydroxyindole-3-acetic acid. A value of 15 pmol/nmol is considered a cutoff value.	Although considered as a 24-hr biomarker for detecting alcohol consumption, in one report the median detection window was 9.8 hr (range, 8–11 hr). In general, consumption of a small amount of alcohol (<10 g) may not be detected, but consumption of 50 g should be detected by this marker. Specificity is almost 100%, and sensitivity is 77%.

sialic acid family is N-acetylated to form *N*-acetylneuraminic acids, which are the most common form of sialic acids. In addition, these *N*-acetylneuraminic acids are almost the only forms of sialic acid found in humans. Sialic acids are rarely found in free form; they are commonly present as components of oligosaccharide chains of mucins, glycoproteins, and glycolipids, where they influence glycoprotein conformation either to serve as recognition sites or to mask recognition sites. In general, the central nervous system has the highest concentration of sialic acid, mostly in gangliosides and glycoproteins [1]. In human serum, the majority of sialic acid is *N*-acetylneuraminic acid, which bears an acetyl group at the fifth carbon atom. Other forms are found only in very trace amounts. Total sialic acid in serum is the sum of protein-bound sialic acid, lipid-bound sialic acid, and free sialic acid, although free sialic acid represents only a very small fraction of total sialic acid. The molecular weights of sialic acids vary with their substitutions, but in general the average molecular weight is assumed to be 328.2 for sialic acid in human serum or plasma. The normal total sialic acid in serum or plasma is 1.58–2.22 mmol/L (52–73 mg/dL), and that of free sialic acid in serum or plasma is 0.5–3 μmol/L (164–985 ng/mL) [2]. Gopaul and Crook reported that total sialic acid level in serum varied from 1.6 to 2.3 mmol/L [3]. Priego-Capote *et al.* reported that normal levels of sialic acid in serum of healthy volunteers ranged from 1.0 to 1.7 mmol/L [4].

Chronic alcohol consumption inhibits glycosylation of many proteins, such as transferrin, fibrinogen, and complement proteins. CDT is used as an alcohol biomarker (see Chapter 6). Because alcohol interferes with glycosylation, it is expected that total sialic acid in serum should increase in individuals who consume alcohol on a regular basis. Therefore, total sialic acid in serum or plasma is an indirect alcohol biomarker. Romppanen *et al.* studied serum sialic acid concentration in 51 alcoholics and 20 healthy individuals. The mean sialic acid concentration was significantly elevated in alcoholics (1.449 mmol/L; 47.5 mg/dL) compared to healthy controls, in whom the mean value was 1.154 mmol/L (37.8 mg/dL). The authors proposed a cutoff value of 1.425 mmol/L (46.7 mg/dL); at this cutoff, sensitivity and specificity of serum sialic acid were 51 and 100%, respectively [5].

Sillanaukee *et al.* studied 38 social drinkers and 77 alcoholics and reported that mean sialic acid concentration in female social drinkers ($n = 22$) was 60 mg/dL, whereas in female alcoholics ($n = 26$) it was significantly elevated to 82 mg/dL. Similarly, sialic acid levels were also significantly elevated in male alcoholics compared to male social drinkers (mean in social drinkers: 64 mg/dL, $n = 16$; mean in alcoholics: 85 mg/dL, $n = 23$). Female social drinkers had self-reported alcohol consumption of less than 30 g per week, whereas female alcoholics consumed more than 800 g of alcohol per week. The cutoff concentration suggested by the authors was 77.8 mg/dL in females and 80 mg/dL in males. Male social drinkers consumed less than 50 g of alcohol per week, whereas male alcoholics consumed more than 1000 g of alcohol per week. During follow-up of 28 alcoholics participating in inpatient treatment, serum sialic acid levels were decreased after 3 weeks of treatment in female alcoholics, but in male alcoholics, such decreases were less remarkable. The sensitivity and specificity of sialic acid as an alcohol biomarker were 57.7 and 95.5%, respectively, in women and 47.8 and 81.3%, respectively, in men [6]. In another study, which involved 75 male alcoholics with mean alcohol consumption of 920 g per week (range, 120–3160 g/week) and mean alcohol dependency of 13 years (range, 1–33 years), the plasma total sialic acid levels were significantly elevated in alcoholics (mean in controls, 67.7 mg/dL; mean in alcoholics, 73.2 mg/dL). Although values of other alcohol biomarkers, such as %CDT, ALT, AST, GGT, and MCV, were reduced statistically with abstinence, no such correlation was observed with total plasma sialic acid concentration in male alcoholics. The authors concluded that %CDT had the highest accuracy among the alcohol biomarkers tested [7].

9.2.1 Other Causes of Elevated Plasma Sialic Acid Concentrations

Total sialic acid in plasma may be increased in a variety of diseases [2,8], which are listed in Box 9.1. Crook *et al.* reported elevated serum total sialic acid concentrations in elderly (mean age, 80.1 years) compared to younger subjects (mean age, 40.3 years). The mean total sialic acid level in serum of elderly subjects was 2.41 mmol/L (79.0 mg/dL), and that in younger subjects was 2.04 mmol/L (66.9 mg/dL) [9].

BOX 9.1 POPULATIONS AND DISEASES IN WHICH PLASMA OR SERUM TOTAL SIALIC ACID CONCENTRATIONS ARE INCREASED

- Elderly
- Various tissue cancers (oral, breast, gastrointestinal tract, lung, stomach, colorectal, gallbladder, thyroid, adrenal, pancreatic, prostate, endometrial, and ovarian) and certain leukemias such as chronic lymphocytic leukemia
- Bacterial infection
- Chronic glomerulonephritis
- Chronic renal failure
- Crohn's disease
- Type 1 and type 2 diabetes
- Cardiovascular diseases
- Atherosclerosis and myocardial infarction
- Sympathetic ophthalmitis
- Inflammatory disease such as rheumatoid arthritis
- Inherited disorders of sialic acid. (Measuring urinary bound and free sialic acid is useful as a screening test for inherited disorders of lysosomal metabolism.)

More than four decades ago, specific tumor characteristics were ascribed to the increased expression of sialic acid sugars on the surface of the cancer cells, which led to the definition of sialic acid as a potential therapeutic target [10]. Narayanan commented that either total sialic acid in serum or plasma or lipid-bound sialic acid can be used as a tumor marker [11]. Dwivedi *et al.* studied plasma lipid-bound sialic acid concentrations in patients with different types of cancer, including breast cancer, lung cancer, colon cancer, ovarian cancer, prostate cancer, leukemia, gastrointestinal cancer, thyroid cancer, pancreatic cancer, and adrenal cancer, as well as normal volunteers and patients with nonmalignant disease. Mean plasma lipid-bound sialic acid concentration was 17.7 mg/dL, and the highest mean lipid-bound sialic acid concentration was observed in patients with adrenal cancer (119.5 mg/dL). Only 2 of 114 cancer patients showed lipid-bound plasma sialic acid levels within the normal range. The authors reported that plasma lipid-bound sialic acid had a sensitivity of 98.2% as a cancer biomarker [12]. Serum sialic acid is also increased in patients with endometrial cancer compared to controls (mean value in cancer patients: 2.38 mmol/L, 78.1 mg/dL; mean value in controls: 1.52 mmol/L, 49.8 mg/dL) [13].

Increased plasma total sialic acid levels are a marker of cardiovascular diseases in patients receiving dialysis. In one study, the authors observed that mean plasma total sialic acid was 91.2 mg/dL in dialysis patients with cardiovascular disease compared to a mean of 82.0 mg/dL in stable renal dialysis patients [14]. Serum sialic acid concentration is elevated in patients with insulin-dependent diabetes mellitus (type 1 diabetes) with normal albumin concentration in urine compared to healthy controls. Serum sialic acid concentrations were further increased in patients with incipient nephropathy and clinical nephropathy [15]. Serum sialic acid concentration is also increased in non-insulin-dependent diabetes mellitus (type 2 diabetes), and this may be related to diabetic complications. Increased serum sialic acid is also an independent risk factor for cardiovascular mortality. Serum or plasma sialic acid levels are increased in myocardial infarction [8]. Like C-reactive protein, sialic acid is an inflammatory marker that is increased in various inflammatory diseases. Von Versen-Hoeynck *et al.* reported elevated plasma levels of sialic acid in pregnancy [16]. Moreover, in liver cirrhosis, free sialic acid in serum is also increased [17].

9.2.2 Laboratory Determination of Total Sialic Acid

Methods such as colorimetric, enzymatic, fluorescence, and chromatographic methods including thin-layer chromatography (TLC) can be used to determine total sialic acid concentrations in serum or plasma. One of the longest-used colorimetric assays for sialic acid is the thiobarbituric acid assay, in which sialic acid is oxidized with sodium periodate in concentrated phosphoric acid, the periodate oxidation product then being coupled with thiobarbituric acid; the chromophore can be extracted in cyclohexanone and measured spectrophotometrically at 549 nm [18]. Orcinol reacts with sialic acid in the presence of ferric ions and hydrochloric acid. The resultant chromophore formed after heating at 100°C can be extracted after cooling in isoamyl alcohol and then measured spectrophotometrically at 572 nm. However, this method suffers from interference (hexoses, pentoses, and uronic acid) [8]. Resorcinol can also be used to determine sialic acid levels in serum colorimetrically, where the chromophore formed in the presence of cupric ion and resorcinol can be measured at 580-nm wavelength. Crook *et al.* compared a commercially available enzyme assay of sialic acid with thiobarbituric acid and resorcinol methods and found that the performances of these three assays were similar. This enzymatic assay can be easily adapted to an automated analyzer [19].

Şimpson *et al.* described an adaptation of an enzymatic assay on microtiter plates for determination of sialic acid in serum. The assay was based on the release of sialic acid from glycoconjugates by neuraminidase, cleavage of sialic acid by *N*-acetyl neuraminic acid aldolase to release acyl-mannosamine and pyruvate, and finally oxidation of pyruvate to generate hydrogen peroxide using pyruvate oxidase. Hydrogen peroxide was determined colorimetrically by the red product formed in the presence of peroxidase, 4-aminoantipyrine, and *N*-ethyl-*N*-2-hydroxyethyl-3-toluidine [20]. Alternatively, pyruvate generated can also be measured using lactate dehydrogenase enzyme and 340-nm wavelength. Many fluorescence methods for the determination of sialic acid have been reported. One of the earliest assays utilized heating of sialic acid in the presence of diluted hydrochloric acid with 3,5-diaminobenzoic acid, which produced an intensely green fluorescent compound. Other fluorescence assays involved an adaptation of the periodic acid/thiobarbituric acid colorimetric assay in which the chromophore was excited at 550 nm and emission measured at 570 nm. The formation of fluorescent product using pyridoxine has also been discussed [8].

Sialic acid in serum or plasma can also be determined by chromatographic methods such as gas chromatography combined with mass spectrometry (GC–MS), high-performance liquid chromatography (HPLC), and liquid chromatography combined with mass spectrometry (LC–MS) or tandem mass spectrometry (LC–MS/MS). Sugawara *et al.* measured the concentration of *N*-acetylneuraminic acid, the major sialic acid in serum, using *N*-[$^{2}H_{3}$]-acetylneuraminic acid (stable isotope of *N*-acetylneuraminic acid) as the internal standard and GC–MS. Both *N*-acetylneuraminic acid and the internal standard were converted into trimethylsilyl derivative prior to analysis. The molecular weight minus the COOTMS fragment at m/z 624 was used for monitoring the trimethylsilyl derivative of *N*-acetylneuraminic acid, and the m/z 627 fragment was used for the internal standard [21]. Romppanen *et al.*

used high-performance anion exchange chromatography and pulsed amperometric detection to determine serum sialic acid levels in healthy volunteers. Total sialic acid in serum was measured after liberating sialic acid by acid hydrolysis [22]. Various derivatization methods have been described for analysis of sialic acid using chromatography, including 1,2-diamino-4,6-dimethoxybenzene, 1,2-diamino-4,5-methylenedioxybenzene, *p*-toluenesulfonyl chloride, and *ortho*-phenylenediamine. In one report, the authors described the quantitative determination of *N*-acetylneuraminic acid and *N*-glycolylneuraminic acid in human serum and urine using LC–MS/MS [4]. However, *N*-acetylneuraminic acid in serum can also be analyzed using chemiluminescence detection. After releasing *N*-acetylneuraminic acid using acid hydrolysis with hydrochloric acid, Ishida *et al.* derivatized *N*-acetylneuraminic acid along with the internal standard *N*-glycolylneuraminic acid using 4,5-diaminophthalhydrazide dichloride, a chemiluminescent derivatizing agent for α-keto acids. The authors used a reverse-phase column for chromatographic separation; for detection, chemiluminescence was produced by the reaction with hydrogen peroxide in the presence of potassium hexacyanoferrate [23]. After releasing sialic acids in serum by hydrolysis, Li converted them to highly fluorescent compounds in borate buffer with malononitrile. The reaction mixture was separated with an octadecyl-bonded silica column using a mobile phase of methanol and ammonium acetate buffer (15:85 by volume). Measurement of fluorescence activity was achieved at 434 nm (extinction wavelength, 357 nm). The linearity of the assay was 30–1000 ng/mL of serum sialic acid with a detection limit of 2 ng/mL [24].

9.3 SIALIC ACID INDEX OF APOLIPOPROTEIN J AS ALCOHOL BIOMARKER

Apolipoprotein J (clusterin) is a disulfide-linked heterodimeric, highly sialylated glycoprotein expressed in a wide variety of tissue and found in all human fluids, including serum and plasma. The physiological function of Apo J is still under investigation, but it has been implicated in several diverse processes, including lipid transport, complement inhibition, tissue remodeling, membrane recycling, cell–cell interaction, sperm maturation, and promotion or inhibition of apoptosis [25]. Apo J has a molecular weight of approximately 70 kDa (Apo Jα, 34–36 kDa; Apo Jβ, 36–39 k Da) and is associated with high-density lipoprotein complex (HDL_2 and HDL_3). Approximately 30% of the Apo J molecule is carbohydrate, with 28 moles of sialic acid associated with one mole of Apo J [26]. Therefore, Apo J contains a much higher amount of sialic acid compared to transferrin (each molecule of Apo J contains 26–28 sialic acid molecules, whereas each transferrin molecule contains only 4–6 sialic acid molecules).

Long-term exposure to alcohol inhibits glycosylation of various proteins, such as incorporation of sialic acid in the transferrin molecule. Therefore, CDT can be used as an alcohol biomarker. Because Apo J contains a higher amount of sialic acid, it can be hypothesized that sialylation of the Apo J molecule should be vulnerable to the deleterious effect of alcohol. Synthesis of the mature Apo J molecule requires the addition of sugars to the molecule in a sequential manner and termination of the attachment with sialic acid molecules. Therefore, in alcoholics, activities of enzymes such as sialyltransferase may be reduced, thus inhibiting incorporation of sialic acid molecules in the mature Apo J molecule. Gong *et al.* reported downregulation of the liver sialyltransferase gene in alcoholics that caused defective glycosylation of a number of proteins, including Apo E and Apo J [27]. As a result, the sialic acid index of plasma Apo J can be used as an alcohol biomarker, where mole of sialic acid per mole of Apo J should be reduced in alcoholics compared to healthy individuals. Value is expressed as moles of sialic acid per mole of Apo J; this is a number with no unit because it is a ratio. Therefore, this value represents the approximate number of sialic acid residues in 1 molecule of Apo J; for example, a value of 14 indicates that approximately 14 molecules of sialic acid are associated with 1 molecule of Apo J. However, the sialic index of plasma Apo J is an indirect alcohol biomarker.

Using a rat model, Ghosh *et al.* observed that following treatment of rats with alcohol, no significant difference in sialylation index of plasma Apo J was observed between rats exposed to alcohol and control rats after 4 weeks. However, a 24% reduction in sialic acid in Apo J was observed after 6 weeks of alcohol treatment (18.6 in rats exposed to ethanol vs. 25.2 in controls). After 8 weeks of treatment, a more significant decrease of 44% was observed (14.0 in rats treated with alcohol vs. 25.2 in controls). Furthermore, a significant recovery was observed following alcohol withdrawal, with 38, 78, 84, and 96% recovery, respectively, after 1, 2, 3, and 4 weeks of alcohol abstinence. These changes in the sialic aid index of Apo J were accompanied by a similar pattern of changes in the enzymatic activities of hepatic sialyltransferase and plasma sialidase in rats undergoing chronic alcohol treatment, withdrawal, and an abstinence period. For example, in rats, the activities of liver sialyltransferase, which catalyzes the incorporation of sialic acid in Apo J, were reduced from 17.4 nmol/g of liver/hr (mean value) on day 0 to 8.4 nmol/g of liver/hr after 8 weeks of chronic alcohol exposure. In contrast, plasma sialidase activities were increased from 92.4 to 212.6 nmol/mL plasma/hr after 8 weeks of chronic exposure of alcohol. The plasma sialidase enzyme catalyzes the removal of sialic acid from the Apo J molecule. The authors concluded that the sialylation index of Apo J is suitable to evaluate ethanol exposure [28].

Because Apo J contains 26–28 sialic acid residues per mole, it is expected that the sialylation index of Apo J should also be a good alcohol biomarker of alcohol in humans. In one study, the authors reported that in human subjects, intake of alcohol for 30 days led to an almost 50% decrease in sialic acid index in Apo J. Patients who consumed 50–60 g of alcohol per day showed a mean sialic index of Apo J (moles of sialic acid per mole of Apo J) of 14 ($n = 15$), whereas in controls, the value was 28 ($n = 15$). However, when such subjects underwent detoxification, the mean value was 21 after 6 weeks and 26 after 8 weeks , indicating 85.7% recovery of sialic acid in the Apo J molecule after 8 weeks. The authors also measured CDT in these subjects and noted that the plasma sialic acid index of Apo J responded to changes in alcohol consumption with higher sensitivity than observed with CDT [29]. The specificity of the sialic acid index of Apo J is approximately 100%, the sensitivity is approximately 90–92%, and the half-life is 4 or 5 weeks [30].

Wurst *et al.* reported that in five male alcohol-dependent patients, the sialic acid indexes of plasma Apo J and phosphatidylethanol were positive; determinations of urine ethyl glucuronide and ethyl sulfate were positive, but serum GGT and MCV were positive in only three of five patients. The individual sialic acid index of Apo J increased during detoxification, and a 44.7% increase was observed after 28 days. Both urinary ethyl glucuronide and ethyl sulfate levels were positive during initiation of the detoxification process, but both ethyl glucuronide and ethyl sulfate concentrations were reduced significantly within 3 days of detoxification, with some values reduced below the cutoff levels. The authors used a cutoff of 100 ng/mL for ethyl glucuronide and 110 ng/mL for ethyl sulfate. They did not establish a cutoff value for the sialic acid index of plasma Apo J, but they observed a significant trend, with average subjects' sialic acid index of plasma Apo J increasing by 1.22 moles of sialic acid per mole of Apo J per week. The authors concluded that the sialic acid indexes of both Apo J and phosphatidylethanol hold potential as markers of heavy alcohol consumption [31].

9.3.1 Laboratory Methods for the Determination of the Sialic Acid Index of Plasma Apolipoprotein J

Measurement of the sialic acid index of plasma Apo J is a multistep process in which the HDL fraction is passed through an immunoaffinity column containing Sepharose 4B matrix with a covalently attached antibody to Apo J. Purified Apo J is eluted from the column using solvent with differential ionic strength. Then, purified Apo J is subjected to sulfuric acid-induced hydrolysis (90 min at 80°C) to release sialic acid from Apo J, followed by quantification of sialic acid using a colorimetric assay such as the thiobarbituric acid assay. Apo J is also quantified using a standard protein assay. Wurst *et al.* precipitated Apo B-containing lipoproteins using manganese and heparin, followed by precipitation of HDL_2 using dextran sulfate (molecular weight, 15 kDa). The precipitated HDL_2 fraction containing Apo J was further purified by

Sephadex column and then passed through a Sepharose 4B matrix conjugated with anti-Apo J (human clusterin). The bound Apo J was eluted with phosphate buffer containing varyied strengths of sodium chloride (0.5–1.0 molar). All eluted fractions were monitored for protein at 280 nm. Then protein fractions were concentrated and authenticity was verified by polyacrylamide gel electrophoresis against known molecular weight markers. Next, the protein fraction was quantified and the Apo J fraction was hydrolyzed with 0.1 N sulfuric acid, and sialic acid content was measured by thiobarbituric acid colorimetric assay [31]. Ghosh *et al.* observed that sialic acid content was 400 pmol/μg of Apo J in healthy individuals. Assuming the molecular weight of Apo J to be 70 kDa, 1 μg of Apo J is equivalent to 0.0142 nmol or 14.2 pmol. Therefore, 400 pmol of sialic acid was associated with 14.28 pmol of Apo J, which translates into 28 pmol of sialic acid associated with 1 pmol of Apo J or 28 mol of sialic acid associated with 1 mol of Apo J. Therefore, the sialic acid index of plasma Apo J (sialic acid index: moles of sialic acid per mole of Apo J) was 28. In alcoholics, the value was reduced to 14 (average value in alcoholics) [29].

9.4 5-HYDROXYTRYPTOPHOL AS ALCOHOL BIOMARKER

Serotonin (5-hydroxytryptamine), a neuromodulator with both neuroendocrine and neurotransmitter function, is synthesized in serotonergic neurons in the central nervous system and in the enterochromaffin cells throughout the gastrointestinal tract. Normally, serotonin is metabolized to 5-hydroxyindole-3-acetaldehyde by the action of monoamine oxidase enzyme (monoamine oxidase A has the highest affinity for serotonin). Then 5-hydroxyindole-3-acetaldehyde is either oxidized into 5-hydroxyindole-3-acetic acid (5-HIAA) or reduced to 5-hydroxytryptophol (5-HTOL). Oxidation of 5-hydroxyindole-3-acetaldehyde by the action of aldehyde dehydrogenase is the major metabolic pathway, where nicotinamide adenine dinucleotide (NAD) is used as a cofactor. Reduction of 5-hydroxyindole-3-acetaldehyde into 5-HTOL is catalyzed by alcohol dehydrogenase and, to some extent, aldehyde reductase. In urine, 5-HTOL is a minor metabolite representing less than 1% of serotonin turnover, whereas 5-HIAA is the major metabolite. However, alcohol consumption causes a shift in the serotonin metabolism from 5-HIAA to 5-HTOL due to competitive inhibition of aldehyde dehydrogenase by acetaldehyde, the major metabolite of alcohol. In addition, increased levels of NADH due to alcohol metabolism also favor 5-HTOL formation [32]. This shift in serotonin metabolism in the presence of ethanol is presented in Figure 9.1. Therefore, 5-HTOL is an indirect alcohol biomarker. In general, serotonin and 5-HIAA are found in urine predominantly in free form, whereas 5-HTOL is mainly excreted as conjugates with glucuronic acid [33]. However, sulfate conjugate of 5-HTOL is

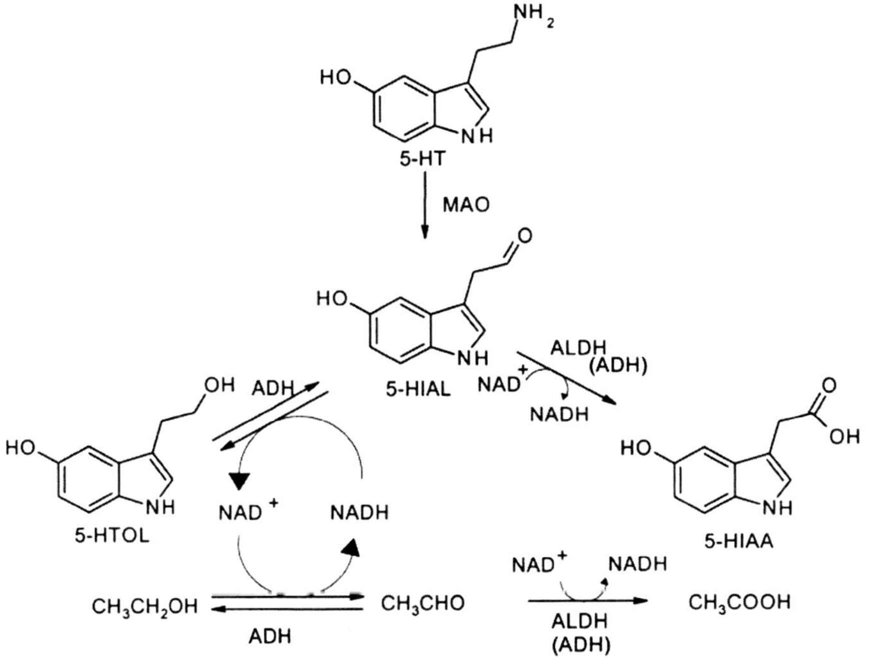

FIGURE 9.1 Metabolic pathway of serotonin in the presence of ethanol.
Serotonin (5-HT) is metabolized in the first step by monoamine oxidase (MAO) into 5-hydroxyindole-3-acetaldehyde (5-HIAL). Normally, 5-HIAL is metabolized to 5-hydroxyindole-3-acetic acid (5-HIAA) by aldehyde dehydrogenase (ALDH). In addition, a minor metabolite, 5-hydroxytryptophol (5-HTOL), is formed by the action of alcohol dehydrogenase (ADH) and to some extent by aldehyde reductase. Conversion of ethanol (CH_3CH_2OH) by ADH to acetaldehyde (CH_3CHO) increases the ratio of NADH:NAD, thus favoring conversion of 5-HIAL to 5-HTOL. In addition to reduction, ADH has the capability to oxidize aldehyde. However, acetaldehyde formed due to ethanol metabolism is converted mostly by ALDH to acetic acid (CH_3COOH). Source: *Svensson et al.*, [32]. © *FEBS/John Wiley. Reprinted with permission.*

also found in urine. Stephanson *et al.* reported that in 13 healthy humans without previous intake of alcohol, urinary 5-hydroxytryptophol glucuronide (GTOL) levels varied from 38 to 327 nmol/L. The authors also determined that the ratio of GTOL with 5-HIAA (pmol/nmol) in these subjects varied from 2.6 to 12.0 [34]. In another study by the same group, urinary GTOL levels varied between 14 and 197 nmol/L, urinary 5-HIAA concentrations varied between 2 and 45 μmol/L, and the GTOL/5-HIAA ratio (nmol/μmol) varied from 2.5 to 10.5 (mean, 5.1). However, in drinkers, the ratio (nmol/μmol) varied from 15.3 to 382 (median, 33.2) [35]. Although the authors used the nanomole (GTOL) to micromole (5-HIAA) ratio, other investigators have used the ratio of picomoles of 5-HTOL to nanomoles of 5-HIAA to investigate the effect of alcohol on urinary secretion of 5-HTOL and 5-HIAA. For the determination of 5-HTOL in urine, glucuronide conjugate must be hydrolyzed to liberate free 5-HTOL prior to analysis. Alternatively, GTOL in urine can be determined directly by LC–MS/MS without hydrolysis if the GTOL/5-HIAA ratio is used as an alcohol biomarker.

The ratio of 5-HTOL and 5-HIAA in urine is usually determined to evaluate alcohol consumption by an individual because the ratio increases significantly in people who consume alcohol. The advantages of using this ratio are that it compensates for urine dilution and accounts for dietary sources of serotonin. In general, it is assumed that the detection window of the 5-HTOL/5-HIAA ratio in urine is approximately 5–15 hr longer than the window of detection of ethanol in urine, and this biomarker is considered as a 24-hr alcohol biomarker. Although consumption of a low amount (<10 g) of ethanol in the evening may not increase the ratio of 5-HTOL and 5-HIAA in urine, consumption of 50 g or ethanol or more should increase the ratio significantly. Higher ratios are indicative of more ethanol consumption almost in a dose-dependent manner. The specificity of this alcohol biomarker is almost 100% [36]. The sensitivity for consuming 50 g of alcohol or more is 77% at a cutoff value of 15 pmol/nmol [37].

Voltaire *et al.* reported that when 30 healthy volunteers (16 males and 14 females) ingested 20, 40, or 60 g of ethanol at 5:00–7:00 PM and urine was collected at 7:00 AM the next morning and also after 9 days of abstinence, the mean ratio of picomoles of 5-HTOL to nanomoles of 5-HIAA was 25 pmol/nmol the morning after alcohol consumption and 8.8 pmol/nmol after 9 days of abstinence. In addition, the authors measured the 5-HTOL/5-HIAA ratio in a group of 69 teetotalers, and the observed mean value was 7.6 pmol/nmol (97% values between 4 and 17 pmol/nmol with no value exceeding 20 pmol/nmol). The authors proposed that a value greater than 20 (picomoles of 5-HTOL to nanomoles of 5-HIAA) can be used to indicate recent alcohol consumption [38]. Whereas Voltaire *et al.* proposed a cutoff value of 20, Helander *et al.* proposed a cutoff of 15 to identify recent alcohol consumption. Moreover, among 47 patients who admitted alcohol consumption (range, 10–230 g; median, 60 g) the night before, breath ethanol testing showed a positive response in only 4 patients. However, 17 patients showed a positive urinary ratio of 5-HTOL/5-HIAA above the cutoff value of 15 pmol/nmol. The values in these 17 patients ranged from 21 to 583 pmol/nmol, with a mean value of 157 pmol/nmol [37].

In a study of 20 social drinkers, Bendtsen *et al.* observed that when volunteers consumed 50 g of alcohol over a 2-hr period during an evening meal, the mean ethanol value at the time of the next morning void collected 6–11 hr after bedtime was only 9 mg/dL, and the corresponding breath alcohol result was negative. The urinary methanol concentration increased from 0.77 mg/dL (pre-drinking value) to 3.06 mg/dL. A very small amount of endogenous methanol is found in blood and urine. After drinking ethanol, the methanol level increases significantly because methanol metabolism by aldehyde dehydrogenase is impaired by ethanol (which has a significantly higher affinity for alcohol dehydrogenase compared to methanol). In addition, methanol is present as congener in alcoholic beverages, further increasing the methanol level in body fluid after consuming ethanol. However, 5-HTOL/5-HIAA increased from a pre-drinking level of 6.36 to 134 pmol/nmol at the time of the first morning void, and such increase was more remarkable than the increase in methanol concentration (6-fold higher methanol level vs. 50-fold higher 5-HTOL/5-HIAA level after drinking). As expected, when other volunteers consumed 80 g of ethanol, the mean value of 5-HTOL/5-HIAA was increased to 296 pmol/nmol from a pre-ethanol value of 6.8 pmol/nmol [39].

Instead of the 5-HTOL/5-HIAA ratio, some authors have used the GTOL/5-HIAA ratio as an alcohol biomarker, which provides similar results to those obtained using the 5-HTOL/5-HIAA ratio. Hoiseth *et al.* reported that when 10 healthy male social drinkers abstained for 1 week before consuming 0.5 g of ethanol per kilogram of bodyweight, median time for observing maximum ethanol concentration in urine was 2.1 hr (range, 1.0–4.8 hr) after drinking. The median detection window was 5.9 hr (range, 4.8–7.0 hr). The median time to reach maximum ethyl glucuronide concentration in urine was 5 hr (range, 4.0–7.0 hr), and the median detection window was 30 hr (25–48 hr). The median time to reach maximum GTOL/5-HIAA level was 4 hr (range, 2.3–6.0 hr), and the median detection window was 9.8 hr (range, 8.0–11.0 hr). The maximum median ethanol level in urine was 0.6 g/L (60 mg/dL; range, 40–70 mg/dL), whereas the maximum median ethyl glucuronide level was 60 mg/L (range, 47–88 mg/L) and the maximum median GTOL/5-HIAA ratio was 275 nmol/μmol/L (range, 199–622 nmol/μmol/L) [40].

CASE REPORT 9.1

During a police investigation of an alleged rape of a 15-year-old girl, blood and urine specimens were collected approximately 15 hr after the incident. No alcohol was detected in blood, whereas urine showed 82 mg/dL of alcohol. The police interpreted the results to indicate that because urine alcohol was 82 mg/dL, the girl was probably drunk at the time of the incident. Because the girl had type 1 diabetes and glycosuria, expert testimony was sought and the possibility of *in vitro* post-collection alcohol formation in urine was suspected because the urine specimen was not preserved with fluoride. A urine analysis of the same specimen after it had been stored for several months at 4°C showed an alcohol level of 550 mg/dL, confirming an ongoing fermentation process and *in vitro* production of alcohol in the specimen. Moreover, the 5-HTOL/5-HIAA ratio was 14 (nmol/μmol), which was below the cutoff value of 15 and inconsistent with urine alcohol level. In fact, for a true urinary alcohol of 82 mg/dL, the expected 5-HTOL/5-HIAA ratio should be approximately 200 (nmol/μmol) or greater. Therefore, post-collection formation of alcohol in urine was confirmed [41].

The 5-HTOL/5-HIAA ratio is used in methadone treatment programs to monitor alcohol intake by patients. Another application is testing for sobriety in disulfiram (Antabuse) treatment programs [37]. However, disulfiram and cyanamide (calcium carbimide) are inhibitors of aldehyde dehydrogenase and may affect serotonin metabolism, thus increasing the 5-HTOL/5-HIAA ratio in urine. Whereas the effect of cyanamide lasts less than 12 hr, the effect of disulfiram may last for several days [42]. Helander commented that apart from alcohol ingestion, treatment with an aldehyde dehydrogenase inhibitor such as disulfiram is the only known cause of an abnormally high 5-HTOL/5-HIAA ratio in urine. The author observed an increased ratio of 5-HTOL/5-HIAA in all patients after initiation of disulfiram therapy. The new higher level reached was stable over time in each patient but varied between patients (mean value varied between 10.0 and 34.4 pmol/nmol). However, during maintenance treatment with 400 mg of disulfiram three times per week, the 5-HTOL/5-HIAA ratio was normally less than 60 pmol/nmol; this level was often exceeded if patients consumed alcohol. For some patients, the ratios were higher when they were not taking medication than when they took medication. The author concluded that during disulfiram therapy, a new dose-related higher 5-HTOL/5-HIAA ratio is observed in urine. However, this alcohol biomarker can still be used to check sobriety because if a patient relapses to drinking while on disulfiram, the ratio is further increased from the new baseline value [43]. The 5-HTOL/5-HIAA ratio in urine is also used in forensic investigations. In one postmortem investigation, the blood ethanol level was 77 mg/dL but the urine ethanol level was zero. In addition,

CASE REPORT 9.2

After attending a banquet, a 25-year-old graduate student attempted to walk home despite an outside temperature of −25°C, and he was later reported missing. Police investigated the case the next day and found his footprints on a frozen river, ending in the middle of it. The fully clothed victim was found downstream 31 days later. Several witnesses from the banquet he had attended told investigators that the victim was a binge drinker and appeared to be drunk the night of his disappearance. At autopsy, water-induced wrinkling of the skin was observed. Blood and vitreous humor were preserved with fluoride/oxalate. His blood alcohol was 260 mg/dL and urine alcohol was 330 mg/dL, indicating acute ethanol intoxication; however, vitreous humor alcohol was only 50 mg/dL, which was not consistent with blood and urine alcohol levels. Although ethanol in blood and urine can form *in vitro* after death by the action of bacteria, postmortem production of ethanol in vitreous humor is rare because the interior of the eye is a sterile medium. Therefore, vitreous humor alcohol is a reliable indicator of body burden of ethanol at the time of death. In this case, it was possible that ethanol from vitreous humor was washed out by water, and due to the low temperature of the river water, postmortem production of alcohol was unlikely. However, the family was adamant that the victim was not intoxicated prior to death. Therefore, urine 5-HTOL and 5-HIAA concentrations were determined: Urine 5-HTOL level was 226 pmol/mL, 5-HIAA level was 0.317 nmol/mL, and the 5-HTOL/5-HIAA ratio was 713 (pmol/nmol), which was consistent with high blood and urine ethanol. The cause of death was determined to be drowning by accident, with acute ethanol intoxication listed as a contributing factor [46].

the 5-HTOL/5-HIAA ratio was within the normal range, which confirmed the suspicion of postmortem ethanol formation in the blood [44]. Using a cutoff value of 15 pmol/nmol for the 5-HTOL/5-HIAA ratio, Johnson *et al.* investigated five possible ethanol-related aviation fatalities and determined that in four of the cases, detected ethanol was present due to postmortem microbial formation but not consumption, based on the 5-HTOL/5-HIAA ratio, despite some indication of possible antemortem ethanol consumption [45].

9.4.1 Laboratory Methods for Determining 5-HTOL and 5-HIAA

Both GC–MS and LC–MS/MS are used for the determination of 5-HTOL and 5-HIAA in urine. Such methods are also applicable for determination of 5-HTOL and 5-HIAA in other biological matrices. In general, 5-HTOL is excreted in urine mostly in conjugated form. Using GC–MS and deuterated analog as the internal standard, Beck *et al.* determined the concentration of 5-HTOL in the cerebrospinal fluid (CSF) of male alcoholics. The procedure involves extraction of 5-HTOL after adding 5 hydroxyindole and internal standard to CSF using chloroform, followed by derivatization using pentafluoropropionic anhydride. During ethanol intoxication in male alcoholics, the mean 5-HTOL value of 10.4 pmol/mL was significantly elevated compared to that of 3.31 pmol/mL observed in controls [47]. Methods based on HPLC have been developed using fluorometric or electrochemical detection after liberation of 5-HTOL glucuronide conjugate in urine using β-glucuronidase. Again, deuterated analog of 5-HTOL can be used as an internal standard. However, the fluorometric method is affected by interference. Immunochemical assay using mouse monoclonal antibody against GTOL has also been reported [36]. Dierkes *et al.* described an enzyme-linked immunosorbent assay (ELISA) for the determination of GTOL in urine [48].

One common approach to measure urinary 5-HTOL and 5-HIAA is to use two different methods. For urinary 5-HTOL, β-glucuronidase enzyme may be used to hydrolyze the glucuronide conjugate, followed by derivatization and analysis using GC–MS and an HPLC method for analysis of urinary 5-HIAA. Because 5-HIAA is not conjugated in urine, no hydrolysis step is needed. Deuterated analog of 5-HTOL as well as deuterated analog of 5-HIAA can be used as internal standards. However, Johnson *et al.* described an LC–MS method for simultaneous analysis of 5-HTOL and 5-HIAA using 5-methoxy-2-methyl-3-indoleacteic acid as the internal standard. Urinary 5-HTOL glucuronide was hydrolyzed first using β-glucuronidase. Then both 5-HTOL and 5-HIAA were converted to trimethylsilyl derivative using *N*,*O*-bis(trimethylsilyl) trifluoroacetamide and analyzed by either LC–MS or LC–MS/MS. The authors applied the method for analysis of postmortem urine specimens using a cutoff of 15 pmol/nmol 5-HTOL/5-HIAA ratio for evaluating antemortem alcohol consumption [49]. Stephanson *et al.* simultaneously analyzed GTOL and 5-HIAA in urine using LC–MS/MS with deuterated GTOL and deuterated 5-HIAA as internal standards. Because GTOL was analyzed directly, no hydrolysis was conducted. The authors used reverse-phase chromatography with gradient elution for chromatographic separation and electrospray ionization and monitoring two product ions per analyte in selected reaction monitoring mode for mass spectrometric analysis. For internal standards, one product ion each was monitored. For GTOL, the parent ion (m/z) was 352.2, with two product ions at m/z 131.0 and 175.9. For deuterated GTOL (internal standard), the parent ion was 356.2, with the product ion at m/z 131.0. For 5-HIAA, the parent ion was at m/z 190.0, with two product ions at m/z 146.2 and 116.2. For deuterated 5-HIAA (internal standard), the parent ion was at m/z 192.0, and the product ion was at m/z 148.2. The analytical measuring range was 6.7–1000 nmol/L for GTOL and 0.07–100 μmol/L for 5-HIAA [35].

9.5 OTHER ALCOHOL BIOMARKERS

Plasma cholesteryl ester transfer protein (CETP) plays an important role in reverse cholesterol transport, the process in which cholesterol is transported from peripheral tissue back to the liver. Alcohol consumption lowers the activity of CETP. Hannuksela *et al.* observed that mean CETP activity was 26% lower in alcoholics (mean daily alcohol consumption of 180 g) compared to controls (10 g of ethanol per day). The authors further commented that sensitivities and specificities of GGT, ALT, AST, and MCV were similar to those of CETP. The authors concluded that CETP activity is not sufficient as a single alcohol biomarker but may be used with other alcohol biomarkers to identify alcohol misuse [50].

Homocysteine is an excitatory amino acid that markedly enhances vulnerability of neural cells to oxidative injury. Chronic alcoholism can increase the serum or plasma concentration of homocysteine. The assumed reason is an impaired metabolism of homocysteine due to dysfunction of methionine synthase secondary to ethanol-induced folate and vitamin B_{12} deficiency. In addition, acetaldehyde, the metabolite of ethanol, can also inhibit methionine synthase. Bayerlein *et al.* reported that alcoholics with a history of withdrawal seizures had a significantly higher homocysteine level (mean, 42.0 μmol/L) than did actively drinking patients without a history of seizures (mean, 22.5 μmol/L). The authors concluded that a high homocysteine level during admission may be a useful screening method to identify actively drinking patients with a higher risk of alcohol withdrawal seizures [51].

Cytokines are proteins implicated in cellular communication and activation. Cytokines play an important role in regulating various processes, including inflammation, cell death, cell proliferation, and cell mitigation. Circulating cytokines such as tumor necrosis factor-α (TNF-α), interleukin-1 (IL-1), and IL-6 are found to be elevated in both chronic and acute alcohol-induced liver disease. In addition, IL-8, IL-12, and monocyte chemoattractant protein-1 (MCP-1) may have potential as alcohol biomarkers. However, cytokines may also be elevated in many neurological disorders [52]. Serum TNF-α levels are increased in alcoholics, especially in patients with alcoholic liver diseases, compared to moderate drinkers and abstainers. In one study, the authors observed that in the general population, most individuals (77%) had an undetectable level (<4 pg/mL) of TNF-α in serum (median, <4 pg/mL and up to 55.6 pg/mL), whereas most alcoholics (90.5%) admitted to the hospital had detectable levels of TNF-α in serum (median, 6.6 pg/mL; range up to 77.4 pg/mL). The 95th percentile was 8.8 pg/mL in the general population and 16.3 pg/mL in alcoholics [53]. Based on a study of 221 alcohol abstainers, 140 light drinkers (1−140 g ethanol per week), 53 moderate drinkers (141−280 g per week), and 45 heavy drinkers (>280 g per week), as well as 137 alcoholics admitted to the hospital, Gonzalez-Quintela *et al.* reported that the proportion of individuals with abnormally high IL-8 (>10 pg/mL) was 5.9% in abstainers, 10.7% in light drinkers, 13.2% in moderate drinkers, and 17.8% in heavy drinkers, and this proportion was exceedingly high (70.1%) among alcoholics admitted to the hospital. Extremely high levels (>100 pg/mL) were observed only among alcoholics and were more frequently observed in females than in males (23.5 vs. 9.7%) [54].

9.6 PROTEOMICS IN ALCOHOL BIOMARKER DISCOVERY

One of the central physiological processes that occurs in any living organism is the conversion of genetic information encoded in DNA into proteins (gene expression), and these proteins then form cells and their structural components or function as enzymes. Therefore, investigators can study the normal function of the body as well as the body's response to a disease or a condition such as alcohol abuse at both gene and protein levels. The field of proteomics consists of analysis of all proteins encoded by DNA as well as protein complexes composed of multiple interacting proteins. A wide array of technologies that are relatively straightforward and amenable to high-throughput analysis are currently available for the study of genomics, proteomics, and even metabolomics.

Proteomics can be subclassified into three groups. Structural proteomics consists of structural analysis of a protein using methods such as X-ray crystallography and nuclear magnetic resonance spectroscopy. In expression proteomics, patterns of protein expression under different conditions or diseases are studied. Functional proteomics explores protein activities and the interaction of proteins with each other. In alcohol biomarker discovery, functional proteomics is used. One of the aims is to discover new biomarkers related to a condition or disease. The two-dimensional electrophoresis technique has been in use for a long time to study various proteins present in a specimen. Currently, two methods are widely used for protein separation in proteomics: electrophoresis and HPLC. For protein identification, affinity reagents such as antibodies can be used. However, mass spectrometry is a more sophisticated technique for protein identification in proteomics, where intact proteins or polypeptides may be analyzed directly or in a bottom-up method in which proteins are digested with an enzyme to produce short peptides that are analyzed by mass spectrometry. Tryptic digestion is commonly used to generate such peptides. Then data are entered into a computer, and specialized software can be used to generate a list of the masses of all measured peptides that can be matched with a database of known proteins. Two mass spectrometric approaches are used in proteomics alcohol biomarker discovery research: matrix-assisted laser desorption time-of-flight (MALDI-TOF) mass spectrometry and electrospray ionization combined with tandem mass spectrometry [55]. An earlier report of application of proteomics in alcohol biomarker research using two-dimensional protein electrophoresis demonstrated that proteins such as α_1-acid glycoprotein, IgA, α_1-antichymotrypsin, haptoglobin, and Apo A-I lipoprotein were elevated, whereas antithrombin III was decreased, in sera of alcoholics. A later report showed that eight proteins were found to be potential biomarkers of fetal alcohol syndrome. Again, haptoglobin and α_1-antichymotrypsin were included in these eight proteins [56].

9.6.1 Specific Proteins Identified as Alcohol Biomarkers Using the Proteomics Approach

In a study that used surface enhanced laser desorption/ionization time-of-flight mass spectrometry (SELDI-TOF-MS) (ProteinChip, SELDI Technology) and included 16 chronic alcoholic patients hospitalized for a rehabilitation program, Nomura *et al.* observed that two peptides with molecular weights of 5.9 and 7.8 kDa were downregulated on admission in these patients. However, the expression level of these proteins increased after 1 week of abstinence. The authors identified the 5.9-kDa protein as a fragment of the fibrinogen α-E chain and the 7.8-kDa protein as a fragment of Apo A-II [57]. Using SELDI-TOF-MS, Sogawa *et al.* identified three potential markers for alcoholism—two peptides with molecular weights of 5.9 and 7.8 kDa and a protein with molecular weight of 28 kDa. The two peptides were downregulated in alcoholics but increased with abstinence. In contrast, the 28-kDa protein was elevated in alcoholics but decreased during abstinence.

The 5.9-kDa peptide was a fragment of the fibrinogen α-E chain, the 7.8-kDa protein was a fragment of Apo A-II, and the 28-kDa protein was identified by the authors as intact Apo A-I. The sensitivity and specificity were highest for the 5.9-kDa peptide. When the 5.9-kDa peptide and the 28-kDa protein were combined with GGT, 96.8% of habitual drinkers were successfully screened with a specificity of 60.9% [58].

Using magnetic beads and MALDI-TOF-MS, Sogawa *et al.* identified 22 peaks that were significantly altered in alcoholics who were admitted to hospital for a rehabilitation program. Of the 22 peaks, 3 had an *m/z* of 3000 or less and had substantial peak intensities; these were subjected to tandem mass spectrometric analysis. The 1466- and 1616-Da peptides were upregulated during admission and were identified as fragments of fibrinopeptide A and phosphorylated fibrinopeptide A. In contrast, the 2660-Da peptide peak, which was downregulated during admission, increased during abstinence. This peptide was identified as a fragment of the fibrinogen α-C chain. These peaks were not detectable using the SELDI-TOF-MS ProteinChip system. The alterations in these peaks induced by alcohol abuse were also seen in GGT nonresponders. The authors concluded that these protein fragments may be used as additional alcohol biomarkers [59]. Later, a sandwich ELISA was developed for rapid analysis of the 5.9-kDa protein so that it might be used as a potential diagnostic tool for measuring this alcohol biomarker. The ELISA result correlated with that of stable isotope dilution mass spectrometry using the ClinProt system [60].

Using ProteinChip and ClinProt systems, Sogawa *et al.*, through a three-step proteome analysis, observed that five proteins—α_2-HS glycoprotein, Apo A-I, glutathione peroxidase-3, heparin cofactor II, and pigment epithelium-derived factor—were significantly higher in concentration on admission in alcoholics compared to 8 weeks after abstinence. Western blotting and ELISA confirmed upregulation of pigment epithelium-derived factor in alcoholics. Serum levels of pigment epithelium-derived factor were significantly higher in moderate to heavy drinkers (14.2 ± 7.7 μg/mL) compared to healthy subjects without a history of alcohol consumption (5.5 ± 3.0 μg/mL). Serum levels of pigment epithelium-derived factor in patients with nonalcoholic chronic liver diseases were comparable to values observed in healthy subjects [61]. Using proteomic workflow including LC–MS/MS with enrichment of serum carrier protein-bound biomarker technique for discovery of alcohol biomarkers, Lai *et al.* initially identified 311 candidate proteins that were bound to serum carrier proteins. Further analysis revealed the following proteins not previously described to be associated with alcohol abuse: gelsolin, selenoprotein P, serotransferrin, tetranectin, hemopexin, histidine-rich glycoprotein, plasma kallikrein, and vitronectin. Altered abundance of these proteins suggests that they may be potential novel biomarkers for alcohol abuse [62].

Using two-dimensional difference in gel electrophoresis, Freeman *et al.* observed altered levels of serum amyloid A4, retinol binding protein, inter-α inhibitor H4 (a serine protease inhibitor), clusterin, and fibronectin in plasma of primates exposed to alcohol. Examination of these target plasma proteins in human subjects demonstrated increased levels of serum amyloid A4 and clusterin and decreased levels of fibronectin in serum of alcoholics compared to control subjects. These proteins may serve as a target of future alcohol biomarker development [63]. In rats that were chronically fed alcohol, Yamada *et al.* observed 46 protein spots that were differentially expressed in liver homogenates and cytosol fraction. The most notable change was downregulation of a 29-kDa protein subsequently identified as carbonic anhydrase III. The messenger RNA level of this protein was also decreased. In rat serum, 41 proteins were differentially expressed; of these proteins, betaine-homocysteine methyltransferase was also found to be differentially expressed in rat liver. The expression of this protein was upregulated in both the liver and the serum of alcohol-fed rats [64]. Using a 17-plasma protein panel, Freeman *et al.* correctly classified abusive drinking in the primate model with 100% sensitivity and 88% accuracy [65]. Potential alcohol biomarkers discovered through proteomic analysis of human sera are summarized in Table 9.2.

Table 9.2 Potential Alcohol Biomarkers Discovered through Proteomic Analysis of Human Serum

Upregulated Proteins as Potential Alcohol Biomarkers	Downregulated Proteins as Potential Alcohol Biomarker
■ α_1-Acid glycoprotein	■ Antithrombin III
■ IgA	■ Fibrinogen α-E chain fragment (5.9 kDa)
■ α_1-Antichymotrypsin	■ Fragment of apolipoprotein A-II (7.8 kDa)
■ Haptoglobin	■ Fibronectin
■ Apolipoprotein A-I	■ 2660-Da peptide identified as a fragment of fibrinogen α-C chain
■ Fragment of fibrinopeptide A (1466 Da)	
■ Fragment of phosphorylated fibrinopeptide A (1616 Da)	
■ Pigment epithelial-derived factor	
■ α_2-HS glycoprotein	
■ Glutathione peroxidase 3	
■ Heparin cofactor II	
■ Serum amyloid A4	
■ Clusterin (Apo J)	

9.7 CONCLUSIONS

Plasma total sialic acid, sialic acid index of Apo J, and the 5-HTOL/5-HIAA ratio are useful alcohol biomarkers in both clinical and forensic settings. Salivary glycoproteins have been investigated as potential alcohol biomarkers because glycosylation of proteins is affected by ethanol; however, more studies are needed to establish the clinical utility of such glycoproteins [66]. Circulating cytokines are also potential alcohol biomarkers that require further investigation. The proteomics approach to discover new alcohol biomarkers has identified several serum proteins that show promise as potential alcohol biomarkers in the near future.

References

[1] Wang B, Brand-Miller J. The role and potential of sialic acid in human nutrition. Eur J Clin Nutr 2003;57:1351–69.

[2] Sillanaukee P, Ponnio M, Jaaskelainen IP. Occurrence of sialic acids in healthy humans and different disorders. Eur J Clin Invest 1999;29:413–25.

[3] Gopaul KP, Crook MA. Sialic acid: a novel marker of cardiovascular disease? Clin Biochem 2006;39:667–81.

[4] Priego-Capote F, Orozco-Salono M, Calderon-Santiago M, Lugue de Castro MD. Quantitative determination and confirmatory analysis of N-acetylneuraminic acid and N-glycolylneuraminic acid in serum and urine by solid phase extraction on line coupled to liquid chromatography–tandem mass spectrometry. J Chromatogr A 2014;1346:88–96.

[5] Romppanen J, Punnonen K, Anttila P, Jakobsson T, et al. Serum sialic acid as a marker of alcohol consumption: effect of liver disease and heavy drinking. Alcohol Clin Exp Res 2002;26:1234–8.

[6] Sillanaukee P, Ponnio M, Seppa K. Sialic acid: new potential marker of alcohol abuse. Alcohol Clin Exp Res 1999;23:1039–43.

[7] Chrostek L, Cylwik B, Szmitkowski M, Korcz W. The diagnostic accuracy of carbohydrate deficient transferrin, sialic acid and commonly used markers of alcohol abuse during abstinence. Clin Chim Acta 2006;364:167–71.

[8] Crook M. The determination of plasma or serum sialic acid. Clin Biochem 1993;26:31–8.

[9] Crook MA, Treloar A, Haq M, Tutt P. Serum sialic acid and acute phase proteins in elderly subjects. Eur J Clin Chem Clin Biochem 1994;32:745–7.

[10] Bull C, Stoel MA, den Brok MH, Adema GJ. Sialic acids sweeten a tumor's life. Cancer Res 2014;74:1–6.

[11] Narayanan S. Sialic acid as a tumor marker. Ann Clin Lab Sci 1994;24:376–84.

[12] Dwivedi C, Dixit M, Hardy RE. Plasma lipid-bound sialic acid alterations in neoplastic diseases. Experientia 1990;46:91–4.

[13] Paszkowska A, Berbec H, Semczul A, Cybulski M. Sialic acid concentration in serum and tissue of endometrial cancer patients. Eur J Obstet Gynecol Reprod Biol 1998;76:211–15.

[14] Afzail B, Baki RS, Bharma-Ariza P, Lumb PJ, et al. Raised plasma total sialic acid levels are markers of cardiovascular disease in renal dialysis patients. J Nephrol 2003;16:540–5.

[15] Yokoyama H, Jensen JS, Jensen T, Deckert T. Serum sialic acid concentration is elevated in IDDM especially in early diabetic nephropathy. J Intern Med 1995;237:519–23.

[16] Von Versen-Hoeynck FM, Hubel CA, Gallahaer MJ, Gammill HS, et al. Plasma levels of inflammatory markers neopterin, sialic acid and C-reactive protein in pregnancy and preeclampsia. Am J Hypertens 2009;22:687–92.

[17] Chrostek L, Supronowicz L, Panasiuk A, Cylwik B, et al. Serum sialic acid levels according to the severity of liver cirrhosis. J Clin Lab Anal 2014;28:465–8.

[18] Warren L. The thiobarbituric acid assay of sialic acids. J Biol Chem 1959;234:1971–5.

[19] Crook M, Haq M, Tutt P. Evaluation of three assays for the determination of serum total sialic acid. Clin Biochem 1993;26:449–54.

[20] Simpson H, Chusney GD, Crook MA, Pickup JC. Serum sialic acid enzymatic assay based on microtiter plates: application for measuring capillary serum sialic acid concentrations. Br J Biomed Sci 1993;50:164–7.

[21] Sugawara Y, Iwamori M, Portoukalian J, Nagai Y. Determination of N-acetylneuraminic acid by gas chromatography–mass spectrometry with a stable isotope as internal standard. Anal Biochem 1983;132:147–51.

[22] Romppanen J, Romppanen EL, Halonen T, Eskelinen M, et al. Reference values of serum total sialic acid for healthy adult subjects and comparison of two different calibrators for quantitation of total sialic acid in high-performance anion-exchange chromatography with pulsed amperometric detection. Clin Chem Lab Med 2003;41:1095–8.

[23] Ishida J, Nakahara T, Yamaguchi M. Measurement of N-acetylneuraminic acid in human serum and urine by high-performance liquid chromatography with chemiluminescence detection. Biomed Chromatogr 1992;6:135–40.

[24] Li K. Determination of sialic acid in human serum by reverse-phase liquid chromatography with fluorometric detection. J Chromatogr 1992;579:209–13.

[25] Trougakos P, Gonos ES. Clusterin/apolipoprotein J in human aging and cancer. Int J Biochem Cell Biol 2002;34:1430–48.

[26] De Silva HV, Stuart WD, Park YB, Mao SJ, et al. Purification and characterization of apolipoprotein J. J Biol Chem 1990;265:14292–7.

[27] Gong M, Castillo L, Redman RS, Garige M, et al. Downregulation of liver Galbeta1,4GlcNac alpha 2,6-sialyltransferase gene by ethanol significantly correlates with alcoholic steatosis in humans. Metabolism 2008;57:1663–8.

[28] Ghosh P, Hale EA, Lakshman R. Long-term ethanol exposure alters the sialylation index of plasma apolipoprotein J (Apo J) in rats. Alcohol Clin Exp Res 1999;23:720–5.

[29] Ghosh P, Hale EA, Lakshman R. Plasma sialic acid index of apolipoprotein J (SIJ): a new novel alcohol intake marker. Alcohol 2001;25:173–9.

[30] Waszkiewicz N, Szajda SD, Kepka A, Szulc A, et al. Glycoconjugates in the detection of alcohol abuse. Biochem Soc Trans 2011;39:365–9.

[31] Wurst F, Thon N, Weinmann W, Tippetts S, et al. Characterization of sialic acid index of plasma apolipoprotein J and phosphatidylethanol during alcohol detoxification—a pilot study. Alcohol Clin Exp Res 2012;36:251–7.

[32] Svensson S, Some M, Lundsjo A, Helander A, et al. Activities of human alcohol dehydrogenase in the metabolic pathways of ethanol and serotonin. Eur J Biochem 1999;262:324–9.

[33] Kema IP, de Vries EG, Muskiet FA. Clinical chemistry of serotonin and metabolism. J Chromatogr B Biomed Sci Appl 2000;747:33–48.

[34] Stephanson N, Dahl H, Helander A, Beck O. Determination of 5-hydroxytryptophol glucuronide by liquid chromatography–mass spectrometry. J Chromatogr B Analyt Technol Biomed Life Sci 2005;816:107–12.

[35] Stephanson N, Helander A, Beck O. Alcohol biomarker analysis: simultaneous determination of 5-hydroxytryptophol glucuronide and 5-hydroxyindoleacetic acid by direct injection of urine using ultra-performance liquid chromatography–tandem mass spectrometry. J Mass Spectrom 2007;42:940–9.

[36] Beck O, Helander A. 5-Hydroxytryptophol as a marker for recent alcohol intake. Addiction 2003;98(Suppl. 2):63–72.

[37] Helander A, von Wachenfeldt J, Hiltunen A, Beck O, et al. Comparison of urinary 5-hydroxytryptophol, breath ethanol and self-report for detection of recent alcohol use during outpatient treatment: a study on methadone patients. Drug Alcohol Depend 1999;56:33–8.

[38] Voltaire A, Beck O, Borg S. Urinary 5-hydroxytryptophol: a possible marker of recent alcohol use. Alcohol Clin Exp Res 1992;16:281–5.

[39] Bendtsen P, Jones AW, Helander A. Urinary excretion of methanol and 5-hydroxytryptophol as biochemical markers of recent drinking in the hangover state. Alcohol Alcohol 1998;33:431–8.

[40] Hoiseth G, Bernard JP, Stephanson N, Normann PT, et al. Comparison between urinary alcohol markers EtG, EtS and GTOL/5-HIAA in a controlled drinking experiment. Alcohol Alcohol 2008;43:187–91.

[41] Jones AW, Eklund A, Helander A. Misleading results of ethanol analysis in urine specimens from rape victims suffering from diabetes. J Clin Forensic Med 2000;7:144–6.

[42] Beck O, Helander A, Carlsson S, Borg S. Changes in serotonin metabolism during treatment with aldehyde dehydrogenase inhibitors disulfiram and cyanamide. Pharmacol Toxicol 1995;77:323–6.

[43] Helander A. Monitoring relapse drinking during disulfiram therapy by assay of urinary 5-hydroxytryptophol. Alcohol Clin Exp Res 1998;22:111–14.

[44] Helander A, Beck O, Jones AW. Distinguishing ingested ethanol from microbial formation by analysis of urinary 5-hydroxytryptophol and 5-hydroxyindoleacetic acid. J Forensic Sci 1995;40:95–8.

[45] Johnson RD, Lewis RJ, Canfield FV, Dubowski KM, et al. Utilizing the urinary 5-HTOL/5-HIAA ratio to determine ethanol origin in civil aviation accident victims. J Forensic Sci 2005;50:670–5.

[46] Singer PP, Jones GR, Lewis R, Johnson R. Loss of ethanol from vitreous humor in drowning deaths. J Anal Toxicol 2007;31:522–5.

[47] Beck O, Borg S, Holmstedt B, Stibler H. Levels of 5-hydroxytryptophol in cerebrospinal fluid from alcoholics determined by gas chromatography–mass spectrometry. Biochem Pharmacol 1980;29:693–6.

[48] Dierkes J, Wolfersdorf M, Borucki K, Weinmann W, et al. Determination of glucuronidated 5-hydroxytryptophol (GTOL), a marker of recent alcohol intake, by ELISA technique. Clin Biochem 2007;40:128–31.

[49] Johnson RD, Lewis RJ, Canfield DV, Blank CL. Accurate assignment of ethanol origin in postmortem urine: liquid chromatographic–mass spectrometric determination of serotonin metabolites. J Chromatogr B Analyt Technol Biomed Life Sci 2004;805:223–34.

[50] Hannuksela M, Kesaniemi YA, Savolainen MJ. Evaluation of plasma cholesteryl ester transfer protein (CETP) as a marker of alcoholism. Alcohol Alcohol 1992;27:557–62.

[51] Bayerlein K, Hillemacher T, Reulbach U, Mugele B, et al. Alcoholism associated hyperhomocystinemia and previous withdrawal seizures. Biol Psychiatry 2005;57:1590–3.

[52] Achur RN, Freeman WM, Veana KE. Circulating cytokines as biomarkers of alcohol abuse and alcoholism. J Neuroimmune Pharmacol 2010;5:83–91.

[53] Gonzalez-Quintela A, Campos J, Loidi L, Quinteiro C, et al. Serum TNF-α levels in relation to alcohol consumption and common TNF gene polymorphism. Alcohol 2008;42:513–18.

[54] Gonzalez-Quintela A, Campos J, Gude F, Perez LF, et al. Serum concentrations of interleukin-8 in relation to different levels of alcohol consumption. Cytokine 2007;38:54–60.

[55] Hiller-Sturmhofel S, Sobin J, Mayfield RD. Proteomic approaches for studying alcohol-induced organ damage. Alcohol Res Health 2008;31:36–46.

[56] Torrente MP, Freeman W, Vrana KE. Protein biomarkers of alcohol abuse. Expert Rev Proteomics 2012;9:425–36.

[57] Nomura F, Tomonga T, Sogawa K, Ohashi T, et al. Identification of novel and downregulated biomarkers for alcoholism by surface enhanced laser desorption/ionization mass spectrometry. Proteomics 2004;4:1187–94.

[58] Sogawa K, Itoga S, Tomonaga T, Nomura F. Diagnostic values of surface enhanced laser desorption/ionization technology for screening of habitual drinkers. Alcohol Clin Exp Res 2007;31(1 Suppl):S22–6.

[59] Sogawa K, Satoh M, Kodera Y, Tomonaga T, et al. A search for novel markers of alcohol abuse using magnetic beads and MALDI-TOF/TOF mass spectrometry. Proteomics Clin Appl 2009;3:821–8.

[60] Noda K, Sogawa K, Kikuchi W, Kiyokawa I, et al. Development of a sandwich ELISA for the detection of the 5.9-kDa fibrinogen alpha C chain fragment detected by serum proteome analysis. Proteomics Clin Appl 2011;4:141–6.

[61] Sogawa K, Kodera Y, Satoh M, Kawashima Y, et al. Increased serum levels of pigment epithelium-derived factor by excessive alcohol consumption detection and identification by three-step serum proteome analysis. Alcohol Clin Exp Res 2011;35:211–17.

[62] Lai X, Liangpunsakul S, Crabb DW, Ringham HN, et al. A proteomic workflow for discovery of serum carrier protein-bound biomarker candidates of alcohol abuse using LC/MS/MS. Electrophoresis 2009;30:2207–14.

[63] Freeman WM, Vanguilder HD, Guidone E, Krystal JH, et al. Plasma proteomic alterations in non-human primates and humans after chronic alcohol self-administration. Int J Neuropsychopharmacol 2011;14:899–911.

[64] Yamada M, Satoh M, Seimiya M, Sogawa K, et al. Combined proteomic analysis of liver tissue and serum in chronically alcohol fed rats. Alcohol Clin Exp Res 2013;37:E79–87.

[65] Freeman WM, Salzberg AC, Gonzales SW, Grant KA, et al. Classification of alcohol abuse by plasma protein biomarkers. Biol Psychiatry 2010;68:219–22.

[66] Kratz EM, Waszkiewicz N, Kaluza A, Szajda SD, et al. Glycosylation changes in the salivary glycoproteins of alcohol dependent patients: a pilot study. Alcohol Alcohol 2014;49: 23–30.

CHAPTER 10

Genetic Markers of Alcohol Use Disorder

10.1 INTRODUCTION

Although moderate alcohol (ethanol) consumption has many beneficial effects, excessive alcohol consumption is detrimental to health (see Chapter 1). Sustained alcohol intake can cause functional alcohol tolerance, which enables increased alcohol consumption by an individual with few symptoms of intoxication. Alcohol tolerance leads to the development of alcohol dependence, which refers to physiological addiction in which abstinence may cause withdrawal symptoms. Alcohol abuse and alcohol dependence are maladaptive patterns of drinking. According to the fourth edition of the *Diagnostic and Statistical Manual of Mental Disorders* (*DSM-IV*), individuals must meet three of seven criteria to be diagnosed with alcohol dependence, and alcohol abuse is defined by meeting at least one of four criteria. In the *DSM-5*, both alcohol dependence and alcohol abuse are combined into one psychiatric illness termed "alcohol use disorder" (see Chapters 2 and 4).

Alcohol use disorder is a risk factor for many diseases, including alcoholic cirrhosis, pancreatitis, liver cancer, and cardiovascular disease (see Chapter 1). Men usually tend to drink more heavily and more frequently than women, and this tendency places men at increased risk of disease and alcohol-related mortality. Therefore, early identification of individuals who are at risk of developing alcohol use disorder is important. Alcohol biomarkers are clinically helpful to identify such individuals and also to monitor the progress of therapy of patients with alcohol use disorder who are undergoing alcohol rehabilitation. Alcohol biomarkers can be broadly classified as state biomarkers and trait biomarkers. A state biomarker provides information regarding an individual's drinking habits, whereas a trait biomarker provides information about a person's genetic predisposition toward alcohol dependence. In Chapters 4–9, various state biomarkers of alcohol abuse were discussed in detail. However, it is also well-established that children of alcoholic parents are at higher risk of developing alcohol use disorder, indicating a genetic

CONTENTS

10.1 Introduction 245

10.2 Heredity, Environment, and Alcohol Use Disorder 246

10.2.1 Effect of Nongenetic Factors on the Development of Alcohol Use Disorder 247

10.3 Genes and Alcohol Use Disorder: An Overview 248

10.4 Polymorphisms in Genes Encoding Alcohol Dehydrogenase and Aldehyde Dehydrogenase ... 250

10.4.1 Polymorphisms that Protect from Alcohol Use Disorder 251

10.4.2 Polymorphisms that may Increase the Risk of Alcohol Use Disorder 253

10.5 Neurobiological Basis of Alcohol Use Disorder 253

10.6 Polymorphisms of Genes in Dopamine Pathway

A. Dasgupta: Alcohol and Its Biomarkers. DOI: http://dx.doi.org/10.1016/B978-0-12-800339-8.00010-9

and Alcohol Use Disorder...............254
10.6.1 Dopamine Receptors.....................255
10.6.2 Dopamine Transporters and Dopamine-Metabolizing Enzymes........................258
10.6.3 Monoamine Oxidase.........................259
10.6.4 Catechol-O-Methyltransferase........259
10.7 Polymorphisms of Genes in the Serotonin Pathway and Alcohol Use Disorder...............260
10.8 Polymorphisms of Genes in the GABA Pathway and Alcohol Use Disorder...............264
10.9 Polymorphisms of Genes Encoding Cholinergic Receptors and Alcohol Use Disorder...............267
10.10 Polymorphisms of Genes in the Glutamate Pathway and Alcohol Use Disorder...............269
10.11 Polymorphisms of Genes Encoding Opioid Receptors and Alcohol Use Disorder...............272
10.12 Polymorphisms of

component of addiction. Unfortunately, there is no single gene related to alcohol use disorder. Therefore, alcohol use disorder is a complex multifactorial disease that is influenced by both genetic predisposition (polygenic disease) and environment. In addition, alcohol use disorder is often comorbid with nicotine abuse, substance abuse, and other psychiatric illness. Another possibility is the interaction between genes and environment mediated by an epigenetic mechanism that may be linked to alcohol use disorder. This chapter provides an overview of the genetics of alcohol use disorder and also a discussion of variations in genetic polymorphism as well as the role of epigenetics in understanding alcohol abuse disorder.

10.2 HEREDITY, ENVIRONMENT, AND ALCOHOL USE DISORDER

Family studies have shown that the risk for alcohol dependence is 4- to 10-fold higher in offspring of an alcoholic parent. A meta-analysis consisting of 10,000 twin pairs showed that heritability of alcohol use disorder is approximately 50%, and other similar studies have indicated that the genetic component is between 40 and 60% [1]. Based on studies of 3516 twins from male–male pairs, Prescott and Kendler observed that the prevalence of alcoholism was substantially higher among identical pairs compared to fraternal pairs. Using the liability threshold model, the authors attributed 48–58% of liability to alcoholism to inherited genetic factors. They concluded that genetic factors may play a more important role in the development of alcoholism in men [2]. However, in a study involving 5091 male and 4168 female twins (including identical, same-sex fraternal, and opposite-sex pairs), Prescott *et al.* attributed 55–66% genetic liability to alcoholism in women and 51–56% in men [3]. Kendler *et al.* reported that heritability of liability of alcoholism was consistently higher in monozygotic compared to dizygotic twins. In addition, heritability of liability of alcoholism in women was in the range of 50–60% [4]. Males with a female co-twin are more likely to become alcohol dependent than are males with a male co-twin [5]. Alcohol dependence may also co-occur with other substance abuse, such as that of cocaine, nicotine, and opiates. Therefore, the genetic component for developing alcohol use disorder is far from the expected 100% as observed in a pure genetic disorder transmitted in a Mendelian manner. It is assumed that several coexisting genetic variants in each affected individual, rather than a single variant (allele), are responsible for susceptibility to alcohol, along with other nongenetic and environmental factors. Various nongenetic factors during pregnancy, childhood, adolescence, and adulthood may play a significant role in the development of alcohol use disorder [1].

10.2.1 Effect of Nongenetic Factors on the Development of Alcohol Use Disorder

During pregnancy, the fetus may inherit genetic the susceptibility variant from the alcoholic mother, but there are also several nongenetic factors that may have an impact on the normal growth of the fetus and later development of alcohol use disorder in the child. Exposure to alcohol during gestation has been associated with mental retardation; attention deficit hyperactivity disorder; and later development of substance and alcohol abuse, anxiety, and personality disorders, including antisocial behavior. Exposure to alcohol during pregnancy is also associated with fetal alcohol syndrome, which may cause miscarriage and even stillbirth [1].

Childhood maltreatment including abuse (physical and/or sexual) and neglect is associated with low self-esteem and development difficulties. It is also a risk factor for binge drinking in adolescence and alcohol dependence in adulthood. Children with high novelty-seeking and low harm-avoidance traits may be at greater risk for early onset of alcohol use and alcohol dependence. Schwandt *et al.* reported that childhood trauma, especially emotional abuse, was the primary predictor of severity of alcohol dependence during adulthood, and such effects were mediated by neuroticism [6]. A possible explanation for this observation may be related to impaired brain development of adolescents who consume heavy amounts of alcohol. Drinking during adolescence significantly increases the risk of developing alcohol use disorder later in life. The average age of first drink in the United States is 11 years for boys and 13 years for girls. Drinking initiation is affected by environmental factors such as alcohol availability, parental attitude including monitoring of children, parent–child attachment, alcohol use and tolerance of alcohol use by parents, attitudes of relatives toward drinking, and peer pressure at the school as well as in the community. Heavy drinking during adolescence has a negative impact on brain development, especially on the hippocampus. Dopaminergic and γ-aminobutyric acid (GABA) neurotransmitter systems also undergo important changes during adolescence, and alcohol intake may impair their normal development. Dopamine is implicated in the rewarding effect of alcohol, and GABA is associated with alcohol's sedative effect as well as development of tolerance. Low response to the sedative effect of alcohol is significantly associated with increased risk of development of alcohol use disorder in the future [1]. Based on a survey of 43,093 adults, Hingson *et al.* observed that compared to respondents who began drinking alcohol at 21 years of age or older, those who began drinking before age 14 years were more likely to experience alcohol dependence within 10 years of first drinking (adjusted hazard ratio, 1.78; 95% confidence interval, 1.51–2.11) [7].

Genes Encoding Cannabinoid Receptors and Alcohol Use Disorder 274
10.13 Adenylyl Cyclase and Alcohol Use Disorder 275
10.14 Neuropeptide Y and Alcohol Use Disorder 277
10.15 Possible Association of Polymorphisms of other Genes With Alcohol Use Disorder 278
10.16 Epigenetics and Alcohol Use Disorder 278
10.17 Conclusions 281
References 281

Keyes *et al.* reviewed the effects of four different stressors (catastrophic events, childhood maltreatment, common adult stressful life events in interpersonal, occupational, financial, and legal domains, and minor stressors) on alcohol consumption and alcohol use disorders. The authors commented that studies generally demonstrate an increase in alcohol consumption associated with disasters such as terrorism. However, childhood maltreatment is a consistent risk factor for early onset of drinking in adolescence as well as the development of alcohol use disorder as an adult. In addition, specific genetic polymorphisms may interact with childhood maltreatment to increase the risk of alcohol abuse. Stressful life events such as divorce and job loss increase the risk of alcohol use disorder, but epidemiological consensus on the specificity of such associations across gender has not been reached. Moreover, perception of discrimination as well as objective indicators of discrimination may also be associated with alcohol abuse and alcohol use disorder among ethnic and sexual minorities [8]. Sher *et al.* reported that although self-reported childhood stressors are strongly related to family history of alcoholism, not all children of alcoholics develop alcohol use disorder [9]. Therefore, a complex interaction between genetics and environmental factors is associated with alcohol use disorder.

In general, it has been recognized that two broad bands of personality—impulsive/novelty-seeking personality and neuroticism/negative emotionality—are strongly associated with alcoholism. Although studied have shown that sons of male alcoholics are at higher risk of developing alcoholism later in life than daughters, whether such risk is related to personality trait is unclear. Longitudinal studies have shown that antisocial behavior and hyperactivity are related to the development of alcoholism later in life [10]. Holdcraft *et al.* concluded that antisocial personality disorder was associated with an earlier age of first intoxication, a more severe and chronic course of alcoholism, more social consequences of drinking, and higher risk of drug abuse. However, depression was associated with a less severe course of alcoholism [11]. Peer pressure plays a role in drinking behavior because it is difficult not to drink when peers are drinking [12]. Even alcohol advertisements are positively associated with beer drinking among adolescents [13]. Environmental, mediating, and genetic factors that may predispose to alcohol use disorder are summarized in Table 10.1.

10.3 GENES AND ALCOHOL USE DISORDER: AN OVERVIEW

Currently, it is accepted that the genetic risk of alcoholism is likely due to variations in numerous genes, with each variation having a minor effect; however, some variants may have a major effect. After many years of family

Table 10.1 Environmental, Mediating, and Genetic Factors that may Predispose a Person to the Risk of Alcohol Use Disorder

Environmental Factors	Mediating Factors	Polymorphism of Gene Encoding Enzymes/Receptors[a]
Early Life Events ■ Childhood neglect/rejection ■ Physical or sexual abuse ■ Alcoholic father/mother or both parents Adult Life Events ■ Death in family ■ Divorce ■ Job loss ■ Moving/job change ■ Eating disorder Personality Trait ■ Impulsive/novelty seeking ■ Neuroticism/negative ■ Antisocial behavior	■ Alcohol availability ■ Alcohol advertisement ■ Peer pressure ■ Lack of social/emotional support	■ Alcohol dehydrogenase and aldehyde dehydrogenase (major effect) ■ Dopamine receptors/transporters ■ MAOA ■ COMT ■ γ-Aminobutyrate receptors ■ Serotonin receptors ■ Acetylcholine receptors ■ Glutamate receptors ■ Cannabinoid receptors ■ Adenylyl cyclase ■ Neuropeptide Y receptors

COMT, catechol-O-methyltransferase; MAOA, monoamine oxidase A.
[a]These are examples of common genetic variants associated with alcohol use disorder. Other genes associated with alcohol abuse have also been described.

based linkage studies and case–control candidate gene studies, the focus is now on large-scale genome-wide association studies (GWAS) for detection of novel variants. Genetic vulnerability to alcoholism may originate in personal traits that may predispose alcohol-seeking behavior differential response to alcohol or differential variation in the neurobiology, which may be the underlying cause of addiction and physiological response to stress. Although alcohol use disorder is often comorbid with other psychiatric illnesses, the heritability is mostly disease specific. Genotyping, which includes the identification of polymorphic gene variants such as single nucleotide polymorphisms (SNPs), has experienced rapid improvement with regard to throughput rates, reduction of costs, increased accuracy, and simplicity of operation. Also, it can be applied in routine clinical investigations, although such tests may be available only in reference laboratories. The most robust finding for the genetic influence of alcoholism remains in the genes encoding enzymes such as alcohol dehydrogenase and aldehyde dehydrogenase, which are involved in metabolism of the majority of alcohol consumed [14].

Unlike opioids and nicotine, which have specific receptors in the brain, there is no specific receptor for ethanol. Therefore, it has been assumed that enhanced GABA (the brain's major inhibitory neurotransmitter), glutamate (an excitatory neurotransmitter), dopamine, opioid peptides, and receptors as well as serotonin neurotransmission have been associated with alcohol administration and potentially mediate some of alcohol's reinforcing effects. The development of tolerance may be caused by the $GABA_A$ receptor system. Alcohol's inhibition of the glutamatergic excitatory neurotransmitter pathway, especially at the postsynaptic *N*-methyl-D-aspartate (NMDA) receptor, may be the cause of the neurotoxic effects of alcohol, particularly intoxication, blackout, and withdrawal symptoms [15]. Therefore, it is expected that in addition to polymorphism in alcohol dehydrogenase and aldehyde dehydrogenase genes, genetic vulnerability to alcoholism may be related to the neurotransmitter system (GABA, opioid, serotonin, dopamine, glutamate, and cannabinoid), signal transduction pathways within the mesolimbic dopamine reward pathway, and interaction of the stress response system. The genetic effect on metabolism of alcohol as well as polymorphism of genes encoding alcohol dehydrogenase and aldehyde dehydrogenase was discussed in detail in Chapter 2. Here, this important issue is discussed briefly, but emphasis is placed on genetic markers that may induce vulnerability to alcoholism in individuals.

10.4 POLYMORPHISMS IN GENES ENCODING ALCOHOL DEHYDROGENASE AND ALDEHYDE DEHYDROGENASE

As described in detail in Chapter 2, the majority of ingested alcohol is eliminated by oxidation catalyzed by the liver enzymes alcohol dehydrogenase (ADH) and aldehyde dehydrogenase (ALDH). In the first step, alcohol is converted into acetaldehyde by alcohol dehydrogenase; in the second step, acetaldehyde is further oxidized to acetate by ALDH (mostly by the mitochondrial ALDH2 enzyme). Acetaldehyde is a toxic by-product that may contribute to the addictive process [16].

Seven known genes clustering on a 370-kb region of chromosome 4 (4q21–24) encode five classes of ADH enzymes. The class I ADH enzymes, mainly expressed in the liver, are responsible for approximately 70% of the total metabolism of ethanol. The reference allele for the *ADH1B* gene is *ADH1B*1* (wild type), which encodes the β_1 subunit of the ADH enzyme with an arginine at amino acid positions 48 (Arg48) and 370 (Arg370). However, a common polymorphism, *ADH1B*2*, encodes the β_2 subunit and has a histidine at position 48 (Arg48His, rs1229984) instead of arginine. This polymorphism is found commonly among East Asians. The polymorphism *ADH1B*3*, which encodes the β_3 subunit, has cysteine at position 370 instead of arginine (Arg370Cys, rs2066702). This allele is found primarily in people of African descent and Native Americans. ADH isoenzymes encoded by *ADH1B*2* or *ADH1B*3* alleles are superactive ADH enzymes with a 30- to 40-fold increase in metabolism of ethanol compared to normally functioning enzymes encoded by the *ADH1B*1* wild-type gene. The reference allele of the *ADH1C* gene is *ADH1C*1* (wild type), which encodes the γ_1 subunit with arginine at position 272 and isoleucine at position 350 (Arg272Ile350). However, the enzyme encoded by two *ADH1C*2*-linked SNPs (Arg272Gln, rs1693482, and Ile350Val, rs698; these two SNPs occur together in almost all cases due to very high linkage disequilibrium) shows two amino acid exchanges. Moreover, enzymatic activity is approximately 2.5 times higher when the enzyme is encoded by the *ADH1C*1* reference allele than when it is encoded by the *ADH1C*2* haplotype (Gln272Val350) [17]. The *ADH4* gene encodes the class II ADH enzyme in the liver and, to a lesser extent, in the kidney. This enzyme has a higher Michaelis constant than most ADH and plays an important role in ethanol metabolism, especially at high blood alcohol concentration, by the liver, accounting for approximately 30% of ethanol metabolism.

ADH genes are located on a single chromosome (chromosome 4), but *ALDH* genes are not. Humans have 19 genes and 3 pseudogenes in the *ALDH* gene superfamily encoding the ALDH superfamily of isoenzymes. However, 3 of these genes—*ALDH1A1*, *ALDH1B1*, and *ALDH2*—are relevant to acetaldehyde metabolism. The ALDH1A enzyme is usually found in the cytosol, whereas ALDH1B and ALDH2 enzymes are produced in the nucleus but have leader sequences that direct them to mitochondria, where they exert their function. The genes *ALDH1A* (located on chromosome 9 encoding cytosolic isoenzyme) and *ALDH2* (located on chromosome 12 encoding mitochondrial isoenzyme) are mostly associated with acetaldehyde metabolism. Although the mitochondrial ALDH2 enzyme accounts for the majority of catalytic activity of ALDH, in the absence of enzymatic activity (or low enzymatic activity) due to polymorphism of the gene, the ALDH1A enzyme plays a significant role in the metabolism of acetaldehyde. The wild-type gene *ALDH2*1*, which encodes the ALDH2 enzyme, has normal enzymatic activity, but a polymorphism *ALDH2*2* found commonly among East Asians encodes a defective enzyme with a very low activity (see Chapter 2 for more detail). Polymorphisms of the *ALDH1A1* and *ALDH1B1* genes have also been described, but their role in alcohol-related problems is poorly studied compared to that of polymorphisms of *ADH1B*, *ADH1C*, and *ALDH2* genes. In Chapter 2, the effect of polymorphisms of the *ALDH1A1* and *ALDH1B1* genes on drinking behavior was discussed.

10.4.1 Polymorphisms that Protect from Alcohol Use Disorder

Currently, the strongest link between alcohol use and genetic polymorphism is associated with protection from alcohol abuse in individuals carrying the *ALDH2*2* allele. This allele is common among East Asians, and ALDH encoded by the *ALDH2*2* gene has much lower enzymatic activity compared to the normal enzyme. As a result, acetaldehyde builds up in blood, causing an unpleasant reaction. Therefore, people with this gene are deterred from consuming alcohol. East Asians who are homozygous for the *ALDH2*2* gene have almost zero risk of developing alcohol use disorder [18].

In addition to the *ALDH2*2* allele, other genetic alleles can also provide protection from alcohol use disorder. In general, any polymorphism that leads to acetaldehyde buildup in blood deters the carrier of the genetic polymorphism from consuming alcohol because many unpleasant reactions after consumption, such as facial flushing, hypotension, headache, and nausea, are related to acetaldehyde. *ADH1B*2*, a variant of *ADH1B* (wild type), encodes a superactive ADH enzyme with higher enzymatic activity. Another variant of *ADH1B* is *ADH1B*3*. ADH isoenzymes encoded by *ADH1B*2* or *ADH1B*3* alleles lead to a 30- to 40-fold increase in the conversion of ethanol (superactive enzyme) into acetaldehyde, and such rapid production of acetaldehyde may cause facial flushing and adverse side effects after alcohol use, thus discouraging people who carry such alleles from drinking alcohol. Kang *et al.* commented that higher blood acetaldehyde levels in individuals with *ADH1B*2/*2* may constitute the mechanism of protection against alcoholism by this allele [19]. Based on a study of a Chinese population living in Taiwan, it has been documented that the *ADH1B*2* allele has a protective effect against alcohol dependence, and homozygous individuals are more protected than heterozygous individuals (odds ratio of 0.12 for homozygous and 0.19 for heterozygous individuals compared to individuals carrying the wild-type gene). In European and African populations, the *ADH1B*2* allele is not common, but if present, it also provides protection against alcoholism. However, the protective effect of *ADH1B*2* appears to be weaker in European than in Asian populations, which may be related to different environmental factors or coinheritance of other, as yet uncharacterized, polymorphisms in the gene encoding ADH. In addition, the *ADH1B*3* allele has a protective effect against alcoholism in African Americans and southern California Indians. The *ADH1B*3* allele also protects against fetal alcohol syndrome. Finally, *ADH1C*1* appears to have a protective effect against alcoholism in Asian populations, but this allele is usually coinherited with the protective *ADH1B*2* allele, and the effect of *ADH1C*1* may not be independent [17]. However, in one study, the authors found an association between *ADH1C*1* and protection from alcohol dependence in Native Americans [20].

As mentioned previously, *ALDH2*2* protects from alcohol abuse because this mutation produces nearly inactive enzyme and as a result acetaldehyde accumulates in blood. If *ALDH2*2* is present along with *ADH1B*2*, the combination has further protective effect against alcoholism. However, Lu *et al.* reported that the *ADH1B*2* allele had no protective effect against alcohol dependence in subjects with antisocial traits who were also alcoholics, although the *ALDH2*2* allele can still provide some protection [21].

10.4.2 Polymorphisms that may Increase the Risk of Alcohol Use Disorder

Although the *ALDH2*2* allele has a protective effect against alcoholism, the functional polymorphism of this gene, *ALDH2*1*, may increase the risk of alcohol abuse because the functional enzyme can rapidly convert acetaldehyde into acetate. Similarly, the wild-type *ADH1B*1* allele may also increase the risk of alcoholism because ethanol is metabolized normally to acetaldehyde, with no appreciable increase in acetaldehyde concentration in blood. Several studies have shown an association of wild-type *ADH1B*1* and *ALDH2*1* alleles with alcoholism [22]. Polymorphism of the *ADH4* gene may be associated with alcohol and drug dependence in European Americans [23]. In addition, other rare polymorphisms of ADH genes may lead to alcohol dependence. (See Chapter 2.)

Liver cytosolic aldehyde dehydrogenase (ALDH1A1) can also convert acetaldehyde into acetate, especially when the mitochondrial enzyme (ALDH2) is inactive, although no individual has been identified with a total absence of catalytic activity of the ALDH1A1 enzyme. Polymorphism of the gene encoding ALDH1A1 can be associated with alcohol dependence. Reports indicate that *ALDH1A1*2* is associated with alcohol dependence in Indo-Trinidadians [24,25]. However, other authors have found no association between this polymorphism and alcohol dependence. Otto *et al.* found no significant effect of *ALDH1A*2* on drinking behavior [26]. In contrast, Spencer *et al.* commented that both *ALDH1A1*2* and *ALDH1A1*3* alleles produced a trend in an African American population that may be indicative of association with alcoholism, but more observations are required to validate this observation [27]. Kortunay *et al.* reported that the *ADH1C*2* allele was associated with alcohol dependence in the Turkish population [28]. Konish *et al.* observed an independent effect of *ADH1C*2* (also known as *ADH3*2*) and *CYP2E1* gene polymorphism contributing to alcoholism in Mexican American men [29]. Associations of polymorphisms of genes encoding ADH and ALDH with either protection from alcohol abuse or increasing risk of alcohol abuse are listed in Table 10.2.

10.5 NEUROBIOLOGICAL BASIS OF ALCOHOL USE DISORDER

The essential features of alcohol abuse disorder are loss of control over drinking habit and craving for more alcohol even when intoxicated and continuation of alcohol abuse despite negative health and social consequences. The

Table 10.2 Polymorphisms of Genes Encoding Alcohol Dehydrogenase and Aldehyde Dehydrogenase That Are Associated with the Protective Effect from Alcohol Use or Increased Risk of Alcohol Abuse[a]

Protective Effect from Alcohol Use	May Increase Risk of Alcohol Abuse
*ALDH2*2* *ADH1B*2* *ALDH2*2* and *ALDH1B*2* combination best evidence of protection *AHD1B*3* *ADH1C*1*	*ADH1B*1* *ALDH2*1* *ADH1C*2*

[a]*There are polymorphisms of ADH4, ADH6, and ADH7 that may be associated with increased risk of developing alcohol use disorder but more studies are needed to establish their role in the development of alcohol use disorder.*

rewarding effect of alcohol consumption eventually leads to reinforcement of consuming alcohol on a regular basis, which eventually results in alcohol tolerance and dependence. Alcohol affects various neurotransmitter systems in the brain, which plays a central part in drug and alcohol addiction. Neurotransmitters are endogenous chemicals responsible for transmission of signals from neurons to target cells across synapses. Major neurotransmitter systems affected by alcohol include dopaminergic, serotoninergic, GABA, and glutamate pathways. It is important to note that moderate consumption of alcohol on a regular basis does not make a person alcohol dependent. It is chronic consumption of alcohol for an extended period that produces changes in brain chemistry that result in alcohol use disorder and withdrawal symptoms upon discontinuation of alcohol. Polymorphisms of genes encoding various receptors and transmitters in the neurotransmission system may be associated with alcohol use disorder.

10.6 POLYMORPHISMS OF GENES IN DOPAMINE PATHWAY AND ALCOHOL USE DISORDER

Dopamine is an important neurotransmitter in the brain, controlling various functions such as motor activity, cognition, emotion, motivation, food intake, and endocrine secretion. Dopamine is used by the mesolimbic, nigrostriatal, and tuberoinfundibular systems of the brain. The mesolimbic dopaminergic pathway also functions as a reward pathway. Dysfunctions in the dopaminergic systems are involved in several pathological conditions, including Parkinson's disease, Tourette's syndrome, drug addiction, and hyperactivity disorders. Dopamine is also involved in drug and alcohol addiction processes, especially drug addiction. The dopaminergic system consists of dopamine receptors as well as dopamine transporters.

Polymorphisms of genes encoding dopamine receptors or transporters may be involved in susceptibility to alcohol abuse as well as novelty-seeking behavior, a personality trait that may cause alcohol abuse.

10.6.1 Dopamine Receptors

Initially, it was thought that only two receptor subtypes exist for dopamine (DRD1 and DRD2), and the target for anti-Parkinson drugs as well as antipsychotic drugs was DRD2. Later, three other subtypes of dopamine receptors (DR3, DR4, and DR5) were discovered [30]. The diverse physiological functions of dopamine receptors encoded by the genes *DRD1– DRD5* and polymorphisms of these genes can result in changes in expression levels [31]. The most studied dopamine receptor in relation to alcohol abuse disorder is DRD2.

DRD1 is involved in the rewarding/reinforcing effects of drugs of abuse, and this receptor gene has been considered as a candidate gene in alcohol dependence. The D1 type family of G protein-coupled receptors, such as DRD1, activates adenylyl cyclase and is involved in various brain functions, including motor control and the reward and reinforcement mechanism, as well as being involved in the attention deficit hyperactive syndrome. DRD1 antagonist may reduce cocaine-seeking behavior. Although none of the polymorphisms of the *DRD1* gene is associated with alcoholism, based on genotyping of 11 polymorphisms of the *DRD* gene family in 535 alcohol-dependent subjects in a Korean population, Kim *et al.* reported that polymorphism in the 5′-untranslated region (5′-UTR) of the *DRD1* (48 A>G) gene located on chromosome 5 (q35.1) was associated with alcohol-related problems. However, none of the polymorphisms of the *DRD5* gene had any significant association with alcoholism [32]. Based on an analysis of 134 alcohol-dependent patients, Batel *et al.* observed that the T allele of the rs686 polymorphism of the *DRD1* gene was significantly more frequent in patients with alcohol dependence. A specific haplotype, rs686*T–rs4532*G, associated with the *DRD1* gene was significantly more precisely associated with alcohol dependence among patients studied by Batel *et al.* [33]. Prasad *et al.* showed that rs4532 present in the 5′-UTR of the *DRD1* gene was associated with alcoholism in East Indian men and noted that the 48 A>G SNP could be a predisposing factor for alcohol abuse. In addition, the *– 120 ins/del* polymorphism in the *DR4* gene was also associated with alcoholism [34].

The *DRD2* gene on chromosome 11 (q22–q23) has been found to be associated with increased alcohol consumption through a mechanism involving incentive salience attribution and craving in patients with alcohol use disorder. The DRD2 receptor is a G protein-coupled receptor located on postsynaptic dopaminergic neurons that plays a role in the reward pathway via the mesolimbic system. The *DRD2* gene shows primarily three kinds of polymorphisms: *141c ins/del*, *Taq1A*, and *Taq1B* alleles.

The dopamine receptor D2 (*DRD2*) gene is most extensively studied gene in addiction disorders, with the *Taq1A* polymorphism being the most frequently studied. The *DRD2 Taq1A* polymorphism is located 10 kb downstream from the coding region of the *DRD2* gene at chromosome 11q23. A mutation in this noncoding region should not cause a structural change in the dopamine receptor. Therefore, the functional significance of this polymorphism is unclear. However, *Taq1A* polymorphism may be in linkage disequilibrium with an upstream regulatory element or another functional gene that could be responsible for susceptibility to alcoholism in these individuals. In fact, the *DRD2*-associated polymorphism has been more precisely located within the coding region of the *ANKK1* (ankyrin repeat and kinase domain containing 1) gene, which is a neighboring gene that may confer a change in amino acid sequence. Higher frequency of the A1 allele of *Taq1A* near the *DRD2* gene was observed among alcoholics. Based on a meta-analysis of 42 studies with 9382 participants, Smith *et al.* observed an association between the presence of one or two copies of the A1 allele and alcohol dependency [35]. Another study also reported an association between the A1 allele and susceptibility to early onset of alcoholism [36]. Based on a study of 103 alcohol-dependent males, Ponce *et al.* reported that approximately 39% of the sample carried the A1 allele (*Taq1A* polymorphism), and these individuals showed a higher prevalence of antisocial personality disorder and family history of alcoholism. Moreover, these people showed early onset of alcohol abuse and more drinking problems [37]. However, Shaikh *et al.* did not observe any association between the A1 allele and the development of alcoholism at a young age (<25 years) in an Indian population [38]. The single nucleotide polymorphism *Taq1B* is closer to the regulatory and structural coding region (5′ region) of the *DRD2* gene, and it may be expected that such polymorphism may play a role in alcohol dependence. The few studies that have explored the association between *Taq1B* and alcohol dependence have produced conflicting results. In one study, the authors found no correlation between alcoholism and *Taq1B* in Mexican Americans, whereas another study observed an association between *Taq1B* polymorphism and early onset of alcohol consumption in Mexican Americans. However, other studies have found no association between this polymorphism and alcohol abuse in Indian populations [39]. The effect of the − *141c ins/del* allele on alcohol dependence is also controversial, with some studies reporting no association and other studies observing some correlation between this allele and alcohol dependence. Prasad *et al.* reported an association between the − *141c ins* allele and alcohol dependence in Indian males. Haplotype with a predisposing − *141c ins* and *Taq1A* alleles seems to confer approximately 2.5 times more risk for developing alcohol dependence [40].

Several polymorphisms of the DRD3 receptor gene have been reported to be associated with susceptibility to alcoholism. Although the *DRD3* rs6280 (Ser9Gly; substitution of a glycine for a serine residue in the extracellular receptor N-terminal domain) polymorphism, which makes the receptor more sensitive to dopamine, is associated with various psychiatric disorders, its association with alcoholism is inconsistent. However, Kang *et al.* observed an association of this polymorphism with alcohol dependence in Korean subjects [41]. In a case–control study of 108 French alcoholic subjects and 71 healthy controls, Limosin *et al.* observed increased homozygosity for *Bal I* polymorphism (Ser9Gly is also known as *Bal I* polymorphism due to a restriction site produced by the variant) in alcohol-dependent patients with low cognitive impulsiveness. However, heterozygous subjects showed higher impulsiveness than homozygous subjects, although this did not reach statistical significance [42]. In contrast, Prasad *et al.* found no association between rs6280 mutation and alcoholism in Indian males [34].

DRD4 receptors, which are G-coupled receptors encoded by the *DRD4* gene, have been linked to several psychiatric disorders (e.g., schizophrenia and bipolar disorder), neurological disorders (e.g., Parkinson's disease), and addictive behavior (e.g., novelty seeking). *DRD4* has a variable number of tandem repeat (VNTR) polymorphisms on exon 3 that may affect gene expression by binding nuclear factors. The *DRD4* VNTR can vary from 2 to 10 with 48-bp repeats, where alleles with 6 or fewer repeats are termed "short alleles" and alleles with 7 or more repeats are termed "long alleles." Compared to short alleles, long alleles may reduce *DRD4* gene expression. The most studied polymorphism of the *DRD4* gene is a 48-bp VNTR in the third exon [43]. In general, *DRD4* VNTR long alleles with 7 or more repeats are associated with novelty-seeking and drinking behavior. In a study of 90 heavy drinking college students, Ray *et al.* observed an association between *DRD4* VNTR long-allele genotype and problem drinking as well as novelty-seeking behavior [44]. Du *et al.* genotyped the VNTR polymorphism of a 48-bp sequence in exon 3 of the *DRD4* gene in 365 alcoholic and 337 nonalcoholic Mexican Americans and observed an association between *DRD4* VNTR and alcoholism [45]. Park *et al.* observed that carriers of the long-allele *DRD4* VTNR polymorphism showed greater susceptibility to environmental effects that may lead to alcohol dependence [46]. α-Synuclein has an important role in regulating dopamine function and dopamine synthesis. The protein α-synuclein is encoded by the *SNCA* gene. α-Synuclein is suspected to play a role in the development of alcohol dependence because a higher level of mRNA expression has been observed in alcohol patients. Foroud *et al.* found no association between *SNCA* polymorphism and alcohol dependence but did observe an association between alcohol craving and polymorphism in the *SNCA* gene [47].

10.6.2 Dopamine Transporters and Dopamine-Metabolizing Enzymes

Dopamine transporter (DAT) belongs to the family of sodium/chloride-dependent transporter proteins and is responsible for the reuptake of extracellular synaptic dopamine into presynaptic neurons, thus terminating the effect of dopamine. Therefore, the dopaminergic reward circuit may function differently when DAT expression level is altered. Neuroimaging studies showed that DAT levels were significantly lower in the striatum of alcohol-dependent patients compared to controls. DAT availability in the brain may be dependent on genetic variation. The *DAT* gene (*DAT1*; locus symbol SLC6A3) is present on chromosome 5q15.3. Expression of DAT is influenced by a VNTR polymorphism located in the 3′-UTR region of the dopamine transporter (*DAT1*) gene. For VNTR, numbers ranging from 3 to 16 have been described, with 9- and 10-repeat alleles being the most common. In general, many studies have found no significant association between polymorphism of the *DAT1* gene and alcohol dependence. However, some studies have shown that the 9-repeat allele (A9 allele) is present in some subgroups of alcoholics with a variety of alcohol withdrawal symptoms, including withdrawal seizure and delirium tremens [48]. Using a group of 102 healthy subjects and 216 alcoholics, Kohnke *et al.* showed an association of allele A9 (9-copy repeat allele in the VNTR polymorphism in the 3′-UTR region of the *DAT1* gene) and alcoholism [49]. Bhaskar *et al.* studied the association between *DAT1* VNTR polymorphism and alcoholism in two culturally different populations (Kota and Badaga) in south India and found that the A10 allele was more common than the A9 allele in both populations. The genotypic distribution was in Hardy–Weinberg equilibrium in both cases and control groups of Kota and Badaga populations. In addition, the *DAT1* VNTR polymorphism was significantly associated with alcoholism in the Badaga population but not the Kota population. The authors concluded that the A9 allele of the *DAT1* gene is involved in vulnerability to alcoholism, but such association could be population specific [50].

Dopamine β-hydroxylase (DβH) catalyzes the conversion of dopamine to norepinephrine, and several polymorphisms in the *DβH* gene have been the focus of addiction research for many years. Several studies have found no association between polymorphism of the *DβH* gene and alcohol dependence. Although DβH may be lower in alcoholics and the *DβH* gene 1012C>T polymorphism is associated with lower plasma activity of the enzyme, no correlation has been observed between *DβH* −1021C>T polymorphism and alcoholism [51]. However, based on a study of 102 healthy controls and 208 alcoholics, Kohnke *et al.* found an association between *DβH* *444 G>A polymorphism, found in exon 2, and alcoholism [52].

10.6.3 Monoamine Oxidase

Monoamine oxidase (MAO), a mitochondrial enzyme in humans, can be subclassified as MAOA and MAOB. MAOA is responsible for metabolism of monoamine neurotransmitters, including dopamine, norepinephrine, and serotonin. Human platelet MAO levels vary considerably, and platelet MAO abnormality has been linked with a number of psychiatric disorders. Low MAO activity probands undergo more psychiatric counseling than high MAO activity probands, and in one study, low MAO probands had a higher incidence of suicide and suicide attempts compared to high MAO probands. Also, individuals with low platelet MAO activity show sensation-seeking behavior, a personality trait related to alcoholism. Although some investigators have observed low levels of platelet MAO activity in alcoholics, indicating that it could be used as a trait marker for alcoholism in males, Farren *et al.* observed no difference in platelet MAO activity between alcoholics and controls [53]. However, Pombo *et al.* observed a significantly low level of platelet MAOB activity in alcoholics compared to controls and commented that MAOB is a potential trait marker for type 1 alcohol-dependent patients [54].

MAOA is encoded by an X chromosome-linked gene (*MAOA* gene; Xp11.23), and the polymorphism of this gene affects its transcriptional activity. The locus termed MAOA-linked polymorphic region (*MAOA-LPR*) is a VNTR that is located approximately 1.2 kb upstream from the *MAOA* start codon and within the gene's transcriptional control region. Alleles of this VTNR (*MAOA-μVTNR*) have different numbers of copies of 30-bp repeated sequence present in 3, 3.5, 4, or 5 copies. The alleles with 3.5 or 4 copies of a repeat sequence are transcribed 2–10 times more efficiently than those with 3 or 5 copies [55]. Studies have shown the association between the 3-repeat allele and a cluster of externalizing behaviors, including alcoholism, antisocial personality, and impulsivity. This is expected because the 3-repeat allele is associated with lower MAOA activity, and low MAO activity has been related to traits associated with a high risk of alcoholism, drug abuse, and impulsive behavior. In a study of Brazilian alcoholics, Contini *et al.* showed an association between the 3-repeat allele and alcohol dependence, earlier onset of alcoholism, comorbid drug abuse among alcoholics, and antisocial symptoms. The authors concluded that the 3-repeat allele of the *MAOA-μVNTR* polymorphism was associated with alcohol and substance abuse as well as impulsive and antisocial behavior [56]. However, Saito *et al.* reported minimal association between the MAOA low-activity promoter allele and alcoholism among male Finnish alcoholics [57].

10.6.4 Catechol-*O*-Methyltransferase

Catechol-O-methyltransferase (COMT) is an enzyme located in the frontal cortex of the brain that metabolizes epinephrine, norepinephrine, and dopamine. The COMT enzyme is encoded by the *COMT* gene, located on chromosome 22 (22q11.21), and the gene produces two versions of this enzyme. The longer form, called membrane-bound catechol-*O*-methyltransferase (MB-COMT), is mainly produced by the nerve cells in the brain. A shorter form of this enzyme, called soluble catechol-*O*-methyltransferase (S-COMT), is found in liver, kidneys, and blood. Polymorphism of the *COMT* gene may cause mental illness such as schizophrenia. As a result of polymorphism of the *COMT* gene, in the longer form of the enzyme, the amino acid variation occurs at position 158 (Val158Met, indicating that the amino acid at position 158 could be either valine or methionine). In the shorter form of this enzyme, the variation occurs at position 108 (Val108Met). The enzyme containing valine is three or four times more active than the enzyme containing methionine (at position 158 or 108, depending on the long form or the short form). Therefore, the *COMT* allele that produces valine-containing enzyme at position 158 should cause a lower dopamine level in the frontal cortex. Homozygosity for the low-activity *COMT* allele (LL genotype) is found in approximately 25% of Caucasians, whereas high-activity homozygosity (HH genotype) is found in approximately 25% of Caucasians. The *COMT* allele (LL genotypes) encodes the low-activity membrane-bound enzyme with methionine at position 158. Homozygous with HH genotype has three or four times higher enzyme activity (valine at position 158) compared to homozygous with LL genotype, whereas heterozygous (LH genotype) shows intermediate enzyme activity. Kauhanen *et al.* reported that men with the LL genotype reported 27% higher weekly alcohol consumption than men with the two other genotypes (LH and HH genotypes). The authors concluded that homozygosity for the low-activity *COMT* allele (LL genotype) is associated with increased alcohol consumption in a population sample of middle-aged Finnish men [58]. However, Ishiguro *et al.* found no association between polymorphism of the *COMT* gene (Val158Met in the membrane-bound form of the enzyme) and alcoholism in their study of 175 Japanese alcoholics and 354 age- and gender-matched Japanese controls [59]. The link between dopamine pathways and alcoholism is summarized in Table 10.3.

10.7 POLYMORPHISMS OF GENES IN THE SEROTONIN PATHWAY AND ALCOHOL USE DISORDER

Apart from the dopamine pathway, a link between the serotonin pathway and alcoholism has also been suggested. Serotonin, also known as 5-hydroxytryptamine (5-HT), is a monoamine neurotransmitter that is derived from tryptophan. Serotonin is found in the central nervous system (CNS), the gastrointestinal tract, and the dense granules of platelets. Some of the functions of serotonin in the CNS involve regulation of mood, appetite, sleep, and

Table 10.3 Link between Polymorphisms of Genes Encoding Receptors, Transporters, or Enzymes Involved in the Dopamine Pathway and Alcoholism

Gene Involved	Chromosome	Polymorphism	Comments
DRD1	5q35:1	Polymorphism in 5′-UTR: *DRDI* (48 A>G)	Associated with alcohol-related problems
		Haplotype rs686*T–rs4532*G	Associated with alcoholism
DRD2	11q22–23	*Taq1A/ANKKI*	Associated with alcoholism
		Taq1B	Conflicting results
		−141c ins/del	Conflicting results
DRD3	3q13.3	*DRD3* rs6280 (Ser9Gly)	Alcohol dependence in Korean subjects
DRD4	11p15.5	*DRD4* VNTR Long allele with seven or more repeats of 48 bp	Associated with problem drinking/alcoholism as well as novelty-seeking behavior
DAT1	5q15.3	*DAT1* gene allele A9	May be associated with alcoholism in certain populations
DβH	9q34.2	*DβH**444 G>A	Associated with alcoholism
MAOA	Xp11.23	Three-repeat allele of *MAOA–μVNTR*	May be associated with alcoholism but particularly associated with substance abuse as well as impulsive and antisocial behavior
COMT	22q11.21	*COMT* LL genotype producing low activity membrane-bound enzyme with methionine at position 158	Associated with high weekly consumption of alcohol, although some studies found no association between alcoholism and *COMT* gene polymorphism

muscle contraction. Serotonin also has cognitive functions, such as memory and learning. Most brain serotonin is not degraded after use but, rather, is collected by serotonin transporters (5-HTT) on the cell surface. The variation in serotonin transporters affects mood with possible interaction with the environment, causing depression. Many drugs that treat depression (selective serotonin reuptake inhibitors) work by increasing serotonin levels in the brain. Serotonin levels are also influenced by many drugs of abuse and alcohol. Pivac *et al.* studied platelet serotonin levels in 148 male and 42 female alcohol-dependent but drug-free subjects for whom the diagnosis of alcohol dependence was based on *DSM-IV* criteria. In addition, the authors studied 110 males and 123 females who were used as controls. The authors observed significantly lower platelet serotonin levels in both male and female alcohol-dependent subjects compared to controls. However, in general, male subjects (both alcoholics and controls) showed higher platelet serotonin levels than corresponding female subjects. Smoking status did not affect platelet serotonin levels. The authors concluded that, in general, alcohol dependence reduced platelet serotonin levels in both male and female alcoholics [60].

Because serotonin is involved in regulating alcohol consumption in both human and animal models, it is assumed that polymorphism of the gene encoding serotonin transporter protein (5-HTT locus SLC6A4) may be associated with alcohol dependence. The human serotonin transporter gene (*5-HTT*), found on chromosome 17 (17q11.1–q12), has a biallelic functional polymorphism in its *5-HTT*-linked promoter region (*5-HTTLPR*), which is composed of a 44-bp insertion (long (L) allele) or deletion (short (S) allele) that regulates the transcription of the serotonin transporter. The S allele is associated with reduced serotonin transporter protein expression. In contrast, the L allele is associated with increased expression of presynaptic serotonin transporter protein, resulting in more efficient serotonin reuptake and reduction of synaptic serotonin levels compared to those of individuals with the S allele.

The S allele has been associated with alcohol dependence; affective disorders; as well as several personality trait, including anxiety, depression, and impulsiveness. Bondy *et al.* found an association between the S allele and violent suicide [61]. Feinn *et al.* conducted a meta-analysis of data from 17 published studies (including 3489 alcoholics and 2325 controls) and concluded that the frequency of the S allele at *5-HTTLPR* was significantly associated with alcohol dependence (odds ratio (OR) 1.18, indicating that the S allele increased the odds by at least 18%). In addition, a greater association with the S allele was seen among individuals with alcohol dependence complicated by either a comorbid psychiatric condition or an early onset of alcohol dependence or more severe alcohol dependence [62]. Pinto *et al.* observed that the S allele of the *5-HTTLPR* polymorphism was associated with relapse in abstinent alcohol-dependent patients [63]. Recently, polymorphism in *5-HTTLPR* with the L allele has received attention as a marker for alcohol craving. In a study of 124 male patients who were admitted to an alcohol detoxification treatment program, Bleich *et al.* found an association between the L allele of *5-HTTLPR* polymorphism and higher compulsive alcohol craving at the beginning of alcohol withdrawal [64]. In a study of 101 alcohol-dependent patients admitted to the hospital, Pombo *et al.* observed that alcohol-dependent patients who were homozygous for the L allele (LL genotype) had self-reported higher scores of alcohol craving compared to patients who were homozygous for the S allele (SS genotype). However, the results were not statistically significant [65].

Another polymorphism of the serotonin transporter gene is intron 2 VNTR, which is associated with three distinct alleles with a 9-, 10-, or 12-bp repeat (*5-HTTVNTR* or *STin2* VNTR, where "ST" stands for serotonin transporter). In a study of 90 Spanish Caucasian alcohol-dependent outpatients, Florez *et al.* observed that *STin2* 12/12 carriers showed poor 6-month treatment outcome during an alcohol rehabilitation program. On the other hand, patients with the 10/10 genotype had a better treatment outcome [66]. However, other studies have failed to confirm this observation.

In addition to serotonin transporter gene polymorphism, serotonin receptor gene polymorphism has been studied in order to identify a potential association with alcohol use disorder. Serotonin receptors can be grouped into seven classes (5-HT_{1-7}), all belonging to the G protein-coupled receptor superfamily except for 5-HT_3, which is a ligand-gated ion channel receptor. The 5-HT_1 receptors can be subgrouped into five categories (1 A, 1B, 1D, 1E, and 1 F), 5-HT_2 receptors into three categories (2 A–2C), and 5-HT_5 into two categories (5 A and 5B). These receptors are encoded by specific genes. Although animal models and imaging studies have suggested a potential role of certain serotonin receptors and polymorphism of genes encoding these receptors (HTR1A, HTR1B, HTR2A, and HTR3), in the pathogenesis of alcohol use disorders, a limited number of genetic association studies have been performed using human subjects, and these studied have produced conflicting results. Himei *et al.* studied the genetic association between alcoholism and alleles of *HTR1A* (coding the 5-HT_{1A} receptor), *HRT2A* (coding the 5-HT_{2A} receptor), and *HRT2C* (coding the 5-HT_{2c} receptor) genes but found no association between alcoholism and these genes. The authors concluded that serotonergic receptor genes may not directly contribute to the etiology of alcoholism [67]. However, Cao *et al.* performed a meta-analysis and observed an association between polymorphism of the *HTR2A* (rs6313; 102 T/C at exon 1) gene and susceptibility to substance abuse disorder, particularly alcohol dependence [68]. In a study of 150 alcohol-dependent patients, Wrzosek *et al.* observed a potential role of *HTR2A* (rs6313; 102 T/C) polymorphism in the development of alcohol dependence. The *HTR2A* gene is located on chromosome 13 (13q14–q21), and this polymorphism is also found in higher frequencies among schizophrenics [69]. In another study, the authors performed a meta-analysis and observed an association between *HTR1B* gene polymorphisms (rs11568817, −261 T>G and rs130058, −161 A>T) and alcohol and drug dependence [70]. Yang and Li reported that one haplotype formed by SNPs rs3891484 and rs3758987 in *HTR3B* was associated with alcohol dependence in African American subjects [71]. Xu *et al.* performed haplotype analysis for two SNPs in the *HTR4* gene (rs17777298 and rs10044881) and observed a protective effect of *HTR4* AG haplotype in a Tibetan population from developing alcohol use disorder [72]. Herman and Balogh noted that polymorphisms of serotonin transporter and serotonin receptor genes (*HTR1B*, *HTR2A*, *HTR2C*, and *HTR3*, including *HTR3A*, *HTR3B*, *HTR3C*, *HTR3D*, and *HTR3R*) likely contribute to substance use disorder, but further research using whole genome sequencing technology and large sample sizes is needed to firmly document the association of certain polymorphisms of these genes with substance abuse disorder, including alcohol abuse disorder [73]. Polymorphisms of genes encoding serotonin receptors and transporters that may be linked with alcohol abuse are listed in Table 10.4.

Table 10.4 Link between Polymorphisms of Genes Encoding Receptors, Transporters, or Enzymes Involved in the Serotonin Pathway and Alcoholism

Gene Involved	Chromosome	Polymorphism	Comments
5-HTT (serotonin transporter gene)	17q11.1–q12	*5-HTTLPR*, short (S) allele, and long (L) allele	S allele results in lower expression of serotonin transporter and is associated with alcohol dependence, depression, and other psychiatric illness. L allele may be associated with alcohol craving, but this information is controversial.
5-HTT	17q11–q12	*5-HTTVNTR* (STin2)	12/12 carriers show poor treatment response during alcohol rehabilitation, but carriers of the 10/10 allele show better response.
HTR2A (serotonin receptor gene)	13q14–q21	SNP (rs6313: 102 T/C)	May be associated with alcohol dependence; also, this polymorphism is found in patients with schizophrenia.
HTR1B (serotonin receptor gene)	6q13	SNPs (rs11568817: 261 T>G and rs130058: 161 A>T)	Associated with alcohol and drug dependence.
HTR3B (serotonin receptor gene)	11q23.1	Haplotype formed by SNPs (rs3891484 and rs3758987)	Associated with alcohol dependence in African Americans.
HTR4	5q31–q33	*HT4* AG haplotype	Protective effect in Tibetan population.

10.8 POLYMORPHISMS OF GENES IN THE GABA PATHWAY AND ALCOHOL USE DISORDER

The main inhibitory neurotransmitter in the CNS is GABA. The neurotransmitter GABA binds to specific receptors (GABA receptors) located in the plasma membrane of neuronal cells. These receptors are heteromeric protein complexes consisting of several homologous membrane spanning glycoprotein subunits. Two main classes of GABA receptors (ionotropic $GABA_A$ and $GABA_C$ receptors), as well as metabotropic $GABA_B$ receptors, have been well-characterized. $GABA_A$ receptors are a family of ligand-activated chloride ion channels and a pentameric assembly out of various possible subunits—α_{1-6}, β_{1-3}, γ_{1-3}, δ, ε, θ, π, and ρ_{1-3}—and their splice variant [74]. $GABA_C$ receptors are composed of the ρ_{1-3} subunits. The composition of the most common $GABA_A$ receptor includes two α_1, two β_2, and one γ_2 subunit. When the neurotransmitter GABA binds to $GABA_A$, the conformational structure of the receptor is changed, and the membrane pore opens to allow the flow of chloride ions. $GABA_B$ receptors, which are widely distributed throughout the human brain, are G protein-coupled receptors that stimulate the opening of potassium channels. $GABA_C$ receptors, which differ in complexity of structure, abundance, distribution, and function from $GABA_A$ and $GABA_B$ receptors, can be found in retina, hippocampus, spinal cord, and pituitary tissues.

Neurotransmitter GABA as well as GABA receptors have long been implicated in mediating at least some pharmacological actions of alcohol. GABA receptors are also a molecular target for benzodiazepines and barbiturates, both of which show some cross-dependence with alcohol. Studies have indicated that $GABA_A$ receptors play an important role in alcohol dependence, and the functional properties of $GABA_A$ receptors are altered following chronic alcohol administration. The mechanism responsible for adaptation of $GABA_A$ receptors to chronic alcohol administration may involve alcohol-induced changes in cell surface expression, subcellular localization, synaptic localization, receptor phosphorylation, and/or changes in $GABA_A$ receptor subunit composition [75]. Therefore, it is expected that polymorphism of GABA genes that encode GABA receptors may be associated with alcohol dependence, alcohol tolerance, and alcohol withdrawal.

$GABA_A$ receptors are encoded by 16 genes, 14 of which are clustered on the following four chromosomes: 4p12–q13 (encoding α_2, α_4, β_1, and γ_1 subunits), 5p31–q35 (encoding α_1, α_6, β_2, and γ_2 subunits), 15q11–q13 (encoding α_5, β_3, and γ_3 subunits), and Xq28 (encoding α_3, α_4, β_4, and ε_1 subunits). Linkage analysis of multiplex families recruited for the Collaborative Study of the Genetics of Alcoholism (COGA) identified a region of chromosome 4p that was linked to alcohol dependence. Edenberg *et al.* performed linkage disequilibrium analysis of 69 SNPs within a cluster of four $GABA_A$ receptor genes (*GABRA2*, *GABRA4*, *GABRB1*, and *GABRG1*), all located on chromosome 4, and observed that the region of strongest association with alcohol dependence extended from intron 3 past the 3′ end of the *GABRA2* gene that encodes the α_2 subunit of the $GABA_A$ receptor. All 43 of the consecutive three-SNP haplotypes in this region of the *GABRA2* gene were significantly associated with alcohol dependence. The authors concluded that the very strong association of *GABRA2* with alcohol dependence and β frequency of the electrogram suggests that *GABRA2* might influence susceptibility to alcohol dependence by modulating the level of neural excitation [76]. This association has been replicated in many different studies involving people of European and African ancestry. Further evidence indicates that the association might extend beyond *GABRA2* to also include the adjacent *GABRG1* gene [77]. Based on a study of 257 German alcohol-dependent patients and 88 healthy controls, Fehr *et al.* observed that *GABRA2* is a susceptible gene for alcohol dependence in a European population. The authors identified rs279836 (T) as the high-risk haplotype. However, rs279871 (T) and rs279845 (A) belong to the nonrisk haplotype [78]. The *GABRGI* and *GABRA2* genes are located in a cluster on chromosome 4p.

Although association of alcohol dependence with markers located at the 3′ region of *GABRA2* has been documented in several studies, some studies suggest that the association signal may be attributable to the adjacent gene, *GABRG1*, located 90 kb distant in the 3′ direction. Because of linkage disequilibrium in European Americans, the origin of the signal is difficult to predict, but decreased linkage equilibrium is observed in African Americans. Therefore, further study of the African American population may be helpful to resolve this issue. Based on a study of 380 African American alcohol-dependent patients and 253 African American controls, Ittiwut *et al.* observed an interrelationship between the *GRBRG1* and *GABRA2* genes and commented that there is a likelihood that risk loci exist in each of these genes [79]. However, based on a study of 547 Finnish Caucasian men (266 alcoholics) and 311 community-derived Plains Indian men and women (181 alcoholics), Enoch *et al.* concluded that *GABRA2* and *GABRG1* haplotypes appear to be independent predictors of alcoholism in both populations [80].

The $GABA_A$ receptor, which contains the combination of α_1, β_2, and γ_2 subunits, is believed to be assembled following the coordinated expression of *GABRA1*, *GABRB2*, and *GABRG2* genes located on chromosome 5 (5q34) and to constitute the major $GABA_A$ receptor subtype present in the adult CNS (50%). The *GABRA6* gene is also present on chromosome 5, and all of these genes on chromosome 5 have been studied in relation to alcohol use. However, results are conflicting. Linkage analyses conducted in the COGA samples found no significant linkage in respect of alcohol dependence to the region of chromosome 5 containing the $GABA_A$ receptors genes (*GABRA1*, *GABRA6*, *GABRB2*, and *GABRG2*). However, in a follow-up study using COGA samples, Dick *et al.* observed an association between polymorphism of the *GABRA1* gene and alcohol dependence, history of blackouts, age at first drunkenness, and level of response to alcohol. The SNP rs980791 was significantly associated with history of alcohol-related blackouts. Another tagSNP rs4263535 showed significant association with age at first drunkenness [81]. Loh *et al.* observed an association between the *GABRG2* gene and alcohol dependence comorbid with antisocial personality disorder in a study based on a Japanese population [82]. Han *et al.* observed an association between the T allele of the 1519 T > C *GABRA6* gene and craving for alcohol and food during treatment of patients with alcohol dependence [83]. Using a Finnish sample of 511 subjects and a Southwestern Native American population of 433 individuals, Radel *et al.* observed an association between the *GABRG2* 1412 T allele and alcohol dependence in both populations. The *GABRA6* 1519 T allele was also associated with alcohol dependence in both populations. The most significant signal was observed at three-locus haplotypes that included one or more *GABRA6* polymorphisms [84]. However, another study found no association between *GABRA1*, *GABRA6*, *GABRB2*, and *GABRG2* genes located on chromosome 5 (5q34–q35) and alcohol dependence or alcohol-dependent comorbid antisocial personality disorder [85].

GABA receptor genes (*GABRA5*, *GABRB3*, and *GABRG3*) present on chromosome 15 (15q11.2–q12) have also been investigated for potential association with alcohol use disorder. These genes are expressed exclusively from the paternal chromosome so that the maternal gene should not influence phenotype. Using COGA samples and family based association tests, Song *et al.* observed an association between the *GABRA5* gene and alcoholism, particularly in the Caucasian population. Evidence of an association was also observed during paternal transmission of the *GABRB3* gene in the Caucasian population. In addition, the authors observed consistent although weak linkage disequilibrium between *GABRB1* (located on chromosome 4) and alcoholism [86]. In a study using both classical trio-based analyses and extended family analyses, the authors found consistent evidence of association between alcohol dependence and polymorphism of *GABRG3*. However, they found no consistent evidence of association between *GABRA5* or *GABRB3* and alcoholism [87].

Compared to $GABA_A$ receptors, few studies have been performed on the link between alcoholism and $GABA_B$ receptors as well as genes encoding $GABA_B$ receptors. Most studies have yielded negative results. However, based on a study of 350 German alcohol-dependent subjects and 234 German controls, Sander *et al.* observed a trend of association of an exonic polymorphism T1974C in the *GABBR1* gene on chromosome 6 (6p21.3) with alcoholism, with the T allele being more frequently associated with alcoholism [88]. GABA transporters are involved in terminating signal transmitted by GABA. The major GABA transporters are GAT-1, GAT-2, and GAR-3 encoded by genes *SLC6A1*, *SLC6A13*, and *SLC6A11*. Although studies to determine an association between polymorphism of genes encoding GABA transporters and alcoholism have mostly produced negative results, polymorphism of these genes may be related to other psychiatric disorders, such as anxiety, schizophrenia, and attention deficit hyperactivity disorder. Some of these psychiatric illnesses may be observed comorbid with alcohol use disorder. Associations of polymorphism of genes that encode GABA receptors with alcohol dependence are listed in Table 10.5.

10.9 POLYMORPHISMS OF GENES ENCODING CHOLINERGIC RECEPTORS AND ALCOHOL USE DISORDER

Acetylcholine, a neurotransmitter in the brain and autonomic nervous system, is the natural agonist of muscarinic and nicotinic receptors. Muscarinic receptors are G protein-coupled acetylcholine receptors in the plasma membrane of

Table 10.5 Link between Polymorphisms of Genes Encoding GABA Receptors and Alcoholism

Gene Involved	Chromosome	Polymorphism	Comments
GABRA2	4p12	SNP (rs279836: T allele)	High-risk allele for alcohol dependence. *GABRA2* polymorphism may also be associated with the adjacent gene *GABRG1* in producing a high-risk allele for alcohol dependence, but some investigators have commented that both genes are independently involved.
GABRA1	5q34	SNP (rs980791)	Associated with a history of alcohol-related blackouts.
		tag SNP (rs263535)	Associated with age at first drunkenness.
GABRA6	5q34	1519 T > C allele	Associated with alcohol dependence in various populations. However, no association between GABA receptor genes on chromosomes *GABRA1*, *GABRA6*, *GABRB2*, and *GABRG2* has been reported.
GABRA5, *GABRB3*	15q11.2–q12	Various polymorphisms	Association of polymorphisms of these $GABA_A$ receptor genes with alcohol dependence is controversial.
GABRG3	15q12	Various SNPs	May be associated with alcoholism.
GABBR1	6p21.3	T1974C	T allele more frequently associated with alcoholism.

neurons throughout the CNS. Nicotinic acetylcholine receptors are receptor ion channels in the autonomic nervous system. Co-occurrence of alcohol and nicotine addiction in humans is well documented, and it is speculated that common genetic polymorphisms may affect the two addictions. Human genetics studies have indicated that nicotinic acetylcholine receptor genes may play a role in mediating early behaviors and risk factors for both alcohol and nicotine dependence. The nicotinic acetylcholine receptors are also an important component of the dopaminergic reward pathway because some of the receptors can activate the release of dopamine. Several genetic studies have shown that common genetic factors may contribute to alcohol and nicotine co-addiction [89].

Genetic studies have indicated an association between two genes that encode the nicotinic acetylcholine receptor α_4 and β_2 subunits (*CHRNA4* and *CHRNB2*) and alcohol and tobacco dependence. In a study of 1068 ethnically diverse young adults, Ehringer *et al.* observed a modest association of six SNPs in the *CHRNA4* gene with 6 preceding months of alcohol use in Caucasians, but they observed no association between tobacco use and *CHRNA4* gene polymorphism. In addition, the SNP (rs2072658) located immediately upstream of *CHRNB2* was associated with the initial subjective response to both alcohol and tobacco [90]. Using a Spanish population of 417 subjects, Landgren *et al.* studied the possible association between 20 tagSNPs in five nicotine subunit receptor genes (*CHRNA3, -4,* and *-6* and

CHRNB2 and *-3*) and observed that a haplotype block in the *CHRNA6* gene consisting of four consecutive SNPs (10–13; rs10087172, rs10109429, rs2196129, and rs16891604) was strongly associated with alcohol consumption. Phasing of this haplotype block resulted in four haplotypes; within this haplotype block, two haplotypes of the *CHRNA6* gene (CCCC and TCGA) showed the strongest association with alcohol consumption. In addition, the haplotype in the *CHRN4* gene (GGTG) associated with increased body mass consists of one synonymous SNP and three intronic SNPs [91].

The cholinergic muscarinic receptor 2, implicated in memory and cognition, is encoded by the *CHRM2* gene, which is located on the long arm of chromosome 7. The *CHRM2* gene contains a single coding exon and a large 5′-UTR region encoded by multiple exons that can be alternatively spliced. Association between the *CHRM2* gene and alcohol dependence has been reported in genome-wide linkage studies. Wang *et al.* examined 11 SNPs spanning the *CHRM2* gene and identified 3 SNPs (one in intron 4 and two in intron 5) that showed a highly significant association with alcoholism. Haplotype analyses revealed that the most common haplotype, TTT (rs1824024–rs2061174–rs324650), was undertransmitted to affected individuals with alcohol dependence and depressive disorder [92]. Luo *et al.* observed that *CHRM2* polymorphism (SNP: rs1824024) was in Hardy–Weinberg disequilibrium in alcohol-dependent and drug-dependent subjects. The authors concluded that polymorphic variation in *CHRM2* predisposes to alcohol and drug dependence as well as affective disorders. The 5′-UTR region of the *CHRM2* gene, which contains one haplotype block harboring SNPs 1–3, may be more important for the development of these disorders than other regions [93]. Based on a study of a Korean population, Jung *et al.* observed that whereas SNP rs324650 showed marginal association with the risk of alcohol dependence, SNP rs1824024 in *CHRM2* was significantly associated with Alcohol Use Disorders Identification Test (AUDIT) score as well as Alcohol Dependence Scale (ADS) scores in patients [94]. Based on a study of 97 female college students, Bauer and Ceballos observed an excess prevalence of minor allele homozygotes (*CHRM2*; SNP rs324650) among frequent binge drinkers. This genotype was previously shown to be associated with substance dependence and major depressive disorder [95]. The Link between polymorphisms of genes encoding cholinergic receptors and alcoholism is summarized in Table 10.6.

10.10 POLYMORPHISMS OF GENES IN THE GLUTAMATE PATHWAY AND ALCOHOL USE DISORDER

Glutamate is the primary excitatory neurotransmitter in the brain acting at ionotropic and metabotropic glutamate receptors. Rapid excitation by

Table 10.6 Link between Polymorphisms of Genes Encoding Cholinergic Receptors and Alcoholism

Gene Involved	Chromosome	Polymorphism	Comments
CHRNA4	20q13.33	Six SNPs	Associated with six month consumption of alcohol in a Caucasian population
CHRNB2	1q21.3	SNP (rs2072658) located immediately upstream of the *CHRNB2* gene	Associated with initial subjective response to both alcohol and tobacco.
CHRNA6	8p11.21	Various polymorphisms	Two haplotypes—CCCC and TCGA—showed strongest association with alcohol consumption
		Haplotype block with four consecutive SNPs (rs10087172, rs10087172, rs10109429, and rs2196129)	Associated with alcoholism
CHRM2	7q31–35	Various SNPs	Associated with alcoholism

glutamate involves action of AMPA (α-amino-3-hydroxy-5-methyl-4-isoxazole-propionic acid receptor), kainate, and NMDA ionotropic glutamate receptors. The NMDA receptor is central to the development and function of the nervous system, and it plays an important role in memory and learning. NMDA receptors are heterotetramers composed of two glycine-binding GluN1 and two glutamate-binding GluN2A–D subunits, whereas the GluN1–GluN2A–GluN2B complex is the predominant receptor on the hippocampal synapses. Glycine and D-serine binding GluN3 subunits are also expressed throughout the nervous system, but their role is less-well-defined. Whereas AMPA and kainate receptors can be activated solely by glutamate, activation of NMDA receptors requires binding of glycine to GluN1 and glutamate to GluN2. Activation of NMDA receptors opens a cation-selective calcium-permeable channel causing further depolarization of cell membrane and influx of calcium, thus relieving the magnesium block to the channel pore [96]. The NMDA receptor genes are located on chromosomes 9q, 16p, 12p, 17q, 19p, and 19q.

A number of studies have suggested that some of the alcohol-induced brain damage associated with chronic alcohol consumption is mediated through changes in NMDA receptors. Increased receptor expression is believed to underlie the development of ethanol tolerance and dependence as well as acute and delayed signs of withdrawal, particularly agitation and delirium tremens [97]. Using human postmortem samples from alcoholics and control subjects, Jin *et al.* observed that mRNA encoding two AMPA receptor subunits (GluA2 and GluA3), three kainate receptor subunits (GluK2, GluK3, and GluK5), and five NMDA receptor subunits (CluN1, GluN2A, GluN2C, GluN2D, and GluN3A) was significantly increased in the hippocampal dentate gyrus region in alcoholics compared to controls. However, no difference was observed in the prefrontal cortex. The authors concluded that excessive long-term alcohol consumption was associated with altered expression of genes encoding glutamate receptors in a brain region in a specific manner. Therefore, genetic predisposition to alcoholism may contribute to these gene expression changes [98].

Genetic polymorphisms of genes encoding NMDA receptors have been studied for potential association with alcohol dependence. Five genes encoding various subunits (NR1, -2A, -2B, -2C, and -2D) of the NMDA receptor have been cloned, and there are eight NR1 splice variants as well as a further group of subunits termed NR3. The function of the NMDA receptor is reduced by alcohol because alcohol is an antagonist. The NR1a splice variants of NMDAR1 and NMDAR2B subunits are most sensitive to alcohol. A second mechanism by which alcohol inhibits the NMDA receptor is through protein phosphorylation. The NMDAR2B subunit has a fyn-kinase phosphorylation site that may rapidly reduce sensitivity of NMDA receptors to alcohol. After chronic ethanol administration, tolerance is developed to the inhibition of NMDA receptor function, and consequently, in the absence of alcohol, NMDA function is increased. This may be related to upregulation of NMDA receptor numbers or altered receptor composition. Peak channel open probability is two- to fivefold higher for NMDA receptors composed of NR1a/NR2A subunits than for those composed of NR1a/NR2B subunits. Therefore, a change in expression of NR2A and NR2B may alter open–close kinetics of NMDA ion channels.

Wernick *et al.* investigated allelic variants of genes encoding most ethanol-sensitive subunits of NMDA receptors (NR1a and NR2B) for possible association with alcoholism using a case–control study of 367 alcoholics and 335 control subjects of German origin. The authors concluded that polymorphisms of genes encoding NMDA receptor 1 and 2B subunits are associated with alcohol-related traits [99]. Although ethanol-inhibited glutamatergic neurotransmission has been shown to mediate pathophysiological mechanisms in the development of alcoholism, including withdrawal symptoms, and NR2B confers a high sensitivity to ethanol-induced inhibition, no association between an SNP (rs1806201) in the *NR2B* gene (*GRIN2B* gene) and alcoholism was observed. Moreover, in another study, no association was found between two polymorphisms (rs1806201 and rs1806191) in the *GRIN2B* gene tested individually or as haplotypes and alcoholism [100]. However, Xia *et al.* observed an association between polymorphism in a metabotropic glutamate receptor 3 (GRM3) subunit gene (7q21.1–q21.2) and alcohol dependence. In on their study of 248 male alcohol-dependent patients and 235 male controls, the authors observed a greater frequency of the A allele of SNP rs6465084 in the alcohol-dependent group compared to the control group [101].

Glycine transporters are endogenous regulators of the dual function of glycine, which acts as an inhibitory neurotransmitter at glycinergic synapses and as a modulator of neuronal excitation mediated by NMDA receptors at glutamatergic synapses. Two genes encoding glycine transporters (*SLC6A5* and *SLC6A9*) have been detected. To clarify the role of glycine transporter gene variants in alcohol dependence, Koller *et al.* studied three SNPs in *SLC6A5* and two SNPs in *SLC6A9* but found no association between alcoholism and glycine transporter gene polymorphism [102]. Despite some negative findings between gene polymorphism coding NMDA receptors and alcohol dependence, it has been well-established that chronic alcohol abuse produces a hyperglutamatergic state characterized by elevated extracellular glutamate and altered glutamate receptors and transporters. Pharmacological manipulation of glutamatergic neurotransmission alters alcohol-related behaviors, including intoxication, withdrawal, and craving for alcohol. Currently, two medicines—topiramate and acamprosate—extensively studied for use as therapy for alcohol rehabilitation appear to have a direct effect on glutamatergic neurotransmission [103].

10.11 POLYMORPHISMS OF GENES ENCODING OPIOID RECEPTORS AND ALCOHOL USE DISORDER

The opioid receptors and endogenous peptide ligands play an important role in neurotransmission and neuromodulation in response to alcohol, heroin, and cocaine. There are three classes of opioid receptors—μ, δ, and κ—and these are widely expressed in the brain. There are three distinct families of opioid peptides: endorphins, enkephalins, and dynorphins. Enkephalins bind to δ receptors, dynorphins bind to κ receptors, and endorphins bind to both μ and δ receptors. Relatively recently, two additional short peptides that display high affinity for μ receptors—endomorphin-1 and endomorphin-2—have been characterized. Both μ and δ receptors have been implicated in the reinforcing effect of alcohol that appears to mediate alcohol-induced dopamine release in the ventral tegmental area. Human and animal experiments have indicated that there is a correlation between opioid peptides, especially β-endorphin, and alcohol abuse. Consumption of alcohol increases β-endorphin levels in specific regions of the brain that are associated with a reward system. However, chronic alcohol consumption may cause β-endorphin deficiency. β-Endorphin is functionally connected with a mesolimbic reward system that plays an important role in alcohol addiction. A decrease in β-endorphin levels may also be responsible for drug craving and withdrawal symptoms. Lowered plasma β-endorphin concentration during alcohol withdrawal might contribute to anxiety, whereas intake of alcohol increases β-endorphin levels. Plasma levels of β-endorphin in subjects genetically at high risk for excessive alcohol consumption show lower basal activity of this peptide [104].

The opioid receptors are encoded by three specific genes: *OPRM1* at 6q24−q25 encodes the μ receptor, *OPRD1* at 1p36.1−p34.3 encodes the δ receptor, and *OPRK1* at 8q11.2 encodes the κ receptor. Association between polymorphisms of these genes and alcohol dependence has been studied with both positive and negative results. Li and Zhang obtained genotype data for 13 *OPRM1* SNPs, 11 *OPRD1* SNPs, and 7 *OPRK1* SNPs among 382 European Americans affected by substance abuse disorders (318 with alcohol dependence, 171 with cocaine dependence, and 91 with opioid dependence). They observed 12 significant patterns in the alcohol dependence data set. In addition, 18 significant patterns in the opioid dependence data set and 4 significant patterns in the cocaine dependence data set were observed. The majority of significance patterns were comprised of marker alleles of *OPRM1* and *OPRD1*, suggesting a greater impact of these two genes on alcohol, cocaine, and opioid dependence compared to *OPRK1*. Interestingly, one *OPRM1* SNP (M2, rs511435) and two *OPRD1* SNPs (D6, rs2298896; D7, rs421300) were consistently present in all three significant data sets. The authors concluded that variations in the opioid genes can jointly influence the vulnerability to alcohol and drug dependence [105].

A coding variation in *OPRM1* (A118G, rs1799971) has been extensively studied because it alters amino acid position 40 in the μ receptor protein from asparagine to aspartate (Asp40). When this substitution occurs, the μ receptor has a threefold increased affinity for β-endorphin, although this finding has been challenged. This allele has been observed among alcohol-dependent subjects. Another polymorphism of the μ receptor gene (C17T), where substitution in the 17th nucleotide position results in change of an alanine residue at position 6 of the receptor protein to valine residue, has also been associated with alcohol dependence [104].

Alcohol-dependent patients with one or two copies of the A118G polymorphism (Asp40) who are treated with the opioid antagonist naltrexone (naltrexone is effective in treating alcohol dependence) have a significantly lower rate of relapse, making naltrexone pharmacotherapy even more effective [106]. However, the association between *OPRM1* A118G polymorphism and alcohol use disorder is controversial. Based on a study of 503 Finnish participants with alcohol dependence and 506 sex- and age-matched controls, Rouvinen-Lagerstrom *et al.* concluded that A118G polymorphism of the *OPRM1* gene may not have any major effect on the development of alcohol use disorder, at least in the Finnish population [107]. Ray *et al.* observed that the contribution of the A118G SNP of the *OPRM1* gene to alcohol-induced stimulation, vigor, and positive mood was moderated by the DAT1 genotype (VNTR; *SLC6A3*). The authors concluded that the purported interaction between opioidergic and dopaminergic systems plays a role in the reinforcing properties of alcohol [108]. Xuei *et al.* performed a comprehensive assessment of 50 different polymorphism genes coding the opioid system to identify any role in alcohol and drug dependence. The authors concluded that there was no evidence of involvement of variations of *OPRM1*, *OPRD1*, *PENK* (encoding endogenous opioid peptide precursor proenkephalins), and *POMC* (encoding pro-opiomelanocortin) genes with alcohol dependence or general illicit drug dependence, although variations in *PENK* and *POMC* appeared to be associated with narrower phenotype of opioid dependence in these families [109].

Studies have indicated that negative feelings and emotions may be related to endogenous opioid dynorphin and the κ opioid receptor system. Dynorphin is a post-translational product of prodynorphin (encoded by the *PDYN* gene). In a study of 1860 European Americans from 219 multiplex alcohol-dependent families, the authors observed an association between multiple SNPs in the promoter 3′ end of *PDYN* and in the intron 2 of *OPRK1*. The variations of these genes coding both the κ opioid receptor and its ligand are associated with the risk of alcohol dependence [110]. Karpyak *et al.* genotyped 23 SNPs in *PDYN* and *OPRK1* genes in 816 alcohol-dependent subjects and observed a significant association of the haplotype-spanning *PDYN* gene (rs6045784, rs910080, rs2235751, and rs2281285) with alcohol dependence and negative craving for alcohol (desire to drink in the context of tension, discomfort, or unpleasant emotion). In addition, a candidate haplotype containing *PDYN* rs2281285–rs1997794 SNPs was also associated with alcohol dependence and negative craving. However, the authors observed no association between variations in the *OPRK1* gene and alcohol dependence [111]. Preuss *et al.* also observed an association between increased risk of drinking to avoid/escape unwanted emotions or somatic events and the minor allele of rs2281285 of the *PDYN* gene [112].

10.12 POLYMORPHISMS OF GENES ENCODING CANNABINOID RECEPTORS AND ALCOHOL USE DISORDER

The endocannabinoid system consists of cannabinoid receptors, endogenous cannabinoids, and enzymes for the synthesis as well as degradation of endocannabinoid. There are two well-characterized G protein-coupled cannabinoid receptors: CB1 and CB2. CB1 receptors are mostly expressed in the CNS, whereas CB2 receptors are found in immune cells and peripheral tissues. Therefore, CB2 is also referred to as a peripheral cannabinoid receptor. Two fatty acid derivatives that are endocannabinoid—namely arachidonylethanolamide and 2-arichidonylglycerol—have been isolated from nerve and peripheral tissue. Both endogenous compounds mimic pharmacological and behavioral effects of marijuana (Δ^9-tetrahydrocannabinol). Rodent experiments have shown that CB1 is involved in the neural circuitry regulating alcohol intake and motivation to consume alcohol. Chronic consumption of alcohol results in downregulation of CB1 receptor binding and its signal transduction capacity, which may be due to increased synthesis of the CB1 receptor agonists arachidonylethanolamide and 2-arichidonylglycerol due to chronic alcohol consumption. Therefore, CB1 receptor gene polymorphism may play a role in the development of alcoholism [113]. Using positron emission tomography, Ceccarini *et al.* showed that chronic heavy drinking resulted in a decrease in CB1 receptor availability in alcoholic patients compared to controls, and such decrease was unaltered during abstinence from alcohol. In contrast, acute alcohol ingestion such as binge drinking resulted in increased CB1 receptor availability [114].

The CB1 receptor is encoded by the cannabinoid receptor gene 1 (*CNR1*) located on chromosome 6 (6q14–q15). *CNR1* appears to be a promising candidate gene for association with alcohol and substance abuse. Zuo *et al.* studied 451 healthy control subjects and 550 substance abuse (alcohol- or drug-dependent) patients and observed that two SNPs in the *CNR1* gene (SNP3, rs6454674; and SNP8, rs806368) were associated with alcohol and/or drug dependence [115]. Marcos *et al.* reported an interaction between the G allele of the rs6454674 SNP and the C allele of the rs806368 SNP in alcohol dependence [116]. In another study, the authors observed that common intragenic *CNR1* polymorphism 1359 G/A exchange confers vulnerability to alcohol withdrawal delirium in alcoholics, especially in homozygous 1359 A/A genotype [117]. Ishiguro *et al.* observed an association between cannabinoid receptor 2 gene polymorphism (Q63R) and alcoholism in a Japanese population [118]. However, Herman *et al.* observed no association between four polymorphisms of the *CNR1* gene and substance dependence (alcohol, cocaine, and opioid) [119].

10.13 ADENYLYL CYCLASE AND ALCOHOL USE DISORDER

Adenylyl cyclase comprises a family of enzymes consisting of nine transmembrane proteins (AC 1–9) that display a distinct response to G protein-coupled receptors and other regulatory factors leading to the generation of cyclic adenosine 3′,5′-monophosphate (cAMP) from adenosine triphosphate (ATP). Adenylyl cyclase enzymes play an important role in memory, synaptic plasticity, and neurodegeneration. In addition, these enzymes are implicated in the development of addiction, including alcohol use disorder. In general, acute alcohol ingestion increases platelet or lymphocyte activity of adenylyl cyclase, whereas chronic ingestion of alcohol reduces such activity. Moreover, platelet or lymphocyte adenylyl cyclase activities of alcohol-dependent individuals are less responsive to various stimulators, such as cesium fluoride, forskolin, or guanine nucleotide analogs (e.g., guanylyl imidodiphosphate), compared to those of controls. The activity of adenylyl cyclase also decreases during alcohol abstinence. However, major depression may also be associated with lower levels of forskolin- or cesium fluoride-stimulated platelet activity of adenylyl cyclase, thus negatively impacting the validity of using adenylyl cyclase as a potential alcohol biomarker. Parsian *et al.* measured adenylyl cyclase activity in membrane preparations of platelets from 51 alcoholic subjects and 54 normal controls. The authors observed lower platelet activity of adenylyl cyclase in male alcoholics compared to male controls [120]. Ikeda *et al.* also observed lower forskolin- and guanylyl imidodiphosphate-stimulated platelet adenylyl cyclase activities in individuals with alcohol use disorder [121].

The adenylyl cyclase genes (*ADCY1–9*) have been investigated in relation to depression, other psychiatric illness, and substance dependence, including alcohol dependence. Adenylyl cyclase isoform 7 is the most sensitive to alcohol. Based on a mouse model, Desrivieres *et al.* observed that female mice in which one copy of the *ADCY7* gene was disrupted showed a higher preference for alcohol compared to wild-type control mice. No such difference was observed with male mice. In addition, the authors studied 1703 patients (1349 men and 354 women) who were alcohol dependent as well as 1347 controls (714 men and 633 women) and observed that SNP rs2302717 was associated with female alcohol-dependent subjects. In addition, TA haplotype formed by the combination of rs2302717 and rs719158 was responsible for the observed association with alcohol dependence in female subjects. The SNP most associated with alcohol dependence is located in intron 5 of the *ADCY7* gene. The authors concluded that polymorphism of *ADCY7* is associated with alcohol dependence in human female subjects [122]. Procopio *et al.* examined nine adenylyl cyclase genes for possible association with comorbid depression and alcoholism in women, and they identified four haplotypes associated with alcoholism in women. One haplotype was in a genomic area in proximity to the *ADCY2* gene. Two haplotypes were in proximity to the *ADCY5* gene, and one haplotype was within the coding region of the *ADCY8* gene. Three of these four haplotypes contributed independently to alcoholism. The authors concluded that polymorphisms in *ADCY2* (in proximity to the *ADCY2* gene but actually within a lincRNA gene) *ACDY5*, and *ACDY8* genes are associated with alcohol-dependent phenotype in females, which is also distinguished by comorbid signs of depression [123].

10.14 NEUROPEPTIDE Y AND ALCOHOL USE DISORDER

Neuropeptide Y (NPY) is a highly conserved 36-amino acid peptide neurotransmitter with abundant expression in the mammalian brain, especially in the hypothalamus (particularly the arcuate and the paraventricular nuclei, the hippocampal formation, the amygdala, and the septum). NPY is a major endogenous regulator of anxiety-related behavior, neuronal excitability, and food intake stimulation. Quantitative trait locus linkage mapping studies in rats as well as research with transgenic and null-mutant mice have indicated that genetic variation in the locus of the *NPY* gene may contribute to heritability of alcoholism in animal models. In addition, mice that were manipulated to overexpress the *NPY* gene showed a low preference for alcohol, whereas mice in which the *NPY* gene was deleted preferred alcohol. Therefore, variation in the NPY receptor or genes encoding such receptors (*NPY1R*, *NPY2R*, and *NPY5T*) may be associated with alcohol dependence as well as alcohol withdrawal symptoms, and hypoactivity of NPY receptors may be related to alcohol dependence.

Studies in humans have shown that alcoholics who consume more than 80 g of alcohol per day for most of their adult life have decreased NPY immunoreactivity in the amygdala as well as reduced gene expression in the frontal and motor cortices. A functional SNP in the neuropeptide receptor gene, Leu7Pro (rs161139), has been extensively studied in association with alcohol dependence. Lappalainen *et al.* reported a higher frequency of Pro7 allele (Leu7Pro) in European American alcohol-dependent subjects compared to controls. The authors concluded that the NPY Pro7 allele is a risk factor for alcohol dependence [124]. Wetherill *et al.* genotyped 39 SNPs across NPY and its three receptor genes in a sample of 1923 subjects from 219 multiplex families of European American descent recruited from the Collaborative Studies of Genetics of Alcoholism (COGA). The authors observed an association of polymorphisms of NPY receptor genes with alcohol dependence, alcohol withdrawal, and comorbid alcohol and cocaine dependence, as well as cocaine dependence. One SNP in the promoter region of the *NPY2R* (rs6857715) gene and two SNPs farther upstream of the gene (rs4333136 and rs4425326) were associated with alcohol dependence. These SNPs were also involved with comorbid alcohol and cocaine dependence. Four SNPs in the *NPY5R* gene (rs4475104, rs4314240, rs4632602, and rs7678265) were associated with the secondary phenotype of withdrawal with seizure. However, neither the *NPY1R* gene nor the *NPY5R* gene were associated with alcohol or cocaine dependence, including the Leu7Pro coding SNP rs161139 in *NPY1R* [125]. In a study of German subjects, Zill *et al.* found no association between Leu7Pro polymorphism as well as two other SNP polymorphisms (G-602 T and T-399C) in the *NPY* gene with alcohol dependence, and they concluded that the analyzed

SNPs as well as the corresponding haplotypes in the *NPY* gene are unlikely to play a major role in the pathophysiology of alcohol dependence [126].

10.15 POSSIBLE ASSOCIATION OF POLYMORPHISMS OF OTHER GENES WITH ALCOHOL USE DISORDER

Genome-wide sibling pair linkage analyses suggest that a gene or genes in chromosome 1 may predispose individuals to alcoholism and other alcohol-induced depression [127]. A variant of the serine incorporator 2 gene (*SERINC2*) is associated with alcohol dependence in individuals of European descent. In addition, a rare variant constellation across the *NKAIN1–SERINC2* region was also associated with alcohol dependence. The *NKAIN1* gene encodes sodium/potassium-transporting ATP subunit β_1 interacting protein. The authors concluded that *SERINC2* is a replicable and significant gene specific for alcohol dependence in individuals of European descent [127]. Zuo *et al.* identified a significant risk region for alcohol and cocaine codependency between the *IPO11* gene (which encodes the importin 11 receptor) and the *HTR1A* gene on chromosome 5q. The authors speculated that the *IPO11–HTR1A* region might harbor a causal variant for alcohol and nicotine dependence [128]. Using GWAS based on data from Study of Addiction: Genetics and Environment (SAFE) and COGA, Han *et al.* identified a subnetwork of 39 genes that were associated with alcohol dependence [129]. In addition, studies have also revealed a possible association between new candidate genes and alcohol dependence, including cadherin 11, cadherin 13, GATA binding protein 4 (GARA4), and the potassium large conductance calcium-activated channel subfamily M, α member 1 gene (*KCNMA1*) [130]. GWAS involving 26,316 individuals showed that SNP rs6943555 in the autism susceptible candidate 2 gene (*AUTS2*) was associated with alcohol consumption with genome-wide significance [131]. Chojnicka *et al.* studied *AUST2* gene polymorphism in 625 individuals who committed suicide and 3861 controls in a Polish Caucasian population and observed an association of the rs6943555 A allele with people who committed suicide under the influence of alcohol. The authors speculated that the rs6943555 allele may be linked to adverse emotional reactions to alcohol [132]. Clarke *et al.* observed that polymorphism of the *KCNJ6* (rs2836016) gene that encodes G protein-coupled inwardly rectifying potassium channel 2 (GIRK2) was associated with alcohol dependence in adults [133].

10.16 EPIGENETICS AND ALCOHOL USE DISORDER

Epigenetics refers to chemical processes that may alter gene activity without changing the sequence of DNA. Chromosomes are composed of chromatin structures, and chromatin is made of nucleosomes consisting of DNA strands wrapped around bead-like histone octamers. In humans, the most stable epigenetic change is methylation of DNA, in which a methyl group is added to cytosine. DNA is methylated at position 5 of the cytosine pyrimidine ring by transfer of a methyl group from 5-adenosyl methionine, a reaction catalyzed by DNA methyltransferase. DNA methylation is a stable chemical mark maintained through the cycle of cell division, and DNA methylation mainly takes place at CpG sites (the region of DNA where a cytosine is present next to a guanine nucleotide). Histone protein can be modified by at least six different pathways: acetylation, methylation, phosphorylation, ubiquitinylation, ADP ribosylation, and sumoylation. Epigenetic changes can regulate gene expression by modulating the structure of the chromatin without changing the DNA code, and aberrant epigenetic mechanisms are implicated in the development of many diseases, including cancer and cardiovascular, immunological, and neuropsychiatric disorders.

Gene–environment interaction can be mediated by epigenetic mechanisms, and alcohol consumption is one environmental factor that may alter epigenetic structure and related gene expression. Alcohol causes site-selective acetylation, methylation, and phosphorylation in histone [134]. Homocysteine plays an important role in DNA methylation because it is metabolized to methionine, which is then activated to *S*-adenosyl methionine by ATP. *S*-adenosyl methionine is the most important methyl group donor in the methylation of DNA. Elevated homocysteine levels have been observed in alcohol-dependent patients, causing an increased level of DNA methylation. However, hypomethylation of DNA has also been reported. Alterations in promoter DNA methylation levels in alcohol-dependent patients have been reported for many genes encoding various proteins, including homocysteine-induced endoplasmic reticulum protein (*HERP*), α-synuclein (*SNCA*), vasopressin (*AVP*), nerve growth factor (*NGF*), atrial natriuretic peptide (*ANP*), NMDA 2b receptor subtype (*NR2B*), *MAOA*, *5-HTT*, μ opioid receptor (*OPRM1*), prodynorphin (*PDYN*), pro-opiomelanocortin (*POMC*), and dopamine transporter (*DAT*) [130].

Increased DNA methylation of the *SNCA* gene in alcoholic patients has been reported. DNA hypermethylation in the promoter region of the *SNCA* gene may cause downregulation of SNCA expression, which may eventually disrupt dopaminergic neurotransmission. Elevated promoter methylation within the *HERP* gene in peripheral blood cells of alcohol-dependent patients has also been reported. Suppressed expression of *HERP* under conditions of alcohol consumption may be related to the elevated rate of seizure and neurological damage. Hypermethylation of the sequence of the *DAT* promoter gene has been observed in alcohol-dependent patients. A negative association between DAT methylation and alcohol craving has also been observed. Human studies based on peripheral blood cells have shown a significant association between high lifetime alcohol consumption as well as high daily consumption and lower DNA methylation in the genomic sequence of the *NR2B* gene in alcohol-dependent patients. It is possible that upregulation of the NR2B receptor in alcohol-dependent patients may be partially explained by DNA methylation of the gene. Epigenetic alteration of the *POMC* gene that encodes POMC in alcohol-dependent patients has also been observed. POMC is modified post-translationally into several active hormones, particularly ACTH, that play an important role in regulation of the HPA axis. Dysfunction of the HPA axis due to alcohol consumption has been reported. DNA methylation status in the gene sequence of *POMC* at single CpG sites differs between patients with alcohol dependence and healthy controls, and a specific CpG cluster shows significant association with alcohol craving. Therefore, epigenetic alteration by alcohol, particularly changes in DNA methylation, may contribute to dysfunction of the HPA axis in alcohol-dependent individuals. However, the epigenetics of alcohol dependence is an emerging field of investigation, and the role of epigenetic alterations by alcohol, including DNA methylation and histone modification in altered transcription factors, in alcohol dependence is still limited. Thus, results should be interpreted with caution [135].

MicroRNAs (miRNAs), a class of noncoding RNAs (approximately 17–22 nucleotides long) transcribed by RNA polymerase II, are post-transcriptional modulators of gene expression. More than 500 miRNAs have been discovered in humans, and they control cellular proliferation, differentiation, and apoptosis [136]. miRNAs are highly abundant in the brain and play an important role in neuronal differentiation, brain development, synapse formation and plasticity, and neurodegeneration. In addition, miRNAs also appear to mediate the cellular adaptation induced by nicotine, cocaine, opioids, and alcohol. Research has shown that expression of 35 miRNAs is upregulated in the brain of alcoholics. Approximately 30% of 35 miRNAs upregulated in human alcoholics (miRNAs 34, 92, 140, 146, 152, 194, 196, 203, 369, and let-7) are modulators of immunity. The miRNA 203 regulates interleukin-6 and interferon-α signaling through targeting of the suppressor of cytokine signaling 3 (SOCS3). The upregulation of miRNA 101 in alcoholics may be involved in modulation of GABAergic transmission in response to alcohol consumption. These results support a role of miRNA-dependent inhibition of gene expression in the prefrontal cortex of human alcoholics. Therefore, discoveries of miRNA signatures in the brains of human alcoholics support the hypothesis that changes in gene expression and regulation of miRNA are responsible for long-term neuroadaptation occurring during development of alcoholism [137]. Lewohl *et al.* also observed upregulation of approximately 35 miRNAs in the prefrontal cortex of 14 alcoholics compared to 13 controls, and they commented that reduced expression of certain genes in human alcoholics may be due to upregulation of miRNAs [138]. Manzardo *et al.* reported an association of alcoholism with upregulation of a cluster of miRNAs located in the genomic imprinted domain of chromosome 14q32, with predicted gene targets involved in oligodendrocyte growth, differentiation, and signaling [139]. A genetic polymorphism of the *miRNA-146a* gene (C.G; rs2910164) is associated with susceptibility to alcohol use disorder [140].

10.17 CONCLUSIONS

Many studies have shown an association between specific genes and alcohol use disorder, and the role of alcohol dehydrogenase and aldehyde dehydrogenase gene polymorphisms that protect from alcohol use disorder has been well documented through many studies, including GWAS studies. It is also important to note that certain polymorphisms in alcohol and/or the aldehyde dehydrogenase gene may increase the risk of alcohol use disorder. Studies continue to reveal other genes in which variants affect the risk of alcoholism or related traits, including *GABRA2* and *CHRM2*, but no single gene is associated with alcoholism. Children of alcoholic parents have a 4- to 10-fold higher risk of developing alcohol use disorder, and it is currently assumed that genetic factors contribute 40–60% to the development of alcohol use disorders. Therefore, complex interaction between environment and genes is also linked with alcohol use disorder. Epigenetics is emerging as an important field in understanding the pathogenesis of alcohol use disorder.

References

[1] Diaz-Anzaldua A, Diaz-Martinez A, Diaz-Martinez LR. The complex interplay of epigenetics and environment in the predisposition to alcohol dependence. Salud Mental 2011;34:157–66.

[2] Prescott CA, Kendler KS. Genetic and environmental contributions to alcohol abuse and dependence in a population-based sample of male twins. Am J Psychiatry 1999;156:34–40.

[3] Prescott CA, Aggen SH, Kendler KS. Sex differences in the sources of genetic liability to alcohol abuse and dependence in a population-based sample in U.S. twins. Alcohol Clin Exp Res 1999;23:1136–44.

[4] Kendler KS, Heath AC, Neale MC, Kessler RC, et al. A population-based twin study of alcoholism in women. JAMA 1992;268:1877–82.

[5] Lenz B, Muller CP, Kornhuber J. Alcohol dependence in same sex and opposite sex twins. J Neural Transm 2012;119:1561–4.

[6] Schwandt ML, Hellig M, Hommer DW, George DT, Ramchandani VA. Childhood trauma exposure and alcohol dependence severity in adulthood: mediation by emotional abuse severity and neuroticism. Alcohol Clin Exp Res 2013;37:984–92.

[7] Hingson RW, Heeren T, Winter MR. Age at drinking onset and alcohol dependence: age at inset, duration, and severity. Arch Pediatr Adolesc Med 2006;160:739–46.

[8] Keyes KM, Harzenbuehler ML, Hasin DS. Stressful life experiences, alcohol consumption, and alcohol use disorders: the epidemiologic evidence for four main types of stressors. Psychopharmacology (Berlin) 2011;218:1–17.

[9] Sher KJ, Gershuny BS, Peterson L, Raskin G. The role of childhood stressors in the intergenerational transmission of alcohol use disorders. J Stud Alcohol 1997;58:414–27.

[10] Mulder RT. Alcoholism and personality. Aust N Z J Psychiatry 2002;36:44–52.

[11] Holdcraft LC, Iacono WG, McGue MK. Antisocial personality disorder and depression in relation to alcoholism: a community-based sample. J Stud Alcohol 1998;59:222–6.

[12] Janssen MM, Mathijssen JJ, van Bon-Martens MJ, van Oers HA, et al. A qualitative exploration of attitudes towards alcohol and the role of parents and peers of two alcohol attitude-based segments of the adolescent population. Subst Abuse Treat Prev Policy 2014;9(1):20.

[13] Faria R, Vendrame A, Silva R, Pinsky I. Association between alcohol advertisement and beer drinking among adolescents. Rav Saude Publica 2011;45:441–7.

[14] Enoch MA. Genetic influence on the development of alcoholism. Curr Psychiatry Rep 2013;15:412.

[15] Enoch MA, Goldman D. The genetics of alcoholism and alcohol abuse. Curr Psychiatry Rep 2001;3:144–51.

[16] Zakhari S. Overview: how is alcohol metabolized by the body? Alcohol Res Health 2006;29:245–54.

[17] Edenberg HJ. The genetics of alcohol metabolism: role of alcohol dehydrogenase and aldehyde dehydrogenase variants. Alcohol Res Health 2007;30(1):5–13.

[18] Wall T. Genetic association of alcohol and aldehyde dehydrogenase with alcohol dependence and their mechanisms of action. Ther Drug Monit 2005;27:700–3.

[19] Kang G, Bae KY, Kim SW, Ki J, et al. Effect of the allelic variant of alcohol dehydrogenase ADH1B*2 on ethanol metabolism. Alcohol Clin Exp Res 2014;38:1502–9.

[20] Osier M, Pakstis AJ, Kidd JR, Lee JF, et al. Linkage disequilibrium at the ADH2 and ADH3 loci and risk of alcoholism. Am J Hum Genet 1999;64(4):1147–57.

[21] Lu RB, Ko HC, Lee JF, Lin WW, et al. No alcoholism protection effects of ADH1B*2 allele in antisocial alcoholics among Han Chinese in Taiwan. Alcohol Clin Exp Res 2005;29:2101–7.

[22] Tolstrup JS, Nordestgaard BG, Rasmussen S, Tybjaerg-Hansen A, et al. Alcoholism and alcohol drinking habits predicted from alcohol dehydrogenase genes. Pharmacogenomics J 2008;8:220–7.

[23] Luo X, Kranzler HR, Zuo L, Lappalainen J, et al. ADH4 gene variation is associated with alcohol dependence and drug dependence in European Americans: results from HWD tests and case–control association study. Neuropsychopharmacology 2006;31:1085–95.

[24] Moore S, Montane-Jaime LK, Carr LG, Ehlers CL. Variations in alcohol-metabolizing enzymes in people of East Indian and African descent from Trinidad and Tobago. Alcohol Res Heath 2007;30(1):28–30.

[25] Moore S, Montane-Jaime K, Shafe S, Joseph R, et al. Association of ALDH1 promoter polymorphisms with alcohol-related phenotypes in Trinidad and Tobago. J Stud Alcohol Drug 2007;68:192–6.

[26] Otto JM, Hendershot CS, Collins SE, Liang T, et al. Association of ALDH1A1*2 promoter polymorphism with alcohol phenotypes in young adults with or without ALDH2*2. Alcohol Clin Exp Res 2013;37:164–9.

[27] Spencer JP, Liang T, Eriksson CJ, Taylor RE, et al. Evaluation of aldehyde dehydrogenase 1 promoter polymorphisms identified in human populations. Alcohol Clin Exp Res 2003;27:1389–94.

[28] Kortunay S, Kosseler A, Ozdemir F, Atalay EO. Association of a genetic polymorphism of the alcohol metabolizing enzyme ADH1C with alcohol dependence: results of a case controlled study. Eur Addict Res 2012;18:161–6.

[29] Konish T, Calvillo M, Leng AS, Feng J, et al. The ADH3*2 and CYP2E1 alleles increase the risk of alcoholism in Mexican American men. Exp Mol Pathol 2003;74:183–9.

[30] Le Foll B, Gallo A, Le Strat Y, Lu L, et al. Genetics of dopamine receptors and drug addiction: a comprehensive review. Behav Pharmacol 2009;20:1–17.

[31] Hoenicka J, Aragues M, Ponce G, Rodriguez-Jimenez R, et al. From dopaminergic genes to psychiatric disorders. Neurotox Res 2007;11:61–72.

[32] Kin DJ, Park BL, Yoon S, Lee HK, et al. 5′-UTR polymorphism of dopamine receptor D1 (DRD1) associated with severity and temperament of alcoholism. Biochem Biophys Res Commun 2007;357:1135–41.

[33] Batel P, Houch H, Daoust M, Ramoz N, et al. A haplotype of the DRD1 gene is associated with alcohol dependence. Alcohol Clin Exp Res 2008;32:567–72.

[34] Prasad P, Ambekar A, Vaswani M. Case–control association of dopamine receptor polymorphisms in alcohol dependence: a pilot study with Indian males. BMC Res Notes 2013;6:418.

[35] Smith L, Watson M, Gates S, Ball D, et al. Meta-analysis of the association of the Taq 1 A polymorphism with the risk of alcohol dependency: a HuGE gene–disease association review. Am J Epidemiol 2008;167:125–38.

[36] Kono Y, Yoneda H, Sakai T, Nonomura Y, et al. Association between early onset alcoholism and the dopamine D2 receptor gene. Am J Med Genet 1997;74:179–82.

[37] Ponce G, Jimenez-Arriero MA, Rubio G, Hoenicka J, et al. The A1 allele of the DRD2 gene (TaqIA polymorphism) is associated with antisocial personality in a sample of alcohol-dependent patients. Eur Psychiatry 2003;18:356–60.

[38] Shaikh KJ, Naveen D, Sherrin T, Murthy A, et al. Polymorphism at the DRD2 locus in early onset alcohol dependence in the Indian population. Addict Biol 2001;6:331–5.

[39] Banerjee N. Neurotransmitters in alcoholism: a review of neurobiological and genetic studies. Indian J Hum Genet 2014;20:20–31.

[40] Prasad P, Amberkar A, Vaswani M. Dopamine D2 receptor polymorphism and susceptibility to alcohol dependence in Indian males: a preliminary study. BMC Med Genet 2010;11:24.

[41] Kang SG, Lee BH, Lee JS, Chai YG, et al. DRD3 gene rs6280 polymorphism may be associated with alcohol dependence and with Lesch type I alcohol dependence in Koreans. Neuropsychobiology 2014;69:140–6.

[42] Limosin F, Romo L, Batel P, Ades J, et al. Association between dopamine receptor D3 gene Bal I polymorphism and cognitive impulsiveness in alcohol-dependent men. Eur J Psychiatry 2005;20:304–6.

[43] Schoots O, Van Tol HH. The human dopamine D4 receptor repeat sequences modulate expression. Pharmacogenomics J 2003;3:343–8.

[44] Ray LA, Bryan A, Mackillop J, McGeary J, et al. The dopamine D receptor (DRD4) gene exon III polymorphism, problematic alcohol use and novelty seeking: direct and mediated genetic effect. Addict Biol 2009;14:238–44.

[45] Du Y, Yang M, Yeh HW, Wan YJ. The association of exon 3 VNTR polymorphism of the dopamine receptor D4 (DRD4) gene with alcoholism in Mexican Americans. Psychiatry Res 2010;177:358–60.

[46] Park A, Sher KJ, Todorov AA, Heath AC. Interaction between the DRD4 VTNR polymorphism and proximal distal environments in alcohol dependence during emerging young adulthood. J Abnorm Psychol 2011;120:585–95.

[47] Foroud T, Wetherill LF, Liang T, Dick DM, et al. Association of alcohol craving with alpha-synuclein (SNCA). Alcohol Clin Exp Res 2007;31:537–45.

[48] Van der Zwaluw CS, Engles RC, Buitelaar J, Verkes RJ, et al. Polymorphism in the dopamine transporter gene (SLC6A3/DAT1) and alcohol dependence in human: a systematic review. Pharmacogenomics 2009;10:853–66.

[49] Kohnke MD, Batra A, Kolb W, Kohnke AM, et al. Association of the dopamine transporter gene with alcoholism. Alcohol Alcohol 2005;40:339–42.

[50] Bhaskar LV, Thangaraj K, Wasnik S, Singh L, et al. Dopamine transporter (DAT1) VNTR polymorphism and alcoholism in two culturally different populations of South India. Am J Addict 2012;21:343–7.

[51] Kohnke MD, Zabetian CP, Anderson GM, Kolb W, et al. A genotype-controlled analysis of plasma dopamine beta-hydroxylase in healthy and alcoholic subjects: evidence for alcohol related differences in noradrenergic function. Biol Psychiatry 2002;52:1151–8.

[52] Kohnke MD, Kolb W, Kohnke AM, Schick S, et al. DBH*444 G/A polymorphism of the dopamine beta-hydroxylase gene is associated with alcoholism but not with severe alcohol withdrawal symptoms. J Neural Transm 2006;113:869–76.

[53] Farren CK, Clare AW, Tipton KE, Dinan TG. Platelet MAO activity in subtypes of alcoholics and controls in a homogenous population. J Psychiatr Res 1998;32:49–54.

[54] Pombo S, Levy P, Bicho M, Ismail F, et al. Neuropsychological function and platelet monoamine oxidase activity levels in type 1 alcoholic patients. Alcohol Alcohol 2008;43:423–30.

[55] Sabol SZ, Hu S, Hamer D. A functional polymorphism in the monoamine oxidase A gene promoter. Hum Genet 1998;103:273–9.

[56] Contini V, Marques FZ, Garcia CE, Hutz MH, et al. MAOA–uVNTR polymorphism in a Brazilian sample: further support for the association with impulsive behaviors and alcohol dependence. Am J Med Genet B Neuropsychiatr Genet 2006;141B:305–8.

[57] Saito T, Lachman HZ, Diaz L, Hallikainen T, et al. Analysis of monoamine oxidase A (MAOA) promoter polymorphism in Finnish male alcoholics. Psychiatry Res 2002;109:113–19.

[58] Kauhanen J, Hallikainen T, Tuomainen TP, Koulu M, et al. Association between the functional polymorphism of catechol-O-methyltransferase gene and alcohol consumption among social drinkers. Alcohol Clin Exp Res 2000;24:135–9.

[59] Ishiguro H, Haruo Shibuya T, Toru M, Saito T, et al. Association study between high and low activity polymorphism of catechol-O-methyltransferase gene and alcoholism. Psychiatr Genet 1999;9:135–8.

[60] Pivac N, Muck-Seler D, Mustapic M, Nenadic-Sviglin K, et al. Platelet serotonin concentration in alcoholic subjects. Life Sci 2004;76:521–31.

[61] Bondy B, Erfurth A, de Jonge S, Kruger M, et al. Possible association of the short allele of the serotonin transporter promoter gene polymorphism (5′-HTTLPR) with violent suicide. Mol Psychiatry 2000;5:193–5.

[62] Feinn R, Nellissery M, Kranzler HR. Meta-analysis of the association of a functional serotonin transporter promoter polymorphism with alcohol dependence. Am J Med Genet B Neuropsychiatr Genet 2005;133:79–84.

[63] Pinto E, Reggers J, Gorwood P, Boni C, et al. The short allele of the serotonin transporter promoter polymorphism influences relapse in alcohol dependence. Alcohol Alcohol 2008;43:398–400.

[64] Bleich S, Bonsch D, Rauh J, Bayerlein K, et al. Association of the long allele of the 5-HTTLPR polymorphism with compulsive craving in alcohol dependence. Alcohol Alcohol 2007;42:509–12.

[65] Pombo S, Ferreira J, Cardoso JM, Ismail F, et al. The role of 5-HTTLPR polymorphism in alcohol craving experience. Psychiatry Res 2014;218:174–9.

[66] Florez G, Saiz P, Garcia-Portilla P, Alvarez S, et al. Association between the Stin2 VNTR polymorphism of the serotonin transporter gene and treatment outcome in alcohol dependent patients. Alcohol Alcohol 2008;43:515–22.

[67] Himei A, Kono Y, Yoneda H, Sakai T, et al. An association study between alcoholism and serotonergic receptor genes. Alcohol Clin Exp Res 2000;24:341–2.

[68] Cao J, Liu X, Han S, Zhang CK, et al. Association of the HTR2A gene with alcohol and heroin dependence. Hum Genet 2014;133:357–65.

[69] Wrzosek M, Jakubczyk A, Wrzosek M, Matsumoto H, et al. Serotonin 2 A receptor gene (HTR2A) polymorphism in alcohol-dependent patients. Pharmacol Rep 2012;64:449–53.

[70] Cao J, LaRocque R, Li D. Associations of the 5-hydroxytryptamine (serotonin) receptor 1B gene (HTR1B) with alcohol, cocaine and heroin abuse. Am J Med Genet B Neuropsychiatr Genet 2013;162B:169–76.

[71] Yang J, Li MD. Association and interaction analyses of 5-HT3 receptor and serotonin transporter genes with alcohol, cocaine and nicotine dependence using SAGE data. Hum Genet 2014;133:905–18.

[72] Xu Y, Guo WJ, Wang Q, Lanzi G, et al. Polymorphisms of genes in neurotransmitter system were associated with alcohol use disorders in a Tibetan population. PLOS ONE 2013;8(11): e80206.

[73] Herman A, Balogh KN. Polymorphisms of the serotonin transporter and receptor genes: susceptibility to substance abuse. Subst Abuse Rehabil 2012;3:49–57.

[74] Connelly WM, Errington AC, Di Giovanni C, Crunelli V. Metabotropic regulation of extrasynaptic GABA receptors. Front Neural Circuits 2013;7:171.

[75] Kumar S, Fleming RL, Morrow AL. Ethanol regulation of gamma-aminobutyric acid A receptors: genomic and nongenomic mechanisms. Pharmacol Ther 2004;101:211–28.

[76] Edenberg HJ, Dick DM, Xuei X, Tian H, et al. Variations in GABRA2, encoding the alpha subunit of the GABA (A) receptors are associated with alcohol dependence with brain oscillations. Am J Hum Genet 2004;74:705–14.

[77] Edenberg HJ, Foroud T. Genetics and alcoholism. Nat Rev Gastroenterol Hepatol 2013;10:487–94.

[78] Fehr C, Sander T, Tadic A, Lenzen KP, et al. Confirmation of association of the GABRA2 gene with alcohol dependence by subtype-specific analysis. Psychiatr Genet 2006;16:9–17.

[79] Ittiwut C, Yang BZ, Kranzler HR, Anton RF, et al. GABRG1 and GABRA2 variation associated with alcohol dependence in African Americans. Alcohol Clin Exp Res 2012;36:588–93.

[80] Enoch MA, Hodgkinson CA, Yuan Q, Albaugh B, et al. GABRG1 and GABRA2 as independent predictors for alcoholism in two populations. Neuropsychopharmacology 2009;34:1245–54.

[81] Dick DM, Plunkett J, Wetherill LF, Xuei X, et al. Association between GABRA1 and drinking behaviors in the collaborative study on the genetics of alcoholism sample. Alcohol Clin Exp Res 2006;30:1101–10.

[82] Loh EW, Higuchi S, Matsushita S, Murray R, et al. Association analysis of the GABA(A) receptor subunit genes cluster on 5q33–34 and alcohol dependence in a Japanese population. Mol Psychiatry 2000;5:301–7.

[83] Han DH, Bolo N, Daniels MA, Lyoo IK, et al. Craving for alcohol and food during treatment for alcohol dependence: modulation by T allele of 1519 T>C GABAalpha6. Alcohol Clin Exp Res 2008;32:1593–9.

[84] Radel M, Vallejo RL, Iwata N, Aragon R, et al. Haplotype-based localization of an alcohol dependent gene to the 5q234 γ-aminobutyric acid type A gene cluster. Arch Gen Psychiatry 2005;62:47–55.

[85] Dick DM, Edenberg HJ, Xuei X, Goate A, et al. No association of the GABAA receptor genes on chromosome 5 with alcoholism in the collaborative study on the genetics of alcoholism sample. Am J Med Genet B Neuropsychiatr Genet 2005;132B:24–8.

[86] Song J, Koller DL, Foroud T, Carr K, et al. Association of GABA(A) receptors and alcohol dependence and the effects of genetic imprinting. Am J Med Genet B Neuropsychiatr Genet 2003;117B:39–45.

[87] Dick DM, Edenberg H, Xuei X, Goate A, et al. Association of GABRG3 with alcohol dependence. Alcohol Clin Exp Res 2004;28:4–9.

[88] Sander T, Samochowiec J, Ladehoff M, Smolka M, et al. Association analysis of exonic variants of the gene encoding GABAB receptor and alcohol dependence. Psychiatr Genet 1999;9:69–73.

[89] Schlaepfer IR, Hoft NR, Ehringer MA. The genetic components of alcohol and nicotine co-addiction: from genes to behavior. Curr Drug Abuse Rev 2008;1:124–34.

[90] Ehringer MA, Clegg HV, Collins AC, Corley RP, et al. Association of the neuronal nicotinic receptor beta 2 subunit gene (CHRNB2) with subjective response to alcohol and nicotine. Am J Med Genet B Neuropsychiatr Genet 2007;144B:596–604.

[91] Landgren S, Engel JA, Andersson ME, Gonzalez-Quintela A, et al. Association on nAChR gene haplotypes with heavy alcohol use and body mass. Brain Res 2009;1305(Suppl.):S72–6.

[92] Wang JC, Hinrichs AL, Stock H, Budde J, et al. Evidence of common and specific genetic effects: association of the muscarine acetylcholine receptor M2 (CHRM2) gene with alcohol dependence and major depressive syndrome. Hum Mol Genet 2004;13:1903–11.

[93] Luo X, Kranzler HR, Zuo L, Wang S, et al. CHRM2 gene predisposes to alcohol dependence, drug dependence and affective disorders: results from an extended case–control structured association study. Hum Mol Genet 2005;14:2421–34.

[94] Jung MH, Park BL, Lee BC, Ro Y, et al. Association of CHRM2 polymorphisms with severity of alcohol dependence. Genes Brain Behav 2011;10:253–6.

[95] Bauer LO, Ceballos NA. Neural and genetic correlates of binge drinking among college women. Biol Psychol 2014;97:43–8.

[96] Lee CH, Lu W, Michel JC, Goehring A, et al. NMDA receptor structures reveal subunit arrangement and pore architecture. Nature 2014;511:191–7.

[97] Vengeliene V, Bachteler D, Danysz W, Spanagel R. The role of the NMDA receptor in alcohol relapse: a pharmacological mapping study using the alcohol deprivation effect. Neuropharmacology 2005;48:822–9.

[98] Jin Z, Bhandage AK, Bazov I, Kononeko O, et al. Selective increases of AMPA, NMDA and kainate receptor subunit mRNAs in the hippocampus and orbitofrontal cortex but not in prefrontal cortex of human alcoholics. Front Cell Neurosci 2014;8:11.

[99] Wernicke C, Samochowiec J, Schmidt LG, Winterer G, et al. Polymorphisms in the N-methyl-D-aspartate receptor 1 and 2B subunits are associated with alcoholism-related traits. Biol Psychiatry 2003;54:922–8.

[100] Tadic A, Dahmen N, Szegrdi A, Rujescu D, et al. Polymorphisms in the NMDA subunit 2B are not associated with alcohol dependence and alcohol withdrawal-induced seizures and delirium. Eur Arch Psychiatry Clin Neurosci 2005;255:129–35.

[101] Xia Y, Wu Z, Ma D, Tang C, et al. Association of single nucleotide polymorphism in a metabotropic glutamate receptor GRM3 gene subunit to alcohol dependent male subjects. Alcohol Alcohol 2014;49:256–60.

[102] Koller G, Zill P, Fehr C, Pogarell O, et al. No association between alcohol dependence with SLC6A5 and SLC6A9 glycine transporter polymorphisms. Addict Biol 2009;14:506–8.

[103] Holmes A, Spanagel R, Krystal JH. Glutamatergic targets for new alcohol medications. Psychopharmacology 2013;229:539–54.

[104] Zalewska-Kaszubska J, Czarnecka E. Deficit in beta-endorphin peptide and tendency to alcohol abuse. Peptides 2005;26:701–5.

[105] Li Z, Zhang H. Analyzing interaction of μ-, δ-, and κ-opioid receptor gene variants on alcohol or drug dependence using a pattern discovery-based method. J Addict Res 2013;14 (Suppl. 7):007.

[106] Setiawan E, Pihl RO, Benkelfat C, Leyton M. Influence of OPRM1 A118G polymorphism on alcohol-induced euphoria, risk of alcoholism and the clinical efficacy of naltrexone. Pharmacogenomics 2012;13:1161–72.

[107] Rouvinen-Lagerstrom N, Lahti J, Alho H, Kovanen L, et al. μ-Opioid receptor gene (OPRM1) polymorphism A118G: lack of association in Finnish population with alcohol dependence or alcohol consumption. Alcohol Alcohol 2013;48:519–25.

[108] Ray LA, Bujarski S, Squeglia LM, Ashenhurst JR, et al. Interactive effects of OPRM1 and DAT1 genetic variation on subjective response to alcohol. Alcohol Alcohol 2014;49:261–70.

[109] Xuei X, Flury-Wetherill L, Bierut L, Dick D, et al. The opioid system in alcohol and drug dependence: family-based association study. Am J Med Genet B Neuropsychiatr Genet 2007;144B:877–84.

[110] Xuei X, Dick D, Flury-Wetherill L, Tian HJ, et al. Association of the kappa-opioid system with alcohol dependence. Mol Psychiatry 2006;11 L:1016–24.

[111] Karpyak VM, Winham SJ, Preuss UW, Zill P, et al. Association of the PDYN gene with alcohol dependence and the propensity to drink in negative emotional states. Int J Neuropsychopharmacol 2013;16:975–85.

[112] Preuss UW, Winham SJ, Biernacka JM, Geske JR, et al. PDYN rs2281825 variant associated with drinking to avoid emotional or somatic discomfort. PLOS ONE 2013;8(11):e78688.

[113] Hungund BL, Basavarajappa BS. Role of endocannabinoid and cannabinoid CB1 receptor in alcohol-related behavior. Ann N Y Acad Sci 2004;1025:515–27.

[114] Ceccarini J, Hompes T, Verhaeghen A, Casteels C, et al. Changes in cerebral CB1 receptor availability after acute and chronic alcohol abuse and monitored abstinence. J Neurosci 2014;34:2822–31.

[115] Zuo L, Kranzler HR, Luo X, Covault J, et al. CNR1 variation modulates risk for drug and alcohol dependence. Biol Psychiatry 2007;62:616–26.

[116] Marcos M, Pastor I, de la Calle C, Barrio-Real L, et al. Cannabinoid receptor 1 gene is associated with alcohol dependence. Alcohol Clin Exp Res 2012;36:267–71.

[117] Schmidt LG, Samochowice J, Finckh U, Fiszer-Piosik E, et al. Association of a CB1 cannabinoid receptor gene (CNR1) polymorphism with severe alcohol dependence. Drug Alcohol Dependence 2002;65:221–4.

[118] Ishiguro H, Iwasaki S, Teasenfitz L, Higuchi S, et al. Involvement of cannabinoid CB2 receptor in alcohol preference in mice and alcoholism in humans. Pharmacogenomics 2007;7:380–5.

[119] Herman AL, Kranzler HR, Cubels JF, Gelernter J, et al. Association study of the CNR1 gene exon 3 alternative promoter region polymorphism and substance dependence. Am J Med Genet B Neuropsychiatr Genet 2006;141B:499–503.

[120] Parsian A, Todd RD, Cloninger CR, Hoffman PL, et al. Platelet adenylyl cyclase activity in alcoholics and subtype of alcoholics: WHO/ISBRA study clinical center. Alcohol Clin Exp Res 1996;20:745–61.

[121] Ikeda H, Menninger JA, Tabakoff B. An initial study of the relationship between platelet adenylyl cyclase activity and alcohol use disorders. Alcohol Clin Exp Res 1998;22:1057–64.

[122] Desrivieres S, Pronko SP, Lourdusamy A, Ducci F, et al. Sex-specific role of adenylyl cyclase type 7 in alcohol dependence. Biol Psychiatry 2011;69:1100–8.

[123] Procopio DO, Saba LM, Walter H, Lesch O, Skala K, et al. Genetic markers of comorbid depression and alcoholism in women. Alcohol Clin Exp Res 2013;37:896–904.

[124] Lappalainen J, Kranzler HR, Malison R, Price LH, et al. A functional neuropeptide Y Leu7Pro polymorphism associated with alcohol dependence in a large population sample from the United States. Arch Gen Psychiatry 2002;59:825–31.

[125] Wetherill L, Schuckit MA, Hesselbrock V, Xuei X, et al. Neuropeptide Y receptor genes are associated with alcohol dependence, alcohol withdrawal phenotypes and cocaine dependence. Alcohol Clin Exp Res 2008;32:2031–40.

[126] Zill P, Preuss UW, Koller G, Bondy B, Soyka M. Analysis of single nucleotide polymorphisms and haplotypes in the neuropeptide Y gene; No evidence for association with alcoholics in a German population. Alcohol Clin Exp Res 2008;32:430–4.

[127] Nurnberger JI, Foroud T, Flury L, Su J, et al. Evidence for a locus on chromosome 1 that influences vulnerability to alcoholism and affective disorder. Am J Psychiatry 2001;158:718–24.

[128] Zuo L, Wang KS, Zhang XY, Li CS, et al. Rare SERINC2 variants are specific for alcohol dependence in individuals of European descent. Pharmacogenet Genomics 2013;23:395–402.

[129] Han S, Yang BZ, Kranzler HR, Liu X, et al. Integrating GWASs and human protein interaction network identifies a gene subnetwork underlying alcohol dependence. Am J Hum Genet 2013;93:1027–34.

[130] Nieratschker V, Batra A, Fallgatter AJ. Genetics and epigenetics of alcohol dependence. J Mol Psychiatr 2013;1:11.

[131] Schumann G, Coin LJ, Lourdusamy A, Charoen P, et al. Genome-wide association study and genetic functional studies identify autism susceptible candidate gene 2 (AUTS2) in the regulation of alcohol consumption. Proc Natl Acad Sci USA 2011;108:7119–24.

[132] Chojnicka I, Gajos K, Strawa K, Brofa G, et al. Possible association between suicide committed under influence of ethanol and a variant in the AUTS2 gene. PLOS ONE 2013;e57199.

[133] Clarke T-K, Laucht M, Ridinger M, Wodarz N, et al. *KCNJ6* is associated with adult alcohol dependence and involved in gene × early stress interactions in adolescence alcohol drinking. Neuropsychopharmacology 2011;36:1142–8.

[134] Shukla SD, Velazquez J, French SW, Lu SC, et al. Emerging role of epigenetics in the action of alcohol. Alcohol Clin Exp Res 2008;32:1525–34.

[135] Hillemacher T. Biological mechanisms in alcohol dependence—new perspectives. Alcohol Alcohol 2011;46:224–30.

[136] Pietrzykowski AZ. The role of microRNAs in drug addiction: a big lesson from tiny molecules. Int Rev Neurobiol 2010;91:1–24.

[137] Nunez YO, Mayfield RD. Understanding alcoholism through microRNA signatures in brains of human alcoholics. Front Genet 2012;3:43.

[138] Lewohl JM, Nunez YO, Dodd PR, Tiwari GR, et al. Upregulation of microRNAs in brain of human alcoholics. Alcohol Clin Exp Res 2011;35:1928–37.

[139] Manzardo AM, Gunewardena S, Butler MG. Over-expression of the miRNA cluster at the chromosome 14q32 in the alcoholic brain correlates with suppression of the predicted target mRNA required for oligodendrocyte proliferation. Gene 2013;526:356–63.

[140] Novo-Veleiro I, Gonzalez-Sarmiento R, Cieza-Borrella C, Pastor I, et al. A genetic variant in the microRNA-146a gene is associated with susceptibility to alcohol use disorder. Eur Psychiatry 2014;29:288–92.

Index

Note: Page numbers followed by "*f*", "*t*" and "*b*" refers to figures, tables and boxes respectively.

A

Absorption
of alcohol, food effect on, 37–39
Acetaldehyde, 40–41, 92–94, 141, 172–173, 221
as alcohol biomarker, 110
Acetaldehyde–erythrocyte protein adducts, 174–175
Acetaldehyde–hemoglobin adducts, 173–174
Acetaldehyde-modified bovine serum albumin
IgA antibody against, 175
Acetaldehyde–protein adducts, 181
as alcohol biomarkers, 172–175, 176*t*
acetaldehyde–erythrocyte protein adducts, 174–175
acetaldehyde–hemoglobin adducts, 173–174
IgA antibody against acetaldehyde-modified bovine serum albumin, 175
Acetaminophen, 124
Acetylcholine, 267–268
Acetyl coenzyme A, 41
ADCY1–9, 276
Adenylyl cyclase, AUD and, 275–276
ADH. *See* Alcohol dehydrogenase (ADH)
ADH4, 46, 57, 250–251
ADH5, 46
ADH6, 46
ADH1A, 46, 57
ADH1B, 46, 57, 250–252
alleles of, 47–48, 48*t*
*ADH1B*1*, 56–57, 250–251
*ADH1B*2*, 53–56, 250–252
*ADH1B*3*, 54–56, 250–252
ADH1C, 46, 57, 250–251
alleles of, 48, 48*t*
*ADH1C*1*, 55–56, 250–251
*ADH1C*2*, 250–251
ADH genes, 251
Adolescence, drinking during
nongenetic factors, 247–248
ADS scores. *See* Alcohol Dependence Scale (ADS) scores
Adult respiratory distress syndrome (ARDS), 22
ADVIA 1650 analyzer, 77
Age
liver function tests and, 123–124, 123*t*
Alanine aminotransferase (ALT), 121
as alcohol biomarkers, 92–94
overview, 121–122
Alcohol
absorption, food effect on, 37–39
caloric value, 37
content of alcoholic beverages, 3–4, 3*t*
molecular properties, 37
by volume to proof, 3
Alcohol abuse/dependence, 1, 4–5, 245. *See also* Binge drinking
adverse health effects, 16–26
alcohol poisoning, 26
cancers, increased risk of, 22–23
cardiovascular disease and stroke, increased risk of, 21
case report, 21*b*, 26*b*
endocrine system and bone damage, 22
fetal alcohol syndrome, 23–24
immune system damage, 21–22
liver diseases and cirrhosis of liver, 16–18
neurological damage, 18–20
reduced life span, 24–25
violent behavior/homicide, 25–26
as cause of global morbidity and mortality, 16–17
fatal car accidents, 6
genetic factors, 246–248. *See also* Genetic markers
as serious public health issue, 91–92
Alcohol-containing hand sanitizers, 189–190
glucuronide and/or ethyl sulfate in urine and, 190
Alcohol dehydrogenase (ADH), 39–41, 43–44, 182, 248–249
genes encoding, 45–48, 47*t*
polymorphism in, 46–48, 48*t*, 51–58, 55*t*, 58*t*, 250–253
overview, 45
Alcohol Dependence Scale (ADS) scores, 269
Alcoholic beverages
alcohol content of, 3–4, 3*t*
vs. red wine, in heart protection, 10

Alcoholic hepatitis, 16–17
Alcoholic odor, in breath, 29–30
Alcohol-induced estrogen levels, 22
Alcohol-induced hypersensitivity, 56
Alcohol-induced oxidative stress, 41–42
Alcohol intoxication, 25–26
Alcohol levels measurement, in body fluids, 65
  blood alcohol determination, 74–81
    alcohol stability in blood during storage, 79–80
    case report, 77*b*
    correlation between blood and breath alcohol, 80–81
    enzymatic alcohol assays and limitations, 75–77
    gas chromatography in, 78–79, 79*f*
  breath alcohol determination, 68–73
    breathing pattern effects, 71
    case report, 71–73, 74*b*
    chemical principle of breath analyzers, 69–71
    interferences in breath analyzers, 71–73, 73*b*
  endogenous alcohol production, 81
    case report, 82*b*
  overview, 65–68
  saliva alcohol determination, 84–85
  specimens, 66*t*
  transdermal alcohol sensors, 85–87
  urine alcohol determination, 82–84
    case report, 83*b*
Alcohol metabolism, 40–45
  factors affecting, 43–45, 45*t*
  first-pass metabolism, 39–40
  non-oxidative pathways, 42, 43*t*
  pathways, 40–45, 43*t*. *See also specific types*
Alcohol Monitoring Systems, 85–86
Alcohol poisoning, 26
  case report, 26*b*
Alcohol-related fatal crashes, 6
Alcohol tax, 2
Alcohol tolerance, 42, 51–52, 245
Alcohol use/consumption
  guidelines for, 4–6, 5*b*
  historical perspective, 1–2
  as leading cause of death, 6
  in moderation, health benefits, 6–16, 7*b*
    arthritis, reduced risk of, 15
    cardiovascular disease, reduced risk of, 7–9, 9*b*
    common cold , reduced chance of getting, 15–16
    dementia/Alzheimer's disease, reduced risk of, 13
    metabolic syndrome and type 2 diabetes, reduced risk of developing, 11–13
    prolong life, 14–15
    reduced cancer risk, 13–14
    red wine *vs.* alcoholic beverages in heart protection, 10
    stroke, reduced risk of, 10–11
  self-assessment of, 111–112
Alcohol use disorder (AUD), 111
  adenylyl cyclase and, 275–276
  ADH and *ALDH* genes, polymorphism of, 250–253
    increased risk of development, 56–58, 58*t*, 253
    protective effect, 51–56, 55*t*, 251–253
  association of polymorphisms of other genes with, 278
  cannabinoid receptors, polymorphisms of genes encoding, 274–275
  cholinergic receptors, polymorphisms of genes encoding, 267–269, 270*t*
  diagnosis using *DSM-IV* and *DSM-5*, 111
  dopamine pathway, polymorphisms of genes in, 254–260, 261*t*
    catechol-O-methyltransferase, 259–260
    dopamine receptors, 255–257
    dopamine transporters and dopamine-metabolizing enzymes, 258
    monoamine oxidase, 259
  environment and, 246–248
  epigenetics and, 278–281
  GABA pathway, polymorphisms of genes in, 264–267, 268*t*
  genes and, 248–250
  glutamate pathway, polymorphisms of genes in, 269–272
  heredity and, 246–248
  neurobiological basis, 253–254
  neuropeptide Y and, 277–278
  nongenetic factors and, 247–248, 249*t*
  opioid receptors, polymorphisms of genes encoding, 272–274
  overview, 245–246
  serotonin pathway, polymorphisms of genes in, 260–263, 264*t*
Alcohol Use Disorder Identification Test—Consumption (AUDIT-C), 111–112
Alcohol Use Disorder Identification Test (AUDIT) score, 111–112, 269
  alcohol biomarkers and, 112–113
Alco-Screen saliva dipstick, 84–85
Alco-Sensor III and IV, 70
Alcotest models, 70
Aldehyde dehydrogenase (ALDH), 40–41, 248–249
  genes encoding, 48–58, 50*t*
    polymorphisms in, 51–58, 55*t*, 58*t*, 250–253
  overview, 48–49
ALDH. *See* Aldehyde dehydrogenase (ALDH)
ALDH2, 48–49, 251
ALDH-2, 40–41, 141, 253
*ALDH2*1*, 56–57, 251
*ALDH2*2*, 52–54, 56–57, 251–253
ALDH1A, 251
ALDH1A1, 48–50, 251, 253
*ALDH1A1*1*, 57–58
*ALDH1A1*2*, 57–58, 253
*ALDH1A1*3*, 57–58

ALDH1A enzyme, 48–49, 57–58, 251, 253
ALDH1B1, 48–49, 251
ALDH1B enzyme, 50, 251
ALDH2 enzyme, 48–50, 251
Alkaline phosphatase (ALP), 121
  overview, 121–122
Alleles. *See also specific entries*
  of *ADH1B*, 47–48, 48*t*
  of *ADH1C*, 48, 48*t*
  of *ALDH* genes, 51–56
ALP. *See* Alkaline phosphatase (ALP)
ALT. *See* Alanine aminotransferase (ALT)
Alzheimer's disease
  moderate alcohol consumption and reduced risk of, 13
Anemia, 139–140
Apo A-1, 239
Apolipoprotein J (Apo J), sialic acid index of
  as alcohol biomarker, 109, 227–230
    laboratory testing, 229–230
ARDS (adult respiratory distress syndrome), 22
Arthritis
  moderate alcohol consumption and reduced risk of, 15
Asialo, 142–144
Aspartate aminotransferase (AST), 121
  as alcohol biomarkers, 92–94
  overview, 121–122
AST. *See* Aspartate aminotransferase (AST)
AST/ALT ratio, 130
Atrial natriuretic peptide (ANP), 279
AUD. *See* Alcohol use disorder (AUD)
AUDIT-C (Alcohol Use Disorder Identification Test—Consumption), 111–112
AUDIT score. *See* Alcohol Use Disorder Identification Test (AUDIT) score
Auto-brewery syndrome, 81. *See also* Endogenous alcohol production

B

Barker's yeast
  glucuronide and/or ethyl sulfate in urine and, 190
Beer
  alcohol content of, 3
  consumption, historical perspective, 2
  phytochemicals in, 6–7
β-endorphin
  as alcohol biomarker, 92–94
β-Hexosaminidase, 181
  as alcohol biomarker, 101–102, 164–170
    analytical methods, in serum and urine, 173*b*
    case report, 170*b*
    characteristics, 168*b*
    elevated levels in various conditions, 168–170, 169*t*
    laboratory methods for measurement, 171–172
    limitations, 97*t*
  and binge drinking, 167
  during glomerular filtration, 167
  isoenzymes of, 101
  isoforms, 163–164
  overview, 163
Big GGT (b-GGT), 132–133
Binge drinking, 5–6. *See also* Alcohol abuse/dependence
  alcohol-related brain damage and, 20
  β-hexosaminidase increased level and, 167
  risks associated with, 5–6
Biomarkers, alcohol, 37–38, 91, 245–246. *See also specific types*
  β-hexosaminidase as, 101–102, 163. *See also* β-Hexosaminidase
  carbohydrate-deficient transferrin as, 98–100, 139. *See also* Mean corpuscular volume (MCV)
    case report, 100*b*
    combined CDT–GGT, 100
  clinical application, 104*t*, 111–116
    AUDIT score, 112–113
    case report, 113*b*
    diagnosis using *DSM-IV* and *DSM-5*, 111
    self-assessment of alcohol use, 111–112
  combination, 114–116
  defined, 91
  direct, 181
    limitations of, 104*t*
  discovery, proteomics in, 237–240
  ethyl glucuronide as, 102–104, 181. *See also* Ethyl glucuronide
  ethyl sulfate as, 102–104, 181. *See also* Ethyl sulfate
  fatty acid ethyl ester as, 104–106, 181. *See also* Fatty acid ethyl esters
  5-HTOL/5-HIAA as, 109–110, 230–236
  indirect, 181
  liver enzymes as, 94–96, 97*t*, 121. *See also* Liver enzymes
  MCV as, 97–98, 139. *See also* Mean corpuscular volume (MCV)
  others, 104–106, 110
  overview, 91–92
  phosphatidylethanol as, 106–107, 181. *See also* Phosphatidylethanol
  sialic acid index of apolipoprotein J, 109, 227–230
  state
    characteristics, 95*t*
    long-term, 93*b*
    short-term, 93*b*
    vs. trait, 92–94, 93*t*
  total plasma sialic acid as, 108, 108*t*, 221–227, 222*t*
  trait *vs.* state, 92–94, 93*t*
Blood alcohol analysis, 38–39
Blood alcohol determination, 74–81
  alcohol stability in blood during storage, 79–80
  and breath alcohol, correlation between, 80–81
  case report, 77*b*
  correlation between blood and breath alcohol, 80–81
  enzymatic alcohol assays and limitations, 75–77
  gas chromatography in, 74, 78–79, 79*f*

Blood alcohol level, 27–30, 67–68
alcoholic odor in breath and endogenous alcohol production, 29–30
case report, 28*b*
gender difference in, 39–40
legal limit of, 27
physiological effects, 27*t*
BMI. *See* Body mass index (BMI)
Body fluids, alcohol levels measurement in, 65
blood alcohol determination, 74–81
alcohol stability in blood during storage, 79–80
case report, 77*b*
correlation between blood and breath alcohol, 80–81
enzymatic alcohol assays and limitations, 75–77
gas chromatography in, 78–79, 79*f*
breath alcohol determination, 68–73
breathing pattern effects, 71
case report, 71–73, 74*b*
chemical principle of breath analyzers, 69–71
interferences in breath analyzers, 71–73, 73*b*
endogenous alcohol production, 81
case report, 82*b*
overview, 65–68
saliva alcohol determination, 84–85
specimens, 66*t*
transdermal alcohol sensors, 85–87
urine alcohol determination, 82–84
case report, 83*b*
Body mass index (BMI)
liver function tests and, 123*t*
Bone(s)
adverse effects of alcohol abuse on, 22
Bovine serum albumin, acetaldehyde-modified
IgA antibody against, 175
Bovine serum albumin–acetaldehyde adduct, 172–173
Breast cancer
alcohol abuse and, 22–23
Breath, alcoholic odor in, 29–30
Breath alcohol analyzers, 68
Breathalyzer, 69
breathing pattern effect on test results, 71
calculation, 68
case report, 74*b*
chemical principle of, 69–71
DataMaster cdm, 69–70
fuel cell technology in, 70
interferences in, 71–73, 73*b*
Intoxilyzer, 69–72
IR spectroscopy and, 69–70
types of, 69
Breath alcohol determination, 68–73
and blood alcohol, correlation between, 80–81
breathing pattern effects, 71
case report, 71–73, 74*b*
chemical principle of breath analyzers, 69–71
interferences in breath analyzers, 71–73, 73*b*
Breathalyzer, 69
Breathing pattern
effect on breath alcohol test results, 71
Brief Michigan Alcoholism Screening Test (Brief MAST), 56–57

C

CAGE, 111–112
California Men's Health Study, 13–14
Cancer
alcohol abuse and increased risk of, 22–23
moderate alcohol consumption and reduced risk of, 13–14
Cannabinoid receptor gene 1 (*CNR1*), 275
Cannabinoid receptors
polymorphisms of genes encoding, AUD and, 274–275
Capacity-limited model, 37–38
Capillary electrophoresis method, 157
Capillary zone electrophoresis (CZE), 156–157, 171
Carbohydrate-deficient transferrin (CDT), 181
as alcohol biomarker, 92–94, 98–100, 142–154
application, 148–151
case report, 149*b*, 154*b*
characteristics, 149*b*
cutoff values, sensitivity, and specificity, 144–147, 146*t*
and GGT as combined biomarker, 147–148
laboratory testing, 154–157
limitations, 97*t*, 151–154
mechanism of formation, 144
overview, 139
glycosylation, 142–143, 150
Carboxyl ester lipase, 201–202
Cardiovascular disease
alcohol abuse and increased risk of, 21
liver function tests and, 123*t*
moderate alcohol consumption and reduced risk of, 7–9, 9*b*
Cardiovascular Health Study, 8
Carter, Jimmy, 2
Catalase, 42
Catechol-O-methyltransferase (COMT), 259–260
CDT. *See* Carbohydrate-deficient transferrin (CDT)
CDT and GGT, as combined alcohol biomarker, 147–148
Center for Nutrition Policy and Promotion, 4
Chemiluminescent detection, 171–172
Childbirth
liver function tests and, 123*t*
Childhood
nongenetic factors and AUD development, 247–248
Cholesteryl ester transfer protein (CETP)
as alcohol biomarker, 110, 236
Cholinergic muscarinic receptor 2, 269
Cholinergic receptors
polymorphisms of genes encoding, AUD and, 267–269, 270*t*

CHRM2 gene, 269
CHRNA4 gene, 268–269
CHRNB2 gene, 268–269
Chronic atrophic gastritis (CAG)
moderate alcohol consumption and, 13–14
Cirrhosis of liver
alcohol abuse-associated, 16–18
ClinProt systems, 239
Clusterin, 240
Code of Federal Regulations, 4
"Coding variations" 46
Coffee consumption
liver function tests and, 123*t*, 129
Collaborative Study of the Genetics of Alcoholism (COGA), 265–266, 277–278
Colorimetry, 171–172
Combined CDT–GGT, as alcohol biomarker, 100
Common cold, moderate alcohol consumption and reduced chances of getting, 15–16
COMT gene, 259–260
Copenhagen City Heart Study, 10–11
Coronary heart disease
smoking as risk factor for, 7
Creatine kinase-MB isoenzyme (CK-MB), 123–124
Cushing's syndrome, 22
Cutoff concentrations
of CDT, 144–147, 146*t*
of ethyl glucuronide and ethyl sulfate
in hair, meconium, and other matrix, 192–194
in urine, 191–192
of phosphatidylethanol, 210–211
Cyanamide (calcium carbimide), 234–235
CYP2E1, 18, 41, 43–44
CYP2E1 gene
polymorphism of, 59
Cytokines, 110, 237

D

DAT. *See* Dopamine transporter (DAT)
DataMaster cdm, 69–70
DAT gene, 258
DβH gene, 258
Dementia/Alzheimer's disease
moderate alcohol consumption and reduced risk of, 13
Depression, AUD development and, 248
Diagnostic and Statistical Manual of Mental Disorders (DSM-IV), 245, 260–261
AUD diagnosis using, 111
Diet
liver function tests and, 123*t*
Direct alcohol biomarkers, 221. *See also* Biomarkers, alcohol; *specific types*
limitations of, 104*t*
sensitivity and specificity of, 213–214, 214*t*
Disasters, alcohol consumption associated with, 248
Disialotransferrin, 142–143
Distillation process, 2
Disulfiram (Antabuse), 49, 234–235
Diurnal variation
liver function tests and, 123*t*
DNA methylation, 278–280
Dolichol
as alcohol biomarker, 110, 175–177
Dopamine, 254–255
Dopamine β-hydroxylase (DβH), 258
Dopamine-metabolizing enzymes, 258
Dopamine pathway, polymorphisms of genes in
AUD and, 254–260, 261*t*
catechol-O-methyltransferase, 259–260
dopamine receptors, 255–257
dopamine transporters and dopamine-metabolizing enzymes, 258
monoamine oxidase, 259
mesolimbic, 254–255
Dopamine receptors, 255–257
Dopamine transporter (DAT), 258
DRD1, 255
DRD2, 255–256
DRD3, 257
DRD4, 257
"Driving under the influence" (DUI), 65–67
"Driving while intoxicated"/"driving while impaired" (DWI), 65–67
Drugs
affecting GGT levels, 133*b*
affecting liver enzymes, 125*b*
macrocytosis and, 142*t*
Drunkenness, 2
DSM-5, 245
AUD diagnosis using, 111
Dudley, Robert, 1

E

80-hr test, 196–197
Electrophoresis, 171, 237–238
Electrospray ionization, 237–238
Electrospray mass spectrometry, 157
Elevated GGT, and risk of certain diseases, 130–132, 132*b*
Elimination process, of alcohol, 37–38
EMIT (enzyme multiplied immunoassay technique), 75–76
Endocrine disorders
liver function tests and, 123*t*
Endocrine system
adverse effects of alcohol abuse on, 22
Endogenous alcohol production, 29–30, 81
case report, 82*b*
Environment
and AUD, 246–248
Enzymatic alcohol assays
in blood alcohol determination, 74–77
case report, 77*b*
limitations, 75–77
Enzyme immunoassay, 172
Enzyme-linked immunosorbent assay (ELISA), 172, 174, 235
Enzyme multiplied immunoassay technique (EMIT), 75–76
Epigenetics, AUD and, 278–281
Esophageal cancers
alcohol abuse and increased risk of, 22–23

Ethanol. *See* Alcohol
Ethyl arachidonate, 202–203
Ethyl docosahexaenoate, 202–203
Ethyl glucuronide, 37–38, 92–94, 181
as alcohol biomarker, 102–104, 186–201
application, 196–199, 200*t*
case report, 197*b*, 199*b*
characteristics, 183*t*
cutoff concentrations in hair, meconium, and other matrix, 192–194
cutoff concentrations in urine, 191–192
false-positive/false-negative results, 194–196
incidental alcohol exposure, 188–191
laboratory testing, 199–201
limitations, 104*t*
in urine, causes of, 188–191, 188*t*
overview, 181–185
phosphatidylethanol and, 209–210
sensitivity of, 213–214, 214*t*
specificity of, 213–214, 214*t*
Ethyl laurate, 202–203
Ethyl linoleate, 202–203
Ethyl linolenate, 202–203
Ethyl myristate, 202–203
Ethyl oleate, 202–203
Ethyl palmitate, 202–203
Ethyl palmitoleate, 202–203
Ethyl stearate, 202–203
Ethyl sulfate, 37–38, 181
as alcohol biomarker, 102–104, 186–201
application, 196–199, 200*t*
case report, 197*b*, 199*b*
characteristics, 183*t*
cutoff concentrations in hair, meconium, and other matrix, 192–194
cutoff concentrations in urine, 191–192
incidental alcohol exposure, 188–191
laboratory testing, 199–201
limitations, 104*t*
in urine, causes of, 188–191, 188*t*
overview, 181–185
phosphatidylethanol and, 209–210
sensitivity of, 213–214, 214*t*
specificity of, 213–214, 214*t*

F

False-negative %CDT, 151*b*
False-negative results
with ethyl glucuronide, 194–196
using urine or hair specimens, 200*t*
False-positive %CDT, 151*b*
False-positive results
with ethyl glucuronide, 194–196
using urine or hair specimens, 200*t*
Fatal car accidents, alcohol use-related, 6, 65–67
Fatty acid ethyl esters, 42, 181
as alcohol biomarker, 104–106, 201–207
case report, 106*b*
characteristics, 183*t*
in hair, 202–204
interpretation of hair and meconium, issues regarding, 205*b*
laboratory analysis, 206–207
limitations, 104*t*
in meconium, 204–205
sensitivity of, 213–214, 214*t*
specificity of, 213–214, 214*t*
Fetal alcohol syndrome, 23–24
Fibronectin, 240
First-pass metabolism, of alcohol, 39–40
5-HTOL/5-HIAA
as alcohol biomarker, 109–110, 230–236
case report, 233*b*, 234*b*
laboratory testing, 235–236
5-HTT, 262
5-HTT-linked promoter region (*5-HTTLPR*), 262
5-Hydroxyindole-3-acetaldehyde, 109–110, 230–232
5-Hydroxyindole-3-acetic acid (5-HIAA), 109–110
5-Hydroxytryptophol (5-HTOL)
as alcohol biomarker, 109–110, 230–236
case report, 233*b*, 234*b*
laboratory testing, 235–236
5-Hydroxytryptophol glucuronide (GTOL), 230–232, 235
Fluorometry, 171–172
Food, effect on alcohol absorption, 37–39
Framingham Heart Study, 7
Free GGT (f-GGT), 132–133
"French paradox" 10
Fruit juices
glucuronide and/or ethyl sulfate in urine and, 190–191
Fuel cell technology, 70
Functional proteomics, 237–238

G

GABA. *See* γ-aminobutyric acid (GABA)
$GABA_A$ receptors, 265–267
GABA pathway
polymorphisms of genes in, AUD and, 264–267, 268*t*
GABA receptors, 265–266
GABRA2 gene, 265–266
GABRA5 gene, 267
GABRA6 gene, 266–267
GABRB3 gene, 267
GABRG1 gene, 265–266
GABRG3 gene, 267
Gallstone
alcohol abuse and, 22–23
γ-aminobutyric acid (GABA), 247, 249–250
as alcohol biomarker, 92–94
γ-glutamyl transferase (GGT), as alcohol biomarkers, 92–94, 96, 121, 126–133, 181
case report, 129*b*
characteristics, 128*b*
elevated, and risk of certain diseases, 130–132, 132*b*
fraction, 132–133, 133*b*
half-life of, 127
limitations, 96, 97*t*
overview, 121–122

Gas chromatography (GC)
in blood alcohol determination, 74, 78–79, 79*f*
Gas chromatography–mass spectrometry (GC–MS), 78–79, 199–200, 206, 211–212, 235
Gastric cancer
moderate alcohol consumption and, 13–14
Gel electrophoresis, 154–155
Gelsolin, 239
Gender difference
in blood alcohol level, 39–40
liver function tests and, 123*t*
Gene–environment interaction, 279
Genes. *See also specific entries*
and AUD, 248–250
encoding ADH, 45–48, 47*t*
polymorphism in, 46–48, 48*t*, 51–58, 55*t*, 58*t*, 250–253
encoding ALDH, 48–58, 50*t*
polymorphisms in, 51–58, 55*t*, 58*t*, 250–253
Genetic markers, of AUD, 245
adenylyl cyclase, 275–276
ADH and *ALDH* genes, polymorphism of, 250–253
increased risk of development, 56–58, 58*t*, 253
protective effect, 51–56, 55*t*, 251–253
association of polymorphisms of other genes with, 278
cannabinoid receptors, polymorphisms of genes encoding, 274–275
cholinergic receptors, polymorphisms of genes encoding, 267–269, 270*t*
dopamine pathway, polymorphisms of genes in, 254–260, 261*t*
catechol-O-methyltransferase, 259–260
dopamine receptors, 255–257
dopamine transporters and dopamine-metabolizing enzymes, 258
monoamine oxidase, 259
environment and, 246–248
epigenetics and, 278–281
GABA pathway, polymorphisms of genes in, 264–267, 268*t*
genes and, 248–250
glutamate pathway, polymorphisms of genes in, 269–272
heredity and, 246–248
neurobiological basis, 253–254
neuropeptide Y and, 277–278
nongenetic factors, 247–248, 249*t*
opioid receptors, polymorphisms of genes encoding, 272–274
overview, 245–246
serotonin pathway, polymorphisms of genes in, 260–263, 264*t*
Genome-wide association studies (GWAS), 248–249
Genotyping, 248–249
Germ-X, 189–190
GGT. *See* γ glutamyl transferase (GGT)
GGT fraction, as alcohol biomarker, 132–133
"Global Status Report on Alcohol and Health 2014", WHO, 91–92
Glomerular filtration
β-hexosaminidase during, 167
Glucuronide conjugate (GTOL), 109–110
Glutamate pathway
polymorphisms of genes in, AUD and, 269–272
Glutathione peroxidase-3, 239
GTOL/5-HIAA ratio, 230–233
Guidelines, for alcohol consumption, 4–6, 5*b*
GWAS. *See* Genome-wide association studies (GWAS)

H

Hair
cutoff concentrations of ethyl glucuronide and ethyl sulfate in, 192–194
ethyl glucuronide level in, 198–199
false-positive/false-negative results, 200*t*
fatty acid ethyl esters in, 202–204
Hair tonics, ethyl glucuronide in, 195–196
Hand sanitizers, alcohol-containing, 189–190
glucuronide and/or ethyl sulfate in urine and, 190
propyl alcohol/isopropyl alcohol-based, 194–195
Helicobacter pylori
elimination of, moderate alcohol consumption and, 13–14
Hemoglobin–acetaldehyde adducts, 92–94, 221. *See also* Acetaldehyde–protein adducts
as alcohol biomarker, 110, 173–174
Hemopexin, 239
Henry's law, 68
Heparin cofactor II, 239
Hepatitis C infection
alcohol abuse and, 16–18
Heptasialotransferrin, 142–143
Herbal supplements
affecting liver enzymes, 125*b*
Heredity
and AUD, 246–248
HEX A gene, 163–164
Hexasialotransferrin, 142–143
HEX B gene, 163–164
Hexosaminidase C, 163–164
High-density lipoprotein (HDL) cholesterol levels, 10
in drinkers *vs.* nondrinkers, 8–9
High-performance liquid chromatography (HPLC), 145, 154–155, 157, 173–174, 210–212, 237–238
Histidine-rich glycoprotein, 239
HIV infection, 21–22
Homicide
alcohol abuse and, 25–26
Homocysteine, 110, 236, 279
Homocysteine-induced endoplasmic reticulum protein (HERP), 279
Honolulu Heart Study, 8–9
HRT2A, 263
HRT2C, 263
α_2-HS glycoprotein, 239
HTR3, 263
HTR1A, 263

HTR1B, 263
Hypoalbuminemia, 121
Hypothyroidism
macrocytosis and, 141–142

I

IgA antibody
against acetaldehyde-modified bovine serum albumin, 175
Ignition interlock devices, 85
IL-6, 237
IL-8, 237
IL-12, 237
Immune system
adverse effects of alcohol abuse on, 21–22
Immunofixation electrophoresis (IFE), 151*b*
Impulsive/novelty-seeking personality, 248
Indirect biomarkers, 181, 221. *See also* Biomarkers, alcohol; *specific types*
Infrared (IR) spectroscopy, 69–70
"Insensible perspiration" 85–86
Inter-α inhibitor H4, 240
Interferences, in breath alcohol analyzers, 71–73, 73*b*
Interleukin-1 (IL-1), 237
Intoxilyzer, 69–72
Ion exchange chromatography, 171
Isoelectric focusing, 171
Isopropyl alcohol, 75

J

Jefferson, Thomas, 2

K

Ketogenic diet, 71–72
Ketonemia, 71–72
Korsakoff's syndrome, 19–20

L

Lab on a computer disc (Lab-CD), 172
Laboratory methods/testings
for β-hexosaminidase measurement, 171–172
for CDT, 154–157
for ethyl glucuronide and ethyl sulfate, 199–201
for fatty acid ethyl esters, 206–207
for 5-HTOL/5-HIAA, 235–236
for liver enzymes, 133–134
for phosphatidylethanol, 211–213
for sialic acid index of plasma Apo J, 229–230
for total plasma sialic acid, 225–227
Lactate, 75–77
Lactate dehydrogenase (LDH), 75–76
Lactococcus garvieae, 29–30
Legal limit
of blood alcohol level, 27
for driving, 67, 67*t*
Life expectancy
alcohol abuse and, 24–25
moderate alcohol consumption and, 14–15
Lincoln, Abraham, 2
Linear regression model, 74
Lipoproteins, 201–202
Liquid chromatography combined with electrospray ionization tandem mass spectrometric method (LC/ESI–MS/MS), 200–201
Liquid chromatography combined with tandem mass spectrometry (LC–MS/MS), 187, 199–201, 206–207, 211–212, 235
Liver, 121
Liver disease, alcohol-induced, 16–18
mechanism, 18
Liver enzymes, as alcohol biomarkers, 94–96, 121. *See also specific entries*
drugs and herbal supplements affecting, 125*b*
GGT, 126–133
case report, 129*b*
characteristics, 128*b*
elevated, and risk of certain diseases, 130–132, 132*b*
fraction, 132–133, 133*b*
laboratory testings, 133–134
limitations, 97*t*
liver function tests
case report, 124*b*
factors affecting, 122–124, 123*t*
normal values, 122*t*
moderate alcohol consumption effect on, 125–126
overview, 121–122
sensitivity, 127*t*
specificity, 127*t*
Liver function tests
case report, 124*b*
factors affecting, 122–124, 123*t*
normal values, 122*t*
Long-term state alcohol biomarkers, 93*b*
Low-density lipoprotein (LDL)–acetaldehyde adduct, 172–173
Low-density lipoprotein (LDL) cholesterol, 8–9
Lysosomal enzymes, 170

M

Macrocytic anemia, 139–140
Macrocytosis, 97–98, 139
causes of, 141–142, 142*b*
defined, 140–141
hypothyroidism and, 141–142
increased MCV in alcoholics, 141
medications and, 142*t*
vitamin B_{12} deficiency and, 141–142
Malnutrition
liver function tests and, 123*t*
MAO. *See* Monoamine oxidase (MAO)
MAOA, 258–259
MAOA-linked polymorphic region *(MAOA-LPR)*, 259
MAOB, 258
Mass spectrometry, 237–238
MAST (Michigan Alcoholism Screening Test), 111–112
Mast cells, 21–22
Matrix-assisted laser desorption/ionization mass spectrometry with time-offlight detection (MALDI-TOF), 157, 237–239

MCV. *See* Mean corpuscular volume (MCV)
Mean corpuscular volume (MCV), 181
as alcohol biomarker, 92–94, 97–98, 139–142
limitations, 97*t*
macrocytosis, causes of, 141–142, 142*b*, 142*t*
mechanism of increased MCV, 141
overview, 139
calculation, 139
Meconium
cutoff concentrations of ethyl glucuronide and ethyl sulfate in, 192–194
fatty acid ethyl esters in, 204–205
Medium GGT (m-GGT), 132–133
Melatonin, 6–7
Membrane-bound catechol-O-methyltransferase (MB-COMT), 259–260
Mesolimbic dopaminergic pathway, 254–255
Metabolic syndrome
moderate alcohol consumption and reduced risk of, 11–13
risk factors, 11–12
Michaelis–Menten model, 37–38
Michigan Alcoholism Screening Test (MAST), 111–112
Microcytic anemia, 139–140
MicroRNAs (miRNAs), 280–281
Microsomal ethanol oxidizing pathway, 41
Moderate alcohol consumption. *See also* Alcohol use/consumption
effect on liver enzymes as alcohol biomarkers, 125–126
health benefits, 6–16, 7*b*
and prolong life, 14–15
and reduced risk of
arthritis, 15
cancer, 13–14
cardiovascular disease, 7–9, 9*b*
common cold, 15–16
dementia/Alzheimer's disease, 13
metabolic syndrome and type 2 diabetes, 11–13
stroke, 10–11
red wine *vs.* alcoholic beverages in heart protection, 10
Monoamine oxidase (MAO), 259
Monocyte chemoattractant protein-1 (MCP-1), 237
Monosialotransferrin, 142–143
Musculoskeletal diseases
liver function tests and, 123*t*
Myocardial infarction
moderate alcohol consumption and reduced risk of, 8

N

N-acetyl-β-hexosaminidase. *See* β-hexosaminidase
N-acetylgalactosamine, 101
N-acetylglucosamine, 101
N-acetylneuraminic acids, 108
National Institute of Alcohol Abuse and Alcoholism, 5–6
Natural selection, 1
N^{ε}-ethyl lysine (NEL), 172–173
Nerve growth factor (NGF), 279
Neurobiological basis, of AUD, 253–254
Neurological damage
alcohol abuse and, 18–20
in binge drinkers, 20
case report, 21*b*
Neuropeptide Y (NPY)
AUD and, 277–278
Neuroticism/negative emotionality, 247–248
Neurotransmitters, 253–254
as alcohol biomarkers, 92–94
N-glycosylation, 150, 164
Nicotinamide adenine dinucleotide (NAD^+), 40–41
Nicotinamide adenine dinucleotide phosphate (NADP), 41
N-Latex CDT assay, 154–156
Nonaqueous capillary electrophoresis, 211–212
"Noncoding variations" 46
Nongenetic factors
and AUD development, 247–248, 249*t*
Non-oxidative metabolites, alcohol metabolism and formation of, 185*f*
Non-oxidative pathways, of alcohol metabolism, 42, 43*t*
Normocytic anemia, 139–140
NPY. *See* Neuropeptide Y (NPY)
NPY gene, 277
Nuclear magnetic resonance spectroscopy, 237–238

O

Octasialotransferrin, 142–143
Odds ratio (OR), 13
Odor, alcoholic, in breath, 29–30
Omnibus Transportation Employee Testing Act, 82
Opioid receptors
polymorphisms of genes encoding, AUD and, 272–274
OPRD1, 273
OPRK1, 273
OPRM1, 273

P

Pancreatitis
alcohol abuse and, 22–23
Parkinson's disease, 254–255
Peak alcohol level, 38–39
Peer pressure, AUD development and, 248
Pentafluoropropionic anhydride, 235
Pentasialotransferrin, 142–143
Percentage (%CDT), 143–144, 151. *See also* Carbohydrate-deficient transferrin (CDT)
test, factors affecting, 151*b*
Personality, AUD development and, 248
Phosphatidylethanol, 42, 181
as alcohol biomarker, 106–107, 207–213
characteristics, 183*t*
cutoff concentration, 210–211
interpretation of results, issues regarding, 212*b*
laboratory analysis, 211–213
limitations, 104*t*

Phosphatidylethanol (*Continued*)
antibodies specific to, 211
as marker to ethyl glucuronide and ethyl sulfate, 209–210
during pregnancy, 209
sensitivity of, 213–214, 214*t*
specificity of, 213–214, 214*t*
Phospholipase D, 42, 207
Phospholipase D_1, 207
Phospholipase D_2, 207
Phospholipids, 201–202
Physician's Health Study, 7–8
Phytochemicals, 6–7
Pigment epithelium-derived factor, 239
Plasma kallikrein, 239
Polymorphism. *See also* Genetic markers, of AUD
ADH genes, 46–48, 48*t*, 250–253
case report, 53*b*
and increased risk of AUD development, 56–58, 58*t*, 253
and protective effect against AUD development, 51–56, 55*t*, 251–253
ALDH genes, 250–253
case report, 53*b*
and increased risk of AUD development, 56–58, 58*t*, 253
and protective effect against AUD development, 51–56, 55*t*, 251–253
association of other genes, 278
CYP2E1 gene, 59
of genes encoding cannabinoid receptors, 274–275
of genes encoding cholinergic receptors, 267–269, 270*t*
of genes encoding opioid receptors, 272–274
of genes in dopamine pathway, 254–260, 261*t*
of genes in GABA pathway, 264–267, 268*t*
of genes in glutamate pathway, 269–272
of genes in serotonin pathway, 260–263, 264*t*
Postmortem investigations
ethyl glucuronide and ethyl sulfate in, 197
Pregnancy
nongenetic factors and AUD development, 247–248
phosphatidylethanol and, 209
serum β-hexosaminidase elevated level during, 169–170
Primatene Mist, 71–72
Prolonged life
moderate alcohol consumption and, 14–15
"Proof" of drink, 3–4
Pro-opiomelanocortin (*POMC*) gene, 279–280
Propanol, 75
Propyl alcohol/isopropyl alcohol-based hand sanitizers, 194–195
ProteinChip system, 239
Proteins, as alcohol biomarkers using proteomics approach, 238–240, 240*t*
Proteomics
in alcohol biomarker discovery, 237–240
specific proteins in, 238–240, 240*t*
functional, 237–238
structural, 237–238
Prothrombin time (PT), 121
Pseudo-Cushing's syndrome, 22

Q

Quantitative ethanol detector (QRD) saliva alcohol test, 84–85

R

Race
liver function tests and, 123*t*
Reduced life span
alcohol abuse and, 24–25
Red wine
vs. alcoholic beverages, in heart protection, 10
resveratrol in, 10
Relative risk (RR), 7
Resveratrol
in red wine, 10
Retinol binding protein, 240
Rheumatoid arthritis
moderate alcohol consumption and reduced risk of, 15

S

S-adenosyl methionine, 279
Saliva alcohol determination, 84–85
SAMHSA (Substance Abuse and Mental Health Services Administration), 196
SCRAM2, 86–87
SCRAM (Secure Continuous Remote Alcohol Monitor) bracelet, 85–87
SELDI-TOF-MS (surfaced enhanced laser desorption/ionization time-of-flight mass spectrometry), 238–239
Selenoprotein P, 239
Self-assessment
of alcohol use, 111–112
Semiconductor alcohol sensors, 70–71
Sensitivity
of CDT, 144–147, 146*t*
of direct alcohol biomarkers, 213–214, 214*t*
of liver enzymes as alcohol biomarkers, 127*t*
Serotonin (5-hydroxytryptamine), 109–110, 230–232, 260–261. *See also* 5-HTOL/5-HIAA
metabolic pathway of, 230–232, 231*f*
Serotonin pathway
polymorphisms of genes in, AUD and, 260–263, 264*t*
Serotransferrin, 239
Short-term state alcohol biomarkers, 93*b*
Sialic acid, 108, 142–143
Sialic acid index of apolipoprotein J
as alcohol biomarker, 109, 227–230
laboratory testing, 229–230
Sialyltransferase, 98

Single nucleotide polymorphism (SNP), 46, 248–249. *See also* Polymorphism
Small GGT (s-GGT), 132–133
Smith & Wesson, 69
Smoking
  liver function tests and, 123*t*
  as risk factor for coronary heart disease, 7
Soluble catechol-*O*-methyltransferase (S-COMT), 259–260
Specificity
  of CDT, 144–147, 146*t*
  of direct alcohol biomarkers, 213–214, 214*t*
  of liver enzymes as alcohol biomarkers, 127*t*
Specimens, alcohol testing of, 66*t*. *See also* Alcohol levels measurement, in body fluids
Stability, of alcohol
  in blood during storage, 79–80
Standard drink, 3
State alcohol biomarkers, 245–246. *See also specific types*
  characteristics, 95*t*
  long-term, 93*b*
  short-term, 93*b*
  vs. trait, 92–94, 93*t*
Stearic acid, 202–203
Steatosis, 16–17
Stone Age, 2
Strenuous exercise
  liver function tests and, 123*t*
Stroke
  alcohol abuse and increased risk of, 21
  moderate alcohol consumption and reduced risk of, 10–11
Structural proteomics, 237–238
Substance Abuse and Mental Health Services Administration (SAMHSA), 196
Sudden infant death syndrome, 75–76
SULT, 182–184
SULT1 (phenol sulfotransferases), 182–184
SULT2 (alcohol sulfotransferases), 182–184
SULT1A2, 182–184
SULT1A3, 182–184
SULT1B1, 182–184
SULT1C2, 182–184
"Sura" 2
Surfaced enhanced laser desorption/ionization time-of-flight mass spectrometry (SELDI-TOF-MS), 238–239
Syndrome X. *See* Metabolic syndrome
α-Synuclein (SNCA), 279

T

Taq1A polymorphism, 256
Tetranectin, 239
Tetrasialotransferrin, 98, 142–143
Thiamine deficiency
  alcohol related brain damage and, 18–20
Thin-layer chromatography, 211–212
3′-phosphoadenosine-5′-phosphosulfate, 182–184
Total plasma sialic acid
  as alcohol biomarker, 108, 221–227
    causes of elevated levels, 224–225, 224*b*
    characteristics, 222*t*
    laboratory testing, 225–227
    limitations, 108*t*
Tourette's syndrome, 254–255
Trait alcohol biomarkers, 245–246
  vs. state, 92–94, 93*t*
Transdermal Alcohol Detection device (TAD), 85–87
Transdermal alcohol sensors, 85–87. *See also* Alcohol levels measurement, in body fluids
Transferrin, 139. *See also* Carbohydrate-deficient transferrin (CDT)
  as alcohol biomarker, 98–100
    overview, 142–143
  domains, 142–143
  isoforms of, 142–143
  molecular properties, 142–143
Transferrin C, 142–143
Triglycerides, 201–202
Trisialotransferrin, 142–143
Tumor necrosis factor-α (TNF-α), 237
Two-dimensional electrophoresis technique, 237–238
Type 2 diabetes
  moderate alcohol consumption and reduced risk of, 11–13

U

UDP-glucuronosyltransferase (UGT), 102, 182
UGT1A1, 182
UGT2B7, 182
Uridine-5′-diphospho (UDP)-glucuronosyltransferase superfamily, 182
Urinary tract infection
  false-negative urinary ethyl glucuronide level and, 195
Urine
  elevated level of β-hexosaminidase in, 168–170, 169*t*
  ethyl glucuronide in
    causes of, 188–191, 188*t*
    cutoff concentrations, 191–192
    false-positive/false-negative results, 200*t*
  ethyl sulfate in
    causes of, 188–191, 188*t*
    cutoff concentrations, 191–192
Urine alcohol test, 82–84
  case report, 83*b*
U.S. Department of Agriculture (USDA), 4
USDA. *See* U.S. Department of Agriculture (USDA)

V

Vasopressin (AVP), 279
Very low-density lipoprotein (VLDL)–acetaldehyde adduct, 172–173

Violent behavior
alcohol abuse and, 25–26
Vitamin B_{12} deficiency
macrocytosis and, 141–142
Vitronectin, 239
Vodka
alcohol content of, 3

W

Wernicke–Korsakoff syndrome, 19–20
Wernicke's encephalopathy, 19–20
Whiskey tax, 2
WHO. *See* World Health Organization (WHO)
Widmark, Eric P., 28
Widmark formula, 28–29
Wilson's disease
liver function tests and, 123–124, 123*t*
Wine aroma, 29
Women
alcohol consumption guidelines, 4–6
alcohol-induced estrogen levels in, 22
alcohol-related neurological damage, 19
first-pass metabolism of alcohol and, 39–40. *See also* Alcohol metabolism
World Health Organization (WHO), 4–5
"Global Status Report on Alcohol and Health 2014" 91–92
Wrist Transdermal Alcohol Sensor (WrisTAS), 85–87

X

X-ray crystallography, 237–238

Y

Yeast, 2
barker's, glucuronide and/or ethyl sulfate in urine and, 190

Printed in the United States
By Bookmasters